Milestones in Drug Therapy
MDT

Series Editors

Prof. Dr. Michael J. Parnham PhD
Director of Preclinical Discovery
Centre of Excellence in Macrolide
Drug Discovery
GlaxoSmithKline Research Centre
Zagreb Ltd.
Prilaz baruna Filipovića 29
HR-10000 Zagreb
Croatia

Prof. Dr. J. Bruinvels
Sweelincklaan 75
NL-3723 JC Bilthoven
The Netherlands

PEGylated Protein Drugs: Basic Science and Clinical Applications

Edited by Francesco M. Veronese

Birkhäuser
Basel · Boston · Berlin

Editors

Francesco M. Veronese
Department of Pharmaceutical Sciences
University of Padova
35131 Padova
Italy

Library of Congress Control Number: 2009928445

Bibliographic information published by Die Deutsche Bibliothek
Die Deutsche Bibliothek lists this publication in the Deutsche Nationalbibliografie;
detailed bibliographic data is available in the Internet at <://dnb.ddb.de>.

ISBN 978-3-7643-8678-8 Birkhäuser Verlag, Basel - Boston - Berlin

© 2009 Birkhäuser Verlag, P.O. Box 133, CH-4010 Basel, Switzerland
Part of Springer Sciebnce+Business Media
Printed on acid-free paper produced from chlorine-free pulp. TCF ∞
Cover illustration: see p. 191. Reproduced with kind permission of Taylor and Francis Group LLC.

Printed in Germany

ISBN 978-3-7643-8678-8   e-ISBN: 978-3-7643-8679-5

9 8 7 6 5 4 3 2 1   www.birkhauser.ch

Contents

List of contributors

Jonathan K. Armstrong, Department of Physiology and Biophysics, Keck School of Medicine, University of Southern California, 1333 San Pablo Street, Los Angeles, California 90033, USA; e-mail: jonathan.armstrong@ usc.edu

Sibu Balan, PolyTherics Ltd. London Bioscience Innovation Centre, 2 Royal College Street, London, NW1 0TU, UK; e-mail: sibu.balan@ polytherics.co.uk

Gian Maria Bonora, Department of Chemical Sciences, Via Giorgieri 1, University of Trieste, 34127 Trieste, Italy; e-mail: bonora@units.it

Steve Brocchini, PolyTherics Ltd. London Bioscience Innovation Centre, 2 Royal College Street, London, NW1 0TU, UK; e-mail: steve.brocchini@ polytherics.co.uk

Penny Bryant, PolyTherics Ltd. London Bioscience Innovation Centre, 2 Royal College Street, London, NW1 0TU, UK; e-mail: penny.bryant@ polytherics.co.uk

Francesca Caboi, Bio-Ker S.r.l, Parco Scientifico e Tecnologico della Sardegna, 09010 Pula, Cagliari, Italy

Elliot K. Chartash, Clinical Development, UCB Inc, Atlanta, GA, USA

Ji-Won Choi, PolyTherics Ltd. London Bioscience Innovation Centre, 2 Royal College Street, London, NW1 0TU, UK; e-mail: ji-won.choi@ polytherics.co.uk

Yuehua Cong, PolyTherics Ltd. London Bioscience Innovation Centre, 2 Royal College Street, London, NW1 0TU, UK; e-mail: yuehua.cong@ polytherics.co.uk

Sara Drioli, Department of Chemical Sciences, Via Giorgieri 1, University of Trieste, 34127 Trieste, Italy; e-mail: sdrioli@units.it

Ruth Duncan, Centre for Polymer Therapeutics, Welsh School of Pharmacy, Redwood Building, King Edward VII Avenue, Cardiff, CF10 3NB, UK; e-mail: duncanr@cf.ac.uk

Victoria Elliott, University of Liverpool, MRC Centre for Drug Safety Science, Department of Pharmacology and Therapeutics, Liverpool. L69 3BX, UK; e-mail: velliott@liverpool.ac.uk

Conan J. Fee, Department of Chemical & Process Engineering, University of Canterbury, Private Bag 4800, Christchurch 8040, New Zealand; e-mail: conan.fee@canterbury.ac.nz

Rory F. Finn, Pfizer Inc, 700 Chesterfield Parkway West, Chesterfield, MO 63017, USA; e-mail: rory.f.finn@pfizer.com

Angelo Fontana, CRIBI, Biotechnology Centre, University of Padua, Viale G. Colombo 3, 35121 Padua, Italy; e-mail: angelo.fontana@unipd.it

Nancy J. Ganson, Duke University Medical Center, Durham, NC 27710, USA

Antony Godwin, PolyTherics Ltd. London Bioscience Innovation Centre, 2 Royal College Street, London, NW1 0TU, UK; e-mail: antony.godwin@ polytherics.co.uk

Mark Hankin, DSRD, Pfizer Global Research and Development, Kent, CT13 9NJ, UK; e-mail: mark.hankin@pfizer.com

Michael S. Hershfield, Box 3049, 418 Sands Building, Duke University Medical Center, Durham, NC 27710, USA; e-mail: msh@biochem. duke.edu

Susan J. Kelly, Duke University Medical Center, Durham, NC 27710, USA

Carlo Maullu, Bio-Ker S.r.l, Parco Scientifico e Tecnologico della Sardegna, 09010 Pula, Cagliari, Italy

Anna Mero, Department of Pharmaceutical Sciences, University of Padua, Via F. Marzolo 5, 35131 Padua, Italy; e-mail: anna.mero@unipd.it

Graham Molineux, Amgen Inc., Mailstop 15-2-A, One Amgen Center Drive, Thousand Oaks, California 91320, USA; e-mail: grahamm@amgen.com

Andrew M. Nesbitt, Inflammation Research, UCB Celltech, 208 Bath Road, Slough SL1 3WE, United Kingdom; e-mail: andrew.nesbitt@ucb.com

Gaetano Orsini, Bio-Ker S.r.l, Parco Scientifico e Tecnologico della Sardegna, 09010 Pula, Cagliari, Italy

B. Kevin Park, University of Liverpool, MRC Centre for Drug Safety Science, Department of Pharmacology and Therapeutics, Liverpool. L69 3BX, UK; e-mail: bkpark@liverpool.ac.uk

Gianfranco Pasut, Department of Pharmaceutical Sciences, University of Padua, Via F. Marzolo 5, 35131 Padua, Italy; e-mail: gianfranco.pasut@ unipd.it

Estera Pawlisz, PolyTherics Ltd. London Bioscience Innovation Centre, 2 Royal College Street, London, NW1 0TU, UK; e-mail: estera.pawlisz@ polytherics.co.uk

Manuchehr Porssa, PolyTherics Ltd. London Bioscience Innovation Centre, 2 Royal College Street, London, NW1 0TU, UK; e-mail: manu.porssa@ polytherics.co.uk

Keith Powell, PolyTherics Ltd. London Bioscience Innovation Centre, 2 Royal College Street, London, NW1 0TU, UK; e-mail: keith.powell@ polytherics.co.uk

Norbert Rumpf, PolyTherics Ltd. London Bioscience Innovation Centre, 2 Royal College Street, London, NW1 0TU, UK; e-mail: norbert.rumpf@ polytherics.co.uk

Mauro Sergi, Ablynx nv, Technologiepark 4, 9052 Zwijnaarde, Belgium; e-mail: mauro.sergi@ablynx.com

Ruchi Singh, PolyTherics Ltd. London Bioscience Innovation Centre, 2 Royal College Street, London, NW1 0TU, UK; e-mail: ruchi.singh@ polytherics.co.uk

Barbara Spolaore, CRIBI, Biotechnology Centre, University of Padua, Viale G. Colombo 3, 35121 Padua, Italy; e-mail: barbara.spolaore@unipd.it

Sue Stephens, Non-Clinical Development, UCB Celltech, Slough SL1 3WE, UK

John S. Sundy, Duke University Medical Center, Durham, NC 27710, USA

Philip Taupin, DSRD, Pfizer Global Research and Development, Kent, CT13 9NJ, UK; e-mail: philip.taupin@pfizer.com

Giancarlo Tonon, Bio-Ker S.r.l, Parco Scientifico e Tecnologico della Sardegna, 09010 Pula, Cagliari, Italy

Francesco M. Veronese, Department of Pharmaceutical Sciences, University of Padua, Via F. Marzolo 5, 35131 Padua, Italy; e-mail: francesco.veronese@unipd.it

Tacey X. Viegas, Serina Therapeutics, Inc., 601 Genome Way, Huntsville, AL 35806, USA; e-mail: tviegas@serinatherapeutics.com

Donald Walker, Pharmacokinetics, Dynamics and Metabolism, Pfizer Global Research and Development, Kent. CT13 9NJ, UK; e-mail: don.walker@pfizer.com

Rob Webster, Pharmacokinetics, Dynamics and Metabolism, Pfizer Global Research and Development, Kent. CT13 9NJ, UK; e-mail: rob.webster@pfizer.com

PEGylated Protein Drugs: Basic Science and Clinical Applications
Edited by F.M. Veronese
© 2009 Birkhäuser Verlag/Switzerland

Preface

PEGylated protein conjugates: A new class of therapeutics for the 21st century

Ruth Duncan[1] and Francesco M. Veronese[2]

[1] *Centre for Polymer Therapeutics, Welsh School of Pharmacy, Redwood Building, King Edward VII Avenue, Cardiff, CF10 3NB, UK*
[2] *Department of Pharmaceutical Sciences, University of Padova, 35131 Padova, Italy*

Introduction

The collected Chapters in this volume describe the current status of poly(ethylene glycol) (PEG) modification of proteins, peptides, oligonucleotides and small molecule drugs, the recent advances in conjugation chemistry, and new clinical products. The book provides an excellent update in this rapidly evolving field, and the comprehensive collection of Chapters complements well past reviews/volumes that have documented the evolution of PEGylation. For example, a reader new to this field is encouraged to gain the historical perspective by reading the following reviews [1–8]. Only then is it possible to see just how far this field has come and understand that it has already established a new class of therapeutics as we start the 21st Century!

In 1990, the Regulatory Authority's approval of the first PEGylated enzymes (PEG-adenosine deaminase; ADAGEN® and PEG-L-asparaginase; ONCASPAR®) was an important landmark. This achievement was the culmination of the pioneering research of Davis, Abuchowski and colleagues in the 1970s that led to the development of these first PEG-enzyme products by Enzon Inc., a company still today contributing important new advances in PEGylation technology. These beginnings, together with the parallel research efforts of a relatively small number of academic groups in the 1980s, gave the credibility to this novel class of drugs, viewed with much scepticism by the pharmaceutical industry at the outset. As with many new ideas, PEGylation was rated as interesting science but impractical to commercialise. How wrong could they be! Today there are thousands of researchers worldwide working in the field and many companies have been founded on the back of this technology. The smaller ones offer speciality PEGs, new conjugation chemistries, and/or they are developing PEGylated liposomes/nanoparticles and PEG-based conjugates of proteins, peptides, oligonucleotides and small molecules as new medicines. Today almost all Pharma sell highly profitable, PEGylated products; for example the two PEG-interferon alpha products and PEG-human-GCSF all have an ~1 billion $US market.

We all know that it is relatively easy now to review the literature-and speculate, although sometimes dangerously as to the likely future directions of a scientific field. Due to the vast wealth of emerging literature, most authors are encouraged to limit their review to those studies published over the last 3–5 years. While this is important, and defines the state of the art, it is also wise to remember the historical evolution of any field, acknowledge its roots, the advances made and the challenges/disappointments encountered. This ensures a realistic starting point for any new developments, avoids repeating mistakes of an earlier generation and allows new technologies to be built on firm foundations, and most rightly gives credit to those who came before [1–8]. It is sometimes too easy to reinvent the wheel on the back of hype! Scientific progress is always evolution and rarely revolution, to quote Einstein "... *my life is based to such a large extent on the work of my fellow human beings, and I am aware of my great indebtedness to them...*" (From 'My Credo', a speech by Albert Einstein to the German League of human Rights, Berlin 1932). This short introduction makes some brief comments relating to the 'recent' historical evolution of the fields of drug targeting and drug delivery, polymers as therapeutics, and the strategic importance of PEG-protein conjugates. These topics are meant to provide a link with the other chapters in the textbook which describe almost all recent progress in chemistry and purification of conjugates, potential issues relating to toxicity and immunogenicity, and also the recent extension of PEGylation strategy to oligonuceotide delivery.

Historical perspective

This year we are celebrating the centenary of Paul Ehrlich's Nobel Prize in Physiology and Medicine (awarded 1908). Ehrlich's vision not only gave important new insights into immunological mechanisms, but he also discovered the first synthetic low molecular weight chemical drug. This was arguably the beginning of drug development as we know it today and medicinal chemistry is still the mainstay of the modern pharmaceutical industry. Moreover, Ehrlich coined the term 'magic bullet', still popular today as an embodiment of the dream of effective disease-specific, targeted therapy. The phrase 'magic bullet' has proved easier to 'say' than achieve in practice. However, it is clear as we enter the 21st Century there is a paradigm shift, both in terms of the changing societal healthcare needs (e.g., increased incidence of diseases relating to the aging population, and emergence of drug resistant infectious diseases), and in parallel, the emergence of exciting new tools that have real potential to help tackle more effectively life-threatening and chronic, debilitating diseases in clinical practice.

Whereas the majority of pharmaceuticals are still natural products or synthetic low molecular weight drugs, the last two decades have seen growing commercialisation of biotech macromolecular therapeutics, particularly antibodies, proteins, peptides and oligonucleotides. The small interfering ribonucleic acids (siRNAs) have most recently entered clinical trials with much anticipation of important new therapeutic benefits. Moreover, genomics and proteomics research is bringing remarkable advances in the understanding of molecular mechanisms of many diseases, which together with the identification of new molecular targets, is leading to an ever-increasing number of biotech drugs. Although these advances have brought many exciting new therapeutic opportunities, it is well acknowledged that effective targeting/delivery of such macromolecular drugs both to diseased cells, and, furthermore, to the particular intracellular compartment they must reach for activity, is very difficult to achieve in practice. The issue of effective drug delivery, and, hopefully, targeting is ever more evident and these challenges are stimulating parallel interest in the design of complementary drug delivery systems (DDS) needed to realise the potential of macromolecular therapeutics.

In the DDS field, the explosion of innovative thinking in the 1970s marked a renaissance period for enabling technologies. A number of distinct classes of DDS appeared that were recently extensively described and reviewed [9]. They included antibody-conjugates, reviewed in [10], liposomes reviewed in [11], nanoparticles reviewed in [12] and polymer–protein [1–8] and polymer-drug conjugates [13, 14]. In these early days, each technology was viewed as competing with the others, and it was naively suggested that one would emerge as the 'best' universal platform for all drug delivery applications. However, clearly each technology has individual advantages and disadvantages [9], and there was increasing realisation that 'the' ideal DDS must be designed on a case-by-case basis, being optimised in respect to the nature of the drug payload to be carried and the specific target for pharmacological action. During the 1980s, a sound biological rational for design of DDS emerged and many modern systems are hybrid, nano-sized technologies, (e.g., PEG-coated liposomes) incorporating multiple components that harness the benefits of several of the original technologies. Moreover, they can be viewed as the 'first generation' nanopharmaceuticals and many have become established clinical products as discussed in [9]. Indeed, the number of Regulatory Authority approved products of this type have grown year on year, and in 2002/2003 the FDA approved more macromolecular drugs and drug delivery systems than small molecules as new medicines [15].

In the context of DDS, it is also important to acknowledge the rapidly rising interest in the application of nanotechnology in medicine [16, 17]. The European Science Foundation's Forward Look in Nanomedicine defined 'nanomedicine' (i.e., nanopharmaceuticals) as "*nanometre size scale systems consisting of at least two components one of which being the active ingredient*". This definition embraces the PEG conjugates as described herein, and the convergence of the basic scientific disciplines relating to 'nano' research is bringing a wealth of new opportunities. For example, to apply existing and new technologies to important emerging clinical challenges, e.g., use of stem cells, and promotion of tissue engineering and repair, design of systems that self-assemble in the patient, and to fabrication of hybrid systems combining DDS technologies and miniaturised devices. Real opportunities exist to design nano-sized, bioresponsive systems able to diagnose and then deliver even macromolecular drugs, so-called theranostics, and to design systems able to promote tissue regeneration and repair in disease, trauma, and during ageing so perhaps in the future it will be possible to circumvent the need for chemotherapy. Although many of the ideas circulating today are still science fiction, it is likely that some facets of 'nanotechnology applied to medicine' will become practical reality within the foreseeable future.

What is increasingly clear, however, is the growing role of natural and synthetic polymers as components of complex DDS, as nanopharmaceuticals and to make nanodevices. Those PEG conjugates described in this volume are nanopharmaceuticals according to the above definition, and they were certainly well ahead of time!

Polymer therapeutics

So, let us begin a brief introduction to 'polymer therapeutics'. In the beginning, the idea of using water-soluble polymers as components of innovative polymer-based therapeutics, particularly for parenteral administration, was viewed by the industry with much scepticism as another totally impractical, scientific curiosity that was much too risky. This was a peculiar stance since natural polymers have been active components of herbal remedies for several millennia, and polymers were widely used as biomedical materials, to fabricate medical devices, as pharmaceutical excipients, and for controlled drug delivery in the form of hydrogels, rate-controlling membranes and biodegradable implants for local delivery. However, it is worthy to remember that the many synthetic polymers we use in society everyday in many different forms (from plastics to computers and mobile phones, to consumer products, etc.) do have a relatively short history. From the outset critics were right to point out that most synthetic and natural polymers are not suitable, and moreover never designed for human administration.

It is sometimes forgotten that the efforts of Hermann Staudinger and his contemporaries led to the birth of polymer science only in the 1920s (post Paul Ehrlich!), and moreover, it was not until 1953 that Staudinger was honoured with the first Nobel Prize for 'polymer chemistry' as reviewed in [18]. Nevertheless, even in these early days, biomedical applications of polymers were envisaged. In the Second World War, synthetic water-soluble polymers were widely adopted as plasma expanders, e.g., poly (vinyl pyrolidone), and large amounts of synthetic polymer were safely administered. This encouraged further exploration of polymers as drugs (e.g., radioprotectants and immunomodulators) and began to underline the potential usefulness of water-soluble, biomedical polymers.

Pioneering work began to emerge in the 1960s and 1970s that lay the foundations for a clearly defined chemical and biological rational for the design of polymeric drugs [13, 19, 20], polymer–protein conjugates [1–8], polymer-drug conjugates [21] and block copolymer micelles [19]. Today, we use the umbrella term 'polymer therapeutics' to include all these classes of polymer-based drugs [13, 14]. From the industrial standpoint, these multicomponent nanosized medicines (typically 5–30 nm) are new chemical entities and macromolecular prodrugs rather than conventional 'drug delivery systems or formulations' which simply entrap, solubilise or control drug release without resorting to chemical conjugation.

There has been a growing realisation that the versatility of synthetic polymer chemistry provides a unique opportunity to tailor synthetic, biomimetic, macromolecular carriers of a specific molecular weight (typically 5,000–100,000 g/mole). Polymer structure can be customised to provide the multi-valency so often needed to promote effective receptor-mediated targeting. Moreover, using the flexibility of dendrimer chemistry, we have a tool kit able to build sophisticated three-dimensional architecture into the structure of synthetic macromolecules, as reviewed in [22], and this is increasingly being built into PEG chemistry via use of branched or dendronised PEGs. Importantly, the linking chemistries used for polymer conjugation have been refined over the years such to enable creation of macromolecular prodrugs (e.g., containing drugs, proteins, oligonucleotides) that are able to display sophisticated rate control and site-specific release of the bioactive moiety. The polymer therapeutics are, still today, often misreported as a rather minor contribution to the therapeutic armoury. This is largely because over the years large companies have made a very small investment in this area compared to biotech and medicinal chemistry/high throughput screening. However, review of the current polymer therapeutics market size (>5 billion US$) compared to antibodies (>17 billion US$) show just how wrong this conclusion is, especially taking in account the disparity in the relative historical economic investment in the two fields!

PEG conjugates

So within this complex landscape of drug delivery and polymers, how best can one summarise the current and future contribution of PEGylation? At the outset [4], PEGylation was developed as a tool to improve delivery of protein drugs and rectify their shortcomings. For example, proteins and peptides can have a short plasma half-life, poor stability, poor formulation properties and they can be immunogenic. Although other polymers, such as dextran, had been explored to address these shortcomings, PEG was initially chosen as the polymer for protein modification as it was already used as 'safe' in body-care products and approved for use as excipient in many pharmaceutical formulations. As a further advantage, it could be synthesised to have a molecular weight of narrow polydispersity and also to have one terminal functional group making it ideal for protein modification without risk of crosslinking. Moreover, this highly hydrated polymer chain makes it theoretically ideal to 'mask' sites responsible for the immunogenicity of proteins to which it was bound. 30 years later, PEGylation is now a well-established tool able to address the limitations of proteins, peptides and oligonucleotides and, in addition a number of PEG-drug conjugates have been tested clinically for both parenteral and oral administration.

Undoubtedly, the Regulatory Authority approval of the first PEG-enzyme conjugates, ADAGEN® and ONCASPAR®, in the 1990s was a significant breakthrough. Indeed, this proof of concept immediately gave credibility to all the emerging classes of polymer therapeutics as a whole. However, although ADAGEN® and ONCASPAR® were important first products, they achieved limited clinical use and only a niche market; particularly ADAGEN®, which is used to treat severe combined immunodeficiency syndrome, a rare disease with few patients worldwide, and a disease that has more recently been treated with mixed success by gene therapy. Nevertheless, these beginnings paved the way for the subsequent application of PEGylation to cytokines such as the interferons (PEG-Intron® and PEGASYS®, (see the chapter by Pasut in this book), which have been successfully used to treat hepatitis C, and a granulocyte colony-stimulating factor (Neulasta®, see chapters by Molinex and by Sergi et al.) used as an adjuvant to repair the effects of neutropenia-inducing chemotherapy. These innovative medicines achieved significant therapeutic benefit, improved patient convenience as they need less frequent dosing compared to the free-protein drug, and achieved considerable economic success and they are now featured in the top marketed drugs lists. Recent Regulatory Approval of the PEG-aptamer Macugen® as a treatment for age-related macular degeneration, (reviewed in [23]), the PEG-anti-TNF antibody Fab' fragment (Cimzia®) for treatment of Crohn's disease, (reviewed in [24] and by Nesbitt et al.) also in clinical development for arthritis, as well as the suggestion to use the enzyme urate oxidase for the refractory gout treatment uricase (chapter by Hershfield et al.), are all showing a move towards application of PEG conjugates in the treatment of chronic diseases. It is important to note that such conjugates have not only therapeutic and formulation advantages, but also the potential to be cost-effective and even cost saving [25, 26].

Evolution of PEGylation chemistry over the last 30 years has been well documented [5–8]. Instrumental to the continuing success of the now emerging products has been the increasing degree of sophistication of the conjugation chemistry and methodology developed for product isolation and characterisation, as described and reviewed in detail by Fee. The first PEGylated enzymes contained multiple PEG chains per protein, whereas now a number of conjugation approaches (chemical and enzymatic also described herein by Bonora/Drioli, Sergi et al. and by Fontana et al.), combined with recombinant protein technology, can ensure 1:1 (polymer: protein) site-specific conjugation. The PEGs used vary in molecular weight from low (~3–5,000 g/mole) to high (20–40,000 g/mole) molecular weight chains and both linear and branched PEGs are now being used (see the chapter by Veronese et al. for properties and limitations description). As PEG is not biodegradable, the use of high molecular weight PEGs and chronic administration of all molecular weights of PEG raise questions about fate and long term safety (see the chapter by Webster et al. on toxicity and the chapter by Armstrong on PEG immunogenicity) that may have regulatory implications in the future depending on proposed conjugate use, dose, frequency of dosing and whether the treatment is for an acute or a chronic disease [27, 28]. Additional chapters deal with the use of PEGylation for the improvement of anticancer drug therapy (see chapter by Mero et al.) and of acromegaly (chapter by Finn). As for all polymer therapeutics, a sound biological rational for design has always been applied to PEG-proteins and it has evolved with time as more has become known of the structure activity relationships in respect to the effect of PEG molecular weight and branching on the pharmacokinetic-pharmacodynamic profile.

As more and more polymer therapeutics are being developed, there is a need to continuously review and consider new Regulatory Guidelines for their approval (see the chapter by Viegas and Veronese).

The future?

It should not be forgotten that it was only the turn of the last century when Paul Ehrlich proposed the first synthetic small molecules as chemotherapy and Hermann Staudinger was suggesting that small molecules, monomer units, might be covalently linked to give us polymer chains! Who could have predicted the plastics revolution that followed?

Introduction of the first biotechnology and polymer-based products over the last two decades of the 20th Century was greeted with the same suspicion that Ehrlich encountered when introducing modern chemotherapy in his day. Things are now rapidly moving on. PEGylated proteins are now well established as therapeutics and PEGylated peptides are gaining momentum. Will they be the mainstay of therapy for all diseases within this Century? Probably not, but it seems certain that as we start the 21st Century we are entering a therapeutic era where low molecular weight chemotherapy, macromolecular drugs, including, antibodies, peptides and proteins, polymer therapeutics, and oligonucleotides and cell therapy will all play an important and complementary role in the prevention, control and cure of diseases. It is rapidly becoming apparent that the future is combination therapy. Many of the PEG conjugates already marketed and those in clinical development will increasingly be used in combination with small molecular chemotherapy and/or any of these new classes of therapeutic/nanopharmaceuticals to ensure successful treatment of complex pathologies. This itself will bring new healthcare challenges including treatment cost, the need to foresee and minimise potential new contraindications and/or drug–drug interactions. There is still much interesting/vital research remaining to be done.

To conclude, thanks to the efforts of a relatively small community (academic and industrial), PEGylation and PEG-proteins as polymer therapeutics are already well established. The recent progress documented in this volume shows that there is more, much more, yet to come and that this is just the beginning!

References

1 Fuertges F, Abuchowski A (1990) The clinical efficacy of poly(ethyleneglycol)-modified proteins. *J Controlled Rel* 11: 139–148
2 Nucci ML, Shorr R, Abuchowski A (1991) The therapeutic values of poly(ethylene glycol)-modified proteins. *Adv Drug Delivery Rev* 6: 133–151
3 Francis GE, Delgado C, Fisher D, Malik F, Argrawl AK (1996) Polyethylene glycol modification: Relevance to improved methodology to tumour targeting. *J Drug Targeting* 3: 321–340
4 Davis FF (2002) The origin of pegnology. *Adv Drug Del Rev* 54: 457–458
5 Harris JM, Chess RB (2003) Effect of pegylation on pharmaceuticals. *Nature Rev Drug Discov* 2: 214–221
6 Veronese FM, Harris JM (Eds) (2002) Introduction and overview of peptide and protein pegylation. *Adv Drug Deliv Rev* 54: 453–609
7 Harris JM, Veronese FM (Eds) Pegylation of peptides and proteins II – Clinical Evaluation. *Adv Drug Deliv Rev* 55: 1261–1277
8 Veronese FM, Harris JM (Eds) (2008) Pegylation of peptides and proteins III: Advances in chemistry and clinical applications. *Adv Drug Deliv Rev* 60: 1–87
9 Duncan R (2005) Targeting and intracellular delivery of drugs. In: RA Meyers (ed.): *Encyclopedia of Molecular Cell Biology and Molecular Medicine*. Wiley-VCH Verlag, GmbH & Co. KGaA, Weinheim, Germany, 163–204
10 Allen TM (2002) Ligand-targeted therapeutics in anticancer therapy. *Nature Rev Drug Discov* 2: 750–763
11 Torchilin VP (2005) Recent advances with liposomes as pharmaceutical carriers. *Nature Rev Drug Discov* 4: 145–160
12 Couvreur P, Vauthier C (2006) Nanotechnology: Intelligent design to treat complex disease. *Pharm Res* 23: 1417–1450
13 Duncan R (2003) The dawning era of polymer therapeutics. *Nat Rev Drug Discov* 2: 347–360
14 Duncan R (2006) Polymer conjugates as anticancer nanomedicines. *Nat Rev Cancer* 6: 688–701
15 US Food and Drug Administration accessed at http://www.accessdata.fda.gov/scripts/cder/drugsatfda/
16 Ferrari M (2005) Cancer nanotechnology: Opportunities and challenges. *Nature Rev Cancer* 5: 161–171
17 European Science Foundation Forward Look on Nanomedicine (2005) http://www.esf.org
18 Ringsdorf H (2004) Hermann Staudinger and the Future of Polymer Research: Jubilees – Beloved Occasions for Cultural Piety. *Angew Chem Int Ed* 43: 1064–1076
19 Gros L, Ringsdorf H, Schupp H (1981) Polymeric antitumour agents on a molecular and cellular level. *Angew Chemie Int Ed Eng* 20: 301–323
20 Regelson W, Parker G (1986) The routinization of intraperitoneal (intracavitary) chemotherapy and immunotherapy. *Cancer Invest* 4: 29–42
21 Ringsdorf H (1975) Structure and properties of pharmacologically active polymers. *J Polymer Sci Polymer Symp* 51: 135–153
22 Lee CC, MacKay JA, Fréchet JMJ, Szoka FC (2005) Designing dendrimers for biological applications. *Nature Biotechnol* 23: 1517–1526
23 Ng EWM, Shima DT, Calias P, Cunningham Jr ET, Guyer DR, Adamis AP (2006) Pegaptanib, a targeted anti-VEGF aptamer for ocular vascular disease. *Nature Rev Drug Discov* 5:125–132
24 Sandborn WJ, Colombel JF, Enns R, Feagan BG, Hanauer SB, Lawrance IC, Panaccione R,

Sanders M, Schreiber S, Targan S, Deventer SV, Goldblum R, Despain D, Hogge GS, Rutgeerts P (2005) Natalizumab induction and maintenance therapy for Crohn's disease *N Engl J Med* 353(18): 1912–1935

25 Eldar-Lissai A, Cosler LE, Culakova E, Lyman GH (2008) Economic analysis of prophylactic peg-filgrastim in adult cancer patients receiving chemotherapy. *Value Health* 11: 172–179

26 Gerkens S, Nechelput M, Annemans L, Peraux B, Beguin C, Horsmans Y (2007) A health economic model to assess the cost-effectiveness of pegylated interferon alpha-2a and ribavirin in patients with moderate chronic hepatitis C and persistently normal alanine aminotransferase levels. *Acta Gastroenterol Belg* 70: 177–187

27 Eaton M (2007) Nanomedicine: industry-wise research, *Nature Mater* 6: 251–253

28 Gaspar R (2007) Regulatory issues surrounding nanomedicines: setting the scene for the next generation of nanopharmaceuticals. *Nanomedicine* 2: 143–147

Protein PEGylation, basic science and biological applications

Francesco M. Veronese, Anna Mero and Gianfranco Pasut

Department of Pharmaceutical Sciences, University of Padua, Via F. Marzolo 5, 35131 Padua, Italy

Abstract

A historical overview of protein-polymer conjugation is reported here, demonstrating the superiority of poly(ethylene glycol) (PEG) among other synthetic or natural polymers, thanks to its unique properties like the absence of toxicity and immunogenicity, and a high solubility in water and in organic solvents. Furthermore, PEG is approved by the FDA for human use. Relevant physicochemical and biological properties of PEG and PEG-conjugates, as the basis of the pharmacokinetic and pharmacodynamic improvements, are reported here and discussed in view of successful therapeutic applications. The chapter also highlights that, although PEGylation is well studied and exploited by many researchers from both academia and industry, it remains difficult to forecast its effects on a predetermined bioactive molecule. The use of PEG-enzymes in bioconversion, which is of interest in drug discovery and production, is also briefly reported.

Historical overview of protein-polymer conjugation

The discovery of PEGylation in the 1970s as a strategy to overcome the problems of administration of therapeutic proteins was neither a fortuitous occurrence, nor the result of careful laboratory investigations, but the result of few months of library work on biochemistry and polymer chemistry. This is what J. Davies, the discoverer of PEGylation, reported in a "commentary" to the 2002 ADDR issue dedicated to PEG [1]. He concluded that the hydrophilic polymer link could reduce immunogenicity and increase the half-life of conjugated proteins *in vivo*. However, the great challenge was to find a safe polymer for a general use. PEG, a hydrophilic polymer easy to obtain in large quantities, was already being used in industry for numerous applications including, 1) an additive for paper production, 2) for controlling the viscosity of printing ink, 3) in biology as a precipitating agent for proteins, and 4) as an inducing agent for cell fusion. Thanks to its low or non-toxic properties it has also been used as a food additive and drug excipient [2]. In all these applications, PEG was used in its diol form, but the availability of the methoxy-PEG, (mPEG) with only one terminal hydroxyl group, attracted Davis' interest since he had seen the possibility of preventing formation of dimers or, more critically, preventing cross-linked forms of proteins using mPEG once the PEG hydroxyl groups were activated for protein conjugation.

The literature shows that several natural or synthetic polymers also share properties with PEG such as hydrophilicity, no toxicity and reactivity with proteins. Many polymers have been proposed to improve the therapeutic application of proteins, peptide or simple non-peptide drugs, but for various reasons none have showed the efficacy of PEG. A few polymers obtained by radical polymerization of suitable acrylic monomers were widely studied, among these poly(N-vinyl pyrrolidone) [3, 4], poly(N-acryloyl morpholine) [5], poly(vinyl alcohol) and succinic acid maleic acid anhydride copolymer [6] and the combination of an acrylic backbone and PEG pendants (see PolyPEG®). Up to now, only the last has been used in therapy as a conjugate with neocarcinostatin, a small protein with anticancer activity. Unfortunately, all these polymers share a common problem: great polydispersity which is due to the chemistry of polymerization. Hopefully, this limitation will be overcome by the improved polymer synthesis methods recently developed [7]. An alternative promising polymer is poly(oxazoline) that, although of quite a different structure as compared to PEG, shares some useful properties: it may be obtained with low polydispersity thanks to the easily controlled anionic polymerization, it is also amphiphylic and it may be obtained with only one reactive terminal group [8]. Biologically active long-lasting conjugates have been obtained with model enzymes as well as with small non-peptide drugs [8–10]. Polysaccharides were also used for conjugation and one drug in Russia, Streptodekase®, reached therapeutic application by conjugation of dextran with streptokinase [11]. This product represented a milestone in the therapeutic use of protein conjugates. However, the method, based on a partial random oxidation of the carbohydrate moieties to yield reactive carbonyl groups, was very crude because it gave rise to heterogeneous and cross-linked products. Thanks to improved sugar chemistry, specific methods of single point activation in the polysaccharide chain now allow more defined conjugates with proteins [12–14]. Enzymatic methods for polysaccharide coupling have also been developed which, together with the development of genetic engineering, have yielded new glycosylated or hyperglycosylated proteins. These methods when applied to different model proteins (e.g., enzymes, cytokines and antibodies) lead to an increased retention time in the blood, a decrease in immunogenicity along with a desired minimal loss of biological activity [15–18]. One such hypersialylated protein has already successfully reached the market (Aranesp®). Globular proteins were also investigated for polymer conjugation. The most studied protein was human serum albumin which was initially randomly conjugated using cross-linking reagents that gave heterogeneous although biologically active, long-lasting and less immunogenic products [19]. Later, a more specific conjugation was proposed that took advantage of the lone free thiol residue of human albumin [20].

There has been a loss of interest in many of these non-PEG modification strategies while some still await a successful clinical application. On the other hand, PEGylation has seen a continual development that has never ceased since 1977, when the first two papers on this technique by Davis and

Abuchowski were published [21, 22]. This is also due to a great deal of research conducted by the pharmaceutical industry which has evaluated PEGylation as a possible solution to shortcomings encountered by proteins of potential therapeutic interest [23].

A synthetic historical overview of PEGylation, as reported in Table 1, demonstrates that new results and applications have always paralleled developments in PEG chemistry. Presently, there are eight marketed PEG-proteins and one PEG-aptamer (Tab. 2).

Table 1. History of PEGylation

Decade	PEGs	Observation	Applications
1970–1980	PEG-chlrotriazine PEG-succinimidyl-succinate PEG-tresil	Immunogenic or toxic starting material, highly polydispersed PEG, lack of selectivity	Research studies, enzyme modification for biocatalysts and application on protein therapeutics
1980–1990	PEG-aldehyde PEG- succinimidyl carbonate PEG-p-nitro-phenyl carbonate PEG-AA-NHS PEG-carbonyl-imidazole, etc.	Site-specific conjugation, less polydisperse PEG, absence of diols	Enzyme replacement therapy
1990–2000	Branched PEG PEG-NHS PEG-maleimide PEG-OPSS	Improved selectivity, marketing of PEGylated drug	Cytokines, hormones, anticancer drugs targeting
2000 on	Releasable PEGs Heterobifunctional PEGs Forked PEGs Star PEGs Monodisperse PEGs	Detailed chemical and biological characterization of conjugates, combination of genetic engineering and PEGylation in the design and discovery of new drugs, more stringent regulatory requirements. Developments of enzymatic methods of coupling.	Non-protein drugs PEGylation, oligonucleotide PEGylation Seven PEG-Protein drugs and one PEG-Aptamer on the market.

In this chapter, basic properties of PEG and important aspects of PEGylation will be described. This chapter may therefore be regarded as an introduction to the following chapters of the book that describe more specific applications of the PEGylation. Other PEG applications, not directly related to therapeutic uses but still important for drug development, will be also reported.

Table 2. PEGylated proteins or oligonucleotides, FDA approved or in advanced clinical trials

Brand	Generic name	Active substance	Indication	Approval year
Adagen®	Pegadamase	Adenosine Deaminase	SCID	1990
Oncaspar®	Pegaspargase	Asparaginase	Leukemia	1994
Neulasta®	Pegfilgrastim	G-CSF	Neutropenia	2002
PEG-INTRON®	Peginterferon-α2b	Interferon-α2b	Hepatitis C	2000
PEGASYS®	Peginterferon-α2a	Interferon-α2a	Hepatitis C	2001
Somavert®	Pegvisomant	Growth hormone antagonist	Acromegaly	2003
Macugen®	Pegaptanib	Anti-VEGF aptamer	ADM	2004
Mircera®	PEG-EPO	EPO	Anemia associated with chronic kidney disease	2007 (Europe)
Cimzia®	Certolizumab pegol	Anti-TNF Fab'	Rheumatoid arthritis and Crohn's disease	Expected 2008

Cimzia did get approval in April 2008.
ADM, age-related macular degeneration; EPO, erythropoietin; G-CSF, granulocyte-colony stimulating factor; IFN, interferon; SCID, severe combined immunodeficiency disease; TNF, tumor necrosis factor; VEGF, vascular endothelial growth factor

A number of PEG and PEG conjugate reviews have already been published over the years. The reader may refer to two specific books [2, 24] and three recent collections of reviews [25–27].

PEG physicochemical properties and availability

The repeated ethylene oxide units along the PEG chain convey unique properties to this polymer: the ethylene moiety confers hydrophobicity, while the oxygen allows strong interactions with water. The polymer is therefore very soluble in both water and in many organic solvents. Furthermore, the carbon–carbon and carbon–oxygen bonds give great flexibility to the overall structure and allow repulsion of incoming molecules (Fig. 1).

For several years Shearwater Polymer Inc. was the only commercial source of activated PEGs at a high degree of purity, devoid of diols and with low polydispersity, but the great success of PEGylation prompted the more recent development of several new dedicated producers. The characteristics of PEG depend on its molecular weight and chain shape. Very low molecular weight PEGs < 400 Da are oils but at ~1.5–2 kDa PEG has a waxy appearance. PEG is a solid at higher molecular weight, provided that it is maintained dried.

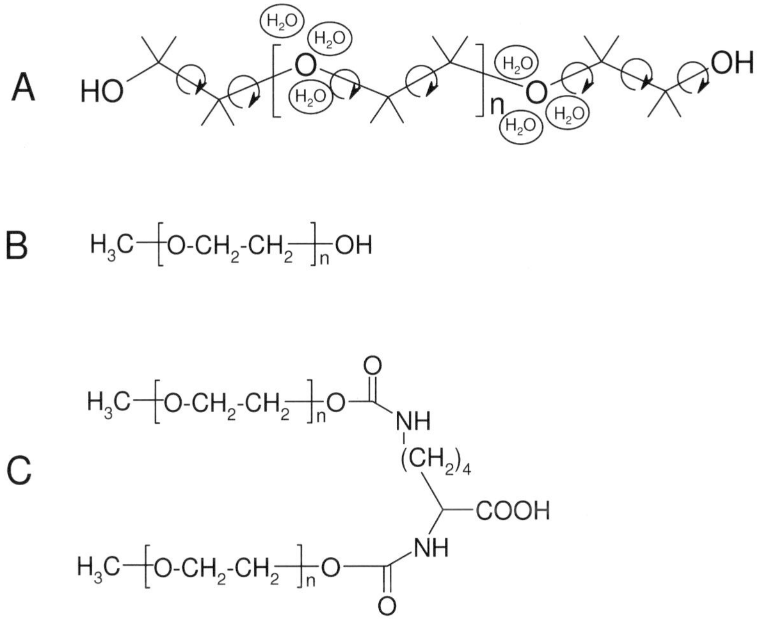

Figure 1. Structure of PEG to show its A) flexibility and hydratation, B) linear methoxy-PEG structure and C) branched PEG structure (PEG2)

Storage under an inert atmosphere is recommended because, although stable towards several chemical reagents, PEG is sensitive to oxidation that may cleave the chain. As with all the synthetic polymers, PEG is polydisperse, the Mw/Mn value is about 1.01 for polymers with molecular weight ranging from 2–10 kDa, while reaching values up to 1.2 for higher molecular weight polymers.

The anionic polymerization for the synthesis of PEG leads to chains with one or two hydroxyl groups at the ends, in the case of methoxy or diol PEG, respectively. These groups must be properly activated to obtain a PEGylating agent suitable for protein conjugation. A variety of reactive PEGs with different molecular weights are commercially available for conjugation to all of the reactive amino and thiol residues found in proteins (see chapter by Bonora and Drioli in this book). Enzymatic methods of PEGylation have also been proposed and are opening a new field of study. Relevant examples are based on transglutaminase (TGase), which catalyzes PEG coupling to glutamine residues [28, 29], or on a double enzyme system that promotes the transfer of a sialic acid PEG to a residue of O-GalNAc which had previously been enzymatically coupled to a serine or a threonine amino acid in an [30] (see also the chapter by Sergi et al. in this book).

Monodisperse PEG has recently become commercially available, but, unfortunately, so far at low molecular weights only, between 500–800 Da [31].

A monodisperse high molecular weight PEG would be very welcome to overcome the subtle differences in biological properties of polydisperse conjugates, but its synthesis and purification using current technologies would be too difficult and expensive for a commercial product. An additional advantage of monodisperse PEGs would be simplification of the analytical problems of conjugate characterization, because electrospray mass spectroscopy could be routinely employed. Currently, MALDI analysis is routinely utilized with polydisperse PEG. A recent enabling application of a monodisperse polymer was the localization of the PEGylation site in G-CSF that had been conjugated by TGase [32].

For many years, only the linear form of the polymer was used, but more recently a branched form, called PEG2, was proposed and has had great success [33, 34] as demonstrated by its use in three commercial drug conjugates, two proteins, α-interferon [35] (see also chapter by Pasut in this book) and anti-TNF-receptor [36] (also see Nesbit chapter of this book) and an anti-VEGF aptamer [37]. The advantage of this special form of PEG resides in the fact that it covers a larger surface of a protein involving only one amino acid residue in the conjugation. Furthermore, branched PEGs may slowly release one of the two PEG lysine linked chains, by cleavage of the carbamate linkage between polymer chains and the branching unit, thus helping its clearance from the body [38].

Advantages of PEGylation

The above reported features of PEG, namely high hydration and flexibility, form the basis of several advantages that PEGylation can attain by polymer coupling to proteins, drugs or surfaces. A short list of these advantages includes: a) the increase in hydrodynamic volume conferred to conjugated molecules, thus reducing their kidney excretion and prolonging *in vivo* half-life, b) the protection of amino acid sequences sensitive to chemical degradation, c) the masking of critical sites sensitive to metabolic enzyme degradation or to antibody recognition, d) the possibility to solubilize proteins in organic solvents allowing new enzyme applications as biocatalysts, e) the solubilization of water insoluble drugs in a physiological medium, f) the reduction of either protein opsonization of liposomes, microparticles or protein adhesion of surfaces coated with PEG, thus increasing their biocompatibility, g) the reduction of protein aggregation. All of these properties conferred by PEGs may have a role in the therapeutic application of proteins and in drug discovery and development.

No studies have been found indicating major disruption of protein conformation using NMR or circular dichroism (CD). On the one hand, the polymer chains in a PEG-protein conjugate prevent the interaction of the protein surface with incoming molecules by steric effects but on the other hand, PEGs also can have direct stabilizing influence on protein conformation. So far, this aspect has not been investigated in mechanistic detail, but some information can be obtained from the studies done on protein glycosylation. In fact PEG and polysaccharides share a common hydrophilic character and preliminary studies carried out on PEG and glycan conjugates, obtained with the same coupling method, demonstrated that PEGylation and glycosylation have a similar effect on proteins [39]. Detailed structural investigations by CD, infrared spectroscopy, and hydrogen/deuterium exchange demonstrated that glycan conjugation increased the rigidity and stability of a protein native structure by increasing electrostatic and Wan der Walls interactions. These results are in agreement with the observation that a high degree of glycosylation reduced the protein B factor, an indication of protein chain flexibility [14, 15, 39]. Recent studies demonstrated that this stabilization also holds for PEGylation, although small differences in rigidity were found which were ascribed to the different structures of the two polymers: PEG being more compact than glycans. Another example on the effects of PEG on protein conformation can be observed in the case N-terminal PEGylation with PEG 20 kDa of brain-derived neurotrophic factor (BDNF). This protein under physiological conditions is a non-covalent dimer while in salt free formulations is present as unstable monomer that undergoes cleavage and aggregation. Unexpectedly it was observed that the rate of protein degradation is accelerated in the PEGylated form. A proper preformulation study demonstrated that when 150 mM of sodium chloride was incorporated into the formulation, improved conformational and thermodynamic stability of both BDNF and PEG-BDNF was achieved [40].

This increased structural stability of PEGylated proteins is of paramount importance for therapeutic applications, ensuring protein stability during drug formulation, storage and *in vivo* circulation. In fact, proteins may undergo modifications by various mechanisms including chemical reactions, such as oxidation, denaturation, hydrolysis, disulfide exchange or conformational changes, resulting in aggregation. These modifications, which in most cases lead to the loss of biological activity, immunogenicity or increased toxicity, are significantly reduced in a more compact PEGylated protein [41].

Effect of PEGylation on absorption, transport, elimination and activity

Most proteins of pharmaceutical interest, such as enzymes, cytokines and hormones, possess a molecular weight between 15–30 kDa. Also falling near this range are monoclonal antibody fragments, which seem exceptionally promising as therapeutics thanks to the reduced risk of immunogenicity and the ability for enhanced tissue mobility when compared to full length antibodies [23, 42]. Almost all proteins within this size range have the shortcoming of a fast *in vivo* clearance that can hamper their therapeutic exploitation. For these biologically active agents, PEGylation represents a suitable solution to increase *in vivo* residence. It is clear that polymer size, amount of PEG coupled and sites of binding on a protein have a role in determining the fate of the conjugate in the body and the rate of excretion (see also the Pasut chapter in this book). Therefore, understanding the *in vivo* behavior of PEG, in its free or conjugated form, is of basic relevance for designing a PEGylation strategy.

It is known that proteins having molecular weights below the kidney filtration threshold, about 60 kDa, are mainly excreted into the urine. However, PEG cannot be compared to a globular protein when related to kidney excretion because this linear and flexible polymer, with a random coil conformation, may cross barriers by a 'reptation' mechanism [43, 44]. Furthermore, each oxyethylene PEG unit is able to coordinate 3–5 water molecules, thus increasing the polymer hydrodynamic volume by an approximate 5–10 fold greater amount than that predicted by the nominal molecular weight, see Figure 1a [24].

Although there is evidence of limited *in vivo* chain degradation for very small PEGs by alcohol dehydrogenase [45], aldehyde dehydrogenase [46] and cytochrome P-450 [47], PEG is considered a non-biodegradable polymer, and for human use it is commonly used at molecular weights below its kidney clearance threshold. So far, the highest PEG molecular weight employed for a conjugate approved for human therapy (i.e., Pegasys®) is 40 kDa. In fact, the threshold for an easy kidney filtration is about 40–60 kDa (a hydrodynamic radius of approximately 45 Å [48]) and over this limit the polymer remains in circulation for longer periods of time and accumulates in the liver. Of interest is the sigmoidal relationship between PEG molecular weights and their *in vivo* half-lives that fits the theoretical models of renal excretion of macromolecules based on the pore sizes of the glomerular capillary wall, in this case with a marked increase of circulation times in the range of 20–30 kDa [49, 50]. This behavior implies that the influence of PEG on the protein conjugates half-lives is not easily predicted and several case by case studies were therefore carried out to determine the half-life increases [51]. Manjula and co-workers reported the effect of different PEG sizes in the case of hemoglobin's (Hb) hydrodynamic volume and conjugate radius. The authors demonstrated that the molecular size of the protein was significantly enhanced after covalent attachment of PEG and exhibited a fairly linear relation with the mass of linked PEG chain. Hb linked with two chains of PEG 10 kDa (total calculated mass 84 kDa) or with two chains of PEG 20 kDa (104 kDa) exhibited a hydrodynamic volume of 712 and 1,436 nm³, respectively, while octamer or dodecamer forms of proteins, without coupled PEG chains, having a molecular weight of 128 and 192 kDa exhibited a hydrodynamic volume three-fold lower than that of the PEGylated proteins [52]. Therefore, oligomerization can also increase the protein half-life by molecular mass augmentation but it is not as effective in the protection against proteolytic enzymes as PEGylation which can effectively mask sensitive amino acid sequences (Fig. 2). Furthermore, computer modeling investigations of PEGylated proteins suggested that PEG chains of 5 and 10 kDa are distributed all around the proteins surface reduc-

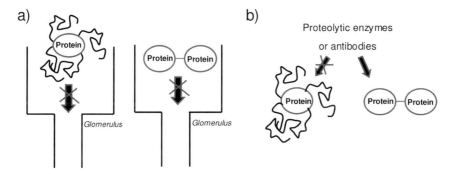

Figure 2. Different behavior of PEGylation and oligomerization of proteins towards kidney excretion and stability; a) both PEGylation and oligomerization increase the hydrodynamic volume of the proteins and decrease the kidney excretion; b) PEG shields the protein surface and masks sensitive sites recognized by proteolytic enzymes and antibodies, whereas oligomerization does not.
The n is chopped off protein in the circles.

ing the protein's exposure, while the 20 kDa polymer may fold up with itself, conferring less protection [52]. Other physical properties influenced by PEG size were studied by comparing the viscosity of different PEG-Hb conjugates: it was found that the hydrodynamic volume of conjugates increased almost linearly with molecular weight of PEG chains while viscosity and colloidal osmotic pressure exhibited an exponential increment with the length of PEG chain [52]. In addition, studies conducted by Fee and co-workers [53] demonstrated that the final molecular size of each PEG-protein specie is determined by the sum of the native globular protein size and the total size amount of conjugated PEG, rather than by the particular PEG molecular weight used or the degree of grafting. Researchers observed that α-lactalbumin linked to a single chain of PEG 20 kDa exhibited a hydrodynamic volume of 53 nm^3, which is very close to the hydrodynamic volume of the same protein linked with four chains of PEG 5 kDa [53]. Further studies by the same authors demonstrated that a protein conjugated with a comparable mass of linear PEGs yields conjugates with similar hydrodynamic volumes but a lower half-life *in vivo* when compared with conjugates obtained using a branched PEG of the same weight [54]. The authors speculated that the polymers form a dynamic, hydrated polymer layer at the protein surface, and the characteristic "umbrella-like" structure of branched polymers (Fig. 1c) masks the protein surface better than the linear form, conferring higher resistance to proteolytic enzymes, antibodies or immunocompetent cells. Indeed, the most recent conjugates that have reached the commercial market contain higher molecular weight PEGs and are coupled site specifically at thiol or are mono PEGylated at lysines, while Adagen® or Oncaspar®, which were first marketed at the beginning of 1990s, were prepared by random conjugation of multiple strands of short 5 kDa PEG. So far, the branched PEG2 has lead to three successful drugs, PEGASYS®, Cimzia® and Macugen®. This PEG allows for a double mass of bound polymers for each site of attachment, thus better improving both pharmacokinetic and pharmacodynamic profiles. These advantages are being pursued with recent conjugates while previous PEGylation aimed at simply extending the half-life of the drug or reducing its adverse reactions, without optimizing the potency.

The most detailed investigation to reach an optimum balance between PD and PK is represented by PEG-interferon α-2a development, where the protein was linked to a single chain of branched 40 kDa PEG resulting in a conjugate that retains only 7% of the native protein activity, but a half-life approximately 11-fold greater [55, 56]. Hence, this favorable balance allowed a weekly dose schedule of the PEGylated form with almost constant drug levels yielding higher rates of viral eradication than the parent interferon α-2a administered three times a week (see also the chapter by Pasut in this book).

Cox and co-workers studied a method to selectively modify human growth hormone (hGH) [57]. The protein was genetically modified by site directed mutagenesis that introduced an unpaired free cysteine into a region not involved in receptor binding. The cysteine was in turn modified with a 20 kDa PEG that was selectively reactive towards thiols. The *in vitro* and *in vivo* properties of mono-PEGylated-hGH were compared with those of a multi-PEGylated-hGH previously described [58]. In the multiply PEGylated protein, eight out of nine lysines and the N-terminal amino acid were modified to various degrees using an activated ester 5 kDa PEG leading to a complex mixture of conjugate isomers. The mixture was composed of multiple PEGylated species containing from 2–7 coupled PEG chains, and each isomer had different values of *in vivo* potency. As expected, the heavy PEGylated species showed reduced or no activity *in vitro* while the monoPEGylated-hGH exhibited an activity 100-fold greater. The mono 20 kDa PEG-hGH conjugate, obtained by genetic engineering, showed an *in vivo* half-life comparable to that of multi-PEGylated hGH, which contained an average of 5–6 PEG chains per protein.

A similar study has been reported for erythropoietin (Epo) [59]. It was found that lysine modification with amine-reactive PEGylating agents leads to bioactivity reduction, whereas attaching a single PEG of 20 kDa to an Epo mutein, in which a cysteine was inserted far from regions important for protein stability and bioactivity, has seven- to eight-fold improved residence in blood with almost complete *in vitro* bioactivity retention. However, the approach to design protein variants containing one free cysteine by genetic engineering is not always optimal because in some case the mutated protein can be less active, unstable, and prone to misfolding or aggregation. An example is IFN-β-1b, where the thiol PEGylation of its mutated forms was particularly challenging due to the instability of the native cysteine bridge under the reaction conditions. In the same work, the authors also studied the expression of muteins where some lysines were depleted. This approach is useful because it can reduce the number of multi-PEGylated isomers when an amino reactive PEG is used. Unfortunately, complications with maintaining the disulfide bonds under conditions favorable for efficient PEGylation preclud-ed the utility of site selective thiol PEGylation. The lysine depleted mutein completely lost antiviral activity or maintained only an unacceptably low residual activity ($<1\%$), even though the PEGylation was far from the active site [60].

Other factors must be taken into consideration as demonstrated by Bowen and co-workers who found that the *in vivo* activities of PEG-NTG-conjugates (granulocyte colony-stimulating factor mutein) increased by increasing the molecular weight of attached PEG [61]. The authors speculated that a lower receptor affinity of these conjugates could also positively affect their *in vivo* half-life, because receptor-mediated endocytosis by mature granulocytes is an important mechanism regulating the levels of hematopoietic growth factors. Actually, there are only a few reported instances of biological activity increase after PEGylation, as in the case of PEG-enzymes where the enhanced activity was ascribed to a positive influence of the polymer in substrate binding [62, 63].

Conjugate size and the amount of linked PEG also have roles in the diffu-sion through tissues, reflected in the volume of distribution and conjugate half-life. These are relevant parameters in choosing the proper administration route for each conjugate. In fact, *i.m.* or *s.c.* administrations may slow down absorption, diffusion and finally the PEGylated drug reaching circulation in blood. In these cases, conjugates may act as a depot, but note also that protein degradation may be favored under these circumstances. For this reason *i.v.* may be the preferred administration route of PEG conjugates. Investigation of the pharmacokinetic profile following different administration routes of free and conjugated forms of superoxide oxidase (SOD), asparaginase adenosine deaminase or glucagon-like peptide-1 are reported [64–66]. For example, the systemic bioavailability of PEG-SOD administered via *i.p.*, *i.m.* or *s.c.* was 71%, 54% or 29% of the value obtained after *i.v.* administration. On the other hand, the maximum protein concentration peak in blood appeared earlier in the case of *i.v.* and *i.p.* than for *i.m.* or *s.c.* Different administration routes were also considered, as for instance the nasal pathway proposed by Youn et al. for PEG-glucagon-like peptide-1 conjugates [67]. Unfortunately, the high mole-cular weights of PEG polymers, commonly used in PEGylation, prevent their bioavailability from oral and dermal routes. In fact, the bioavailability of an orally administered 1,000 Da PEG is only 2%. This value can rise to between 79–100% but only for PEG oligomers with a molecular weight of up to 600 Da [68].

All the above reported data demonstrate that every protein has its specifici-ty and behavior, therefore there is not a single strategy for PEGylation and each new project must be accompanied by a PK and PD evaluation [69]. Summary of PK and PD values in differently modified proteins is reported in Table 3.

Table 3. Comparison between PK and PD PEGylated and the native protein (adapted from [69])

PEGylated-drugs	PK Half-life ($t_{1/2}$) (h)		PD *In vitro* activity		
	Parent drug ($t_{1/2}$) (h)	PEGylated drug ($t_{1/2}$) (h)	% Activity retained	Species	Ref.
PEG-arginine deiminase	2.8	50	48	rats	97
PEG-Catalase	1	10	95	mice	21
PEG-Methioninase	2	38	70	primates	84
PEG-Superoxide dismutase	0.01	38	51	mice	64
PEG-Uricase	3	72	100	human	98
PEG-Tricosanthin	3.3	8.3	14	rats	86
PEG-Lysostaphin	5	>>25	57	mice	85
PEG-GCSF	1.8	7	41	rats	99
PEG-IFN-α2a	0.7	51	7	mice	55
PEG-IFN-β1a	0.98	13	50	rats	100
PEG-IFN-β1b	1.1	16.3	50	rats	60
PEG-IL6	0.05	48	51	rats	101
PEG-TNFα	0.07	0.7	80	rats	36, 102
PEG-Calcitonin	3.31	15.4	50	rats	103, 104
PEG-GLP	0.04	0.56	83	mice	67
PEG-hGH	0.34	10	24	rats	57
Fab'Fragment	0.33	9.05	100	rats	42
PEGvisomant	0.50	100	22	rats	69
Anti-VEGF RNA aptamer	24	94	25	primates	37

Effect of PEGylation on protein recognition, uptake and processing

The observations reported above highlight the difficult task of foreseeing the effect of PEGylation on protein pharmacokinetics and pharmacodynamics. In fact, many parameters such as protein absorption, elimination, degradation and unfolding are affected by PEGylation and in particular by both the amount of linked PEG and the site of coupling. As a consequence the development of a proper chemical strategy for PEGylation of a specific protein is not straight-forward. Such constraints are potentially circumvented with a releasable approach in which a fully active protein may be recovered after *in vivo* administration. Methods involving releasable PEGs are based on special linkers between polymer and protein, which can be cleaved by chemical or enzymatic hydrolysis [70] (see also the chapter by Bonara and Drioli in this book).

Predicting the effects of PEGylation on immunogenicity is even more problematic due to the great numbers of biological variables that can generate neutralizing antibodies or can break self-tolerance. An antibody response involves a complex cascade of events: antigen internalization by antigen processing cells, their processing to peptides followed by presentation to B or T cells, also

T and B cells maturation that results in cell division and antibody release [71, 72]. Furthermore, an immune response may also be strongly elicited by a series of repetitive moieties that together act as an epitope. This is for instance the case of aggregated proteins which are more immunogenic than non-aggregated proteins which cause of breaking self-tolerance, a phenomena that can take place even without T cell involvement [73, 74].

In this complex scenario, PEGylation may act at different levels, among these we may recall: a) the reduced mobility of PEGylated proteins in tissue that may influence the translocation from blood to cells or organs involved in immunogenicity, b) the reduced recognition and uptake by antigen presenting cells (APC), c) the increased stability of PEG conjugates towards proteolytic enzymes inside the APC and finally d) a decreased recognition by B cell receptors of epitopes masked by PEG [75–77]. Quantitative studies of all these phenomena *in vivo* are still missing, although interesting investigations of the effect of PEG on certain non-covalent interactions were reported. It was demonstrated, for instance, that the presence of PEG on cyclic immunogenic peptides does not prevent their binding to the major histocompatibility complex (MHC) but the modified peptides are not immunogenic and do not stimulate the production of MHC-restricted T cells as in the case of free peptides [78]. In another study, the uptake, intracellular transport and degradation of PEG conjugates were all investigated with PEGylated asialofutein, a protein known to bind to the galactose receptors of hepatocytes. It was found that receptor-mediated uptake was decreased by the presence of PEG, due to a reduced rate of formation of the receptor-ligand complex. Furthermore, sub-cellular fractionation by density gradient, demonstrated that PEG-modified asialofutein is transported and degraded intracellulary in the same manner as the native protein, although the rate of proteolysis is reduced [79]. The ability of PEGylation to reduce protein aggregation is of great importance in preventing the immunogenicity of recombinant human proteins that, if administered at high doses, may form aggregates that are potentially immunogenic, thus breaking self tolerance. An interesting case of the effect of PEGylation on protein aggregation has been reported for G-CSF. In this case, the authors determined that the chemistry linking the PEG to protein may have a role in controlling the aggregation rate [80]. Further insights on protein aggregation can be obtained from another work also dealing with G-CSF. In this last case, PEGylation was directed to the lone free cysteine that has a relevant role in forming covalent aggregates through thiol scrambling. The aim was to reduce both non-covalent aggregation, as shown in the above reported study [81], and covalent aggregation mediated by the cysteine. Unexpectedly, the thiol conjugate PEG-G-CSF showed a higher propensity to form non-covalent aggregates with respect to the native protein. This result demonstrated that the site of PEGylation has a role in protein aggregation [81].

Precise rules of the effects of PEGylation on immunogenicity are difficult to draw because of the difficulty of comparing results from different studies. The problem in comparing results is due either to the various methods used to evaluate antigenicity, each one with a distinct sensitivity and specificity, or to the different conjugate administration routes. Units are not standardized and often not reported, so it is difficult for any direct comparison between different laboratories [82]. Furthermore, most of the immunological studies are carried out in animals that are not always predictive of human behavior. However, note that the presence of antibodies in blood does not necessarily preclude therapeutic efficacy because in several cases the antibodies may dissipate over time or do not neutralize the therapeutic protein's biological activity and there are limited studies published addressing neutralizing antibodies. The antigenicity of PEGylated proteins, i.e., recognition by the antibodies elicited against the unmodified protein, was sometimes evaluated and reported [83]. Antibodies against PEG can also be raised. This aspect is reported in detail in the chapter by Armstrong of this book.

It is beyond the scope of this chapter to fully review the literature on immunogenicity of PEGylated proteins. Therefore we have chosen to highlight only a few examples. Several studies have highlighted how the elicitation of an antibody response against a conjugated protein did not completely prevent the protein's biological activity. Examples are enzymes where the active site does not involve a large array of exposed amino acids, but also cytokines in which the binding region is rather extended on the protein surface. PEGylated methioninase, a heterologous enzyme active towards the small substrate methionine, promotes an antibody response but still maintains the ability to degrade its substrate; meanwhile PEG increases its circulation residence time, as demonstrated in monkeys [84]. In other cases, where the substrate is a macromolecule, the decrease in activity may be ascribed more to the steric hindrance of PEG chains that prevent the approach of the substrate than to the effect of neutralizing antibodies. Reduced antibody binding upon PEGylation was found with the non-human enzyme lysostaphin, an endopeptidase that disrupts the cell walls of *Staphylococcus aureus*. In this case, the biological activity is partially reduced, probably due to the effect of PEG steric hindrance on the binding of the large substrate (cell walls): the larger the PEG is, the greater the activity loss. A hindered branched form of PEG was chosen for the modification, rather than a linear one to more efficiently prevent the entry of the polymer inside the active site of the enzyme [85].

Trichosantin, a ribosome inactivating protein extracted from plants, was also recently studied. In this case, different mutants were produced and two of these were PEGylated at one thiol group with a 5 kDa polymer. A marked decrease in immunogenicity was found, but it was accompanied by some reduction of biological activity. The authors concluded that the PEGylated sites in the muteins are at or near the protein's antigenic sites. Also, in this case the decreased activity may be ascribed to the steric hindrance of PEG that prevented the enzyme's approach to the DNA, a large substrate [86].

Certolizumab Pegol, a PEGylated humanized Fab' fragment directed toward anti-tumor necrosis factor α, (TNF-α) produces low incidences of antibodies in patients, thus not affecting drug efficacy. Furthermore, Certolizumab Pegol did not cross react with anti-Infliximab antibodies [36], Infliximab being a different non-human monoclonal antibody against TNF-α that has been approved and marketed as Remicade®. This will allow the use of Certolizumab Pegol when the therapy with Infliximab must be interrupted due to the formation of neutralizing antibodies.

Detectable neutralizing antibodies were occasionally observed for PEG-interferon α-2b administered to mice (one animal out of 10), while they were always raised to a substantial amount by the unmodified cytokine and to even higher titers by its aggregates [55]. Interferon-β-1-b, a protein that tends to aggregate and is difficult to formulate due to its high hydrophobicity, was conjugated to PEGs of various molecular weights and shapes resulting in a series of derivatives with different sites and degrees of modification [60]. The specie that exhibited the best pharmacokinetic properties was the branched 40 kDa-conjugate, which was extensively investigated for immunogenicity in rats. In this case, the authors used a number of tests to assay immunogenicity: direct and indirect ELISA, flow cytometry, Biacore and antiviral activity neutralization. The PEGylated protein demonstrated the formation of much lower titers of IgG when compared to the unmodified IFN-β-1b. Interestingly, the antibodies raised against the native protein did not demonstrate significant binding to PEG-IFN-β-1b, whereas the IgGs against PEG-IFN-β-1b also bind the native protein. Furthermore, PEGylation increased the solubility and stability of IFN-β-1b preventing the formation of aggregates that could be immunogenic [60].

PEGylation also achieved success in the reduction of immunogenicity of glycosylated proteins. One case is represented by recombinant erythropoietin (rhEPO), a protein that during therapeutic protocols for chronic administration for anemia treatment can lead to immunogenicity and EPO resistant anemia. PEGylated rhEPO products were obtained with linked single and multiple chains of the polymer. Only the mono-PEGylated species was assessed for immunogenicity because the multi-PEGylated proteins were less effective in increasing hemoglobin levels. While the unmodified rhEPO was found to raise antibodies in 69% of rats following 12 weeks of treatment, no antibody was found in the PEGylated protein treated animals while the hematopoietic activity was maintained [87].

PEGylated enzymes for biocatalysis in organic solvents

Enzyme catalysis currently represents an interesting alternative to chemical synthesis for production of fine chemicals, in particular when chemo- or regio-selectivity is required. Biotransformation procedures received further application by the so called 'non aqueous' enzymology that allows for the synthesis or modification of compounds insoluble in water as is the case of several drugs or their precursors. Furthermore, the absence of water may in some cases modify the specificity of reactions and reverse reaction equilibrium [88–90].

Generally, when enzymes are employed in organic solvents, they are used as suspended powders and the dispersion degree may be a critical factor in the expression of catalytic activity. The activity of one enzyme depends on the number of productive encounters that occur between the enzyme and a substrate. As expected, due to diffusion limitations, dispersed enzymes in organic solvents were found to exhibit 10–1,000 fold reduced catalytic activities in comparison to enzymes dissolved in aqueous solution [91]. Consequently, every method that may increase dispersion in organic solvents may improve the catalytic performance and those that will allow complete dissolution will be highly desirable. Among the proposed methods to solubilize enzymes, PEGylation is preferred because the amphiphilic polymer conveys its dissolution properties to the proteins. It was found that the solvent may influence enzyme stability, rate of reaction and selectivity. Among the several enzymes that were PEGylated and studied in organic solvents, so far we are describing only a couple of examples that may illustrate the potentials and limitations of this procedure. *Candida rugosa* lipase treated with PEG-p-nitro-phenyl-chlroformiate [92] and PEG-cyanuric chloride [93] were found to exhibit enhanced stability in isooctane where the first conjugate was found to be more active than the second one. In both cases, however, they exhibited decreased lipase and esterase activities, compared to activity in aqueous systems although transesterification activity was improved [94].

Subtilisin Carlsberg exhibited an activity in dioxane 32-fold less than in water [95], but the PEGylated form exhibited complete solubility in dioxane and an enzymatic activity that depended upon the level of PEGylation. Indeed it was demonstrated that increasing the extent of PEGylation increases the enzymatic activity, while the enzyme's enantioselectivity decreased.

A further possible utility of PEGylation is to reduce the diffusion of an enzyme entrapped in a hydrogel matrix by increasing its size. Thus, PEGylated lipase could be permanently entrapped in a polyvinyl alcohol (PVA) hydrogel obtained by freezing and thawing a PVA-enzyme solution. The entrapped enzyme maintained catalytic activity and showed regioselectivity in the hydrolysis of water insoluble acetoxycoumarines [96].

Conclusions

From the short overview of protein PEGylation described in this introductory chapter, one can understand the great potentials of the technique. The applications are only partially disclosed because many uses of poly(ethylene glycol) conjugation are beyond the scope of this book. Some of these unreported applications have already achieved pharmaceutical relevance, such as modification of surface properties and improvement of *in vivo* behavior of liposomes, microparticles or materials. Others remain as interesting biological experiments only, waiting for further developments, for example, the case of the modification of the cell surface for transplantation purposes or the use of PEGylation in two phase partitioning. Thanks to the unique physical–chemical and biological properties of PEG, doubtless further new applications will be discovered.

All of this demonstrates how wrong those biochemists were who, at the Bad Neuenahar 1977 'Enzyme Engineering Meeting', dismissed the Abuchowski report on protein immunogenicity as a dream without future.

References

1 Davis F (2002) The origin of pegnology. *Adv. Drug. Del. Rev* 54: 457–458
2 Working PK, Newman SS, Johnson J, Cornacoff JB (1997) Safety of poly(ethylene glycol) derivatives. In: Harris JM, Zalipsky S (eds): *Poly(ethylene glycol) Chemistry and Biological Applications*. ACS Books, Washington, 45–54
3 Von Spect BH, Seinfeld H, Brendel W (1973) Polyvinylpyrrolidone as a soluble carrier of proteins. *Physiol Chem* 354: 1659–1660
4 Ranucci E, Spagnoli G, Sartore L, Bigotti P, Schiavon O, Caliceti P, Veronese FM (1995) Synthesis and molecular weight characterization of end functionalized poly(N-vinylpyrrolidone) oligomers. *Macrom. Chem. Phys* 196: 763–774
5 Schiavon O, Caliceti P, Ferruti P, Veronese FM (2000) Therapeutic proteins: a comparison of chemical and biological properties of uricase conjugated to linear or branched poly(ethylene glycol) and poly(N-acryloylmorpholine). *Farmaco* 55: 264–269
6 Maeda H (2001) SMANCS and polymer-conjugated macromolecular drugs: advantages in cancer chemotherapy. *Adv. Drug Deliv. Rev* 46: 169–185
7 Braunecker WA, Matyjaszewski K (2007) Controlled/living radical polymerization: Features, developments, and perspectives. *Prog. Polymer Sci* 32: 93–146
8 PCT/US Patent 2008/002626
9 Gaertner FC, Luxenhofer R, Blechert B, Jordan R, Essler M (2007) Synthesis, biodistribution and excretion of radiolabeled poly(2-alkyl-2-oxazoline)s. *J. Contr. Rel.* 119: 291–300
10 Mero A, Pasut G, Dalla Via L, Fijten MWM, Schubert US, Hoogenboom R, Veronese F*M (2008)* Synt*h*esis and characterization of poly(2-ethyl 2-oxazoline)-conjugates with proteins and drugs: Suitable alternatives to PEG-conjugates? *J. Contr. Rel* 125: 87–95
11 Torchilin VP, Mazaev AV, Voronkov I (1982) The use of immobilised streptokinase for the therapy of thromboses. *Ther. Arch* 54: 21–28
12 Davis BG, Lloyd RC, Jones JB (1998) Controlled site-selective glycosylation of proteins by a combined sitedirected mutagenesis and chemical modification approach. *J. Org. Chem* 63: 9614–9615
13 Hang HC, Bertozzi CR (2001) Chemoselective approaches to glycoprotein assembly. *Acc. Chem. Res* 34: 727–736
14 Solá RJ, Griebenow K (2006) Chemical glycosylation: New insights on the interrelation between protein structural mobility, thermodynamic stability, and catalysis. *FEBS Letters* 580: 1685–1690
15 Imperiali B, O'Connor SE (1999) Effect of N-linked glycosylation on glycopeptide and glycoprotein structure. *Curr. Opin. Chem. Biol* 3: 643–649
16 Sinclair AM, Elliott S (2005) Glycoengineering: the effect of glycosylation on the properties of therapeutic proteins. *J. Pharm. Sci* 94: 1626–1635
17 Fernandes AI, Gregoriadis G (2001) The effect of polysialylation on the immunogenicity and antigenicity of asparaginase: implication in its pharmacokinetics. *Int. J. Pharm* 217: 215–224
18 Gregoriadis G, Jain S, Papaioannou I, Laing P (2005) Improving the therapeutic efficacy of peptides and proteins: a role for polysialic acids. *Int. J. Pharm* 300: 125–130
19 Wong K, Cleland LG, Poznanski MJ (1980) Enhanced anti-inflammatory effects and reduced immunogenicity of bovine liver superoxide dismutase by conjugation with homologous albumin. *Agent Actions* 10: 231–239
20 Tao Hu, Zhiguo Su (2002) Bovine serum albumin-bovine hemoglobin conjugate as a candidate blood substitute. *Biotech Lett* 24: 275–278
21 Abuchowski A, McCoy JR, Palczuk NC, van Es T, Davis FF (1977) Effect of covalent attachment of polyethylene glycol on immunogenicity and circulating life of bovine liver catalase. *J. Biol. Chem* 252: 3582–3586

22 Abuchowski A, van Es T, Palczuk NC, Davis FF (1977) Alteration of immunological properties of bovine serum albumin by covalent attachment of polyethylene glycol. *J. Biol. Chem* 252: 3578–3781

23 Leader B, Baca QJ, Golan DE (2008) Protein therapeutics: a summary and pharmacological classification. *Nat. Rev. Drug Discov* 7: 21–39

24 Harris MJ (ed.) (1991) *Poly(Ethylene Glycol) Chemistry: Biotechnical and Biomedical Applications.* Plenum Press, New York

25 Harris JM, Veronese FM (eds): (2002) Peptide and protein PEGylation. *Adv. Drug Del. Rev* 54: 453–610

26 Harris JM, Veronese FM (eds): (2003) Peptide and protein PEGylation II Clinical Evaluation. *Adv. Drug Del. Rev* 55: 1259–1350

27 Harris JM, Veronese FM (eds): (2008) Peptide and protein PEGylation III: Advances in Chemistry and Clinical Applications *Adv. Drug Del. Rev* 60: 1–88

28 Sato H (2002) Enzymatic procedure for site-specific pegylation of proteins. *Adv. Drug Del. Rev* 54: 487–504

29 Fontana A, Spolaore B, Mero A, Veronese FM (2008) Site-specific modification and PEGylation of pharmaceutical proteins mediated by transglutaminase. *Adv. Drug Deliv. Rev* 60: 13–28

30 DeFrees S, Wang ZG, Xing R, Scott AE, Wang J, Zopf D, Gouty DL, Sjoberg ER, Panneerselvam K, Brinkman-Van der Linden EC et al. (2006) GlycoPEGylation of recombinant therapeutic proteins produced in Escherichia coli. *Glycobiology* 16: 833–843

31 Berna M, Dalzoppo D, Pasut G, Manunta M, Izzo L, Jones AT, Duncan R, Veronese FM (2006) Novel monodisperse PEG-Dendrons as new tools for targeted drug delivery: synthesis, characterization and cellular uptake. *Biomacromol* 7: 146–153

32 Mero A, Spolaore B, Veronese FM, Fontana A (2009) Transglutaminase-mediated PEGylation of proteins: direct identification of the sites of protein modification by mass spectrometry using a novel monodisperse PEG. *Bioconjug Chem* 20: 384–389

33 Monfardini C, Schiavon O, Caliceti P, Morpurgo M, Harris JM, Veronese FM (1995) A branched monomethoxypoly(ethylene glycol) for protein modification. *Bioconjug Chem* 6: 62–69

34 Veronese FM, Caliceti P, Schiavon O (1997) Branched and linear Poly(ethyl glycol): influence of the polymers structure on enzymological, pharmacokinetic and immunological properties of protein conjugates. *J Bioac Biocomp Polym* 12: 196–207

35 Foster GR (2004) Pegylated interferons: chemical and clinical differences. *Aliment Pharmacol. Ther* 20: 825–830

36 Blick S, Curran M (2007) Certolizumab pegol. *Biodrugs* 21: 196–201

37 Ng EWM, Shima DT, Calias P, Cunningham ET, Guyer DR, Adamis AP (2006) A targeted anti-VEGF aptamer for ocular vascular disease. *Nat. Rev. Drug Dis* 5: 123–132

38 Guiotto A, Canevari M, Pozzobon M, Moro S, Orsolini P, Veronese FM (2004) Anchimeric assistance effect on regioselective hydrolysis of branched PEGs: a mechanistic investigation. *Bioorg Med Chem* 12: 5031–5037

39 Solá RJ, Rodríguez-Martínez JA, Griebenow K (2007) Modulation of protein biophysical properties by chemical glycosylation: biochemical insights and biomedical implications. *Cell Mol. Life Sci* 64: 2133–2152

40 Callahan W, Narhi L, Kosky A, Treuheit M (2001) Sodium chloride enhances the storage and conformational stability of BDNF and PEG-BDNF. *Pharm. Res* 18: 261–266

41 Frokjaer S, Otzen DE (2005) Protein drug stability: a formulation challenge. *Nat. Rev Drug Discov* 4: 298–306

42 Chapman AP (2002) PEGylated antibodies and antibody fragments for improved therapy: a review. *Adv. Drug Del. Rev* 54: 531–545

43 Russell TP, Deline VR, Dozier WD, Felcher GP, Agrawal G, Wool RP, Mays W (1993) Direct observation of reptation at polymer interfaces *Nature* 365: 235–237

44 Veronese FM (2001) Peptide and protein PEGylation: a review of problems and solutions. *Biomaterials* 22: 405–417

45 Kawai F (2002) Microbial degradations of polyethers. *Appl. Microbiol. Biotechnol* 58: 30–38

46 Friman S, Egestad B, Sjövall J, Svanik J (1993) Hepatic excretion and metabolism of polyethylene glycols and mannitol in the cat. *J. Hepatol* 17: 48–55

47 Beranova M, Wasserbauer R, Vancurova D, Stifter M, Ocenaskova J, Mara M (1990) Effect of cytochrome P-450 inhibition and stimulation on intensity of polyethylene degradation in microsomial fraction of mouse and rat livers. *Biomaterials* 11: 521–524

48 Petrak K, Goddard P (1989) Transport of macromolecules across the capillary walls. *Adv. Drug Del. Rev* 3: 191–214

49 Yamaoka T, Tabata Y, Ikada Y (1994) Distribution and tissue uptake of poly(ethylene glycol) with different molecular weights after intravenous administration to mice. *J. Pharm. Sci* 83: 601–606

50 Yamaoka T, Tabata Y, Ikada Y (1995) Fate of water-soluble polymers administered via different routes. *J. Pharm. Sci* 84: 349–354

51 Hamidi M, Azadi A, Rafiei P (2006) Pharmacokinetic consequences of pegylation. *Drug Delivery* 13: 399–409

52 Manjula BN, Tsai A, Upadhya R, Perumalsamy K, Smith PK, Malavalli A, Vandegriff K, Winslow RM, Intaglietta M, Prabhakaran M et al. (2003) Site-specific PEGylation of hemoglobin at Cys-93: correlation between the colligative properties of the PEGylated protein and the length of the conjugated PEG chain. *Bioconj. Chem* 14: 464–472

53 Fee CJ, Van Alstine JM (2004) Prediction of the viscosity radius and the size exclusion chromatography behavior of PEGylated proteins. *Bioconj. Chem* 15: 1304–1313

54 Fee CJ (2007) Size comparison between proteins PEGylated with branched and linear poly(ethylene glycol) molecules. *Biotechnol. Bioeng* 98: 725–731

55 Bailon P, Palleroni A, Schaffer CA, Spence CL, Fung WJ, Porter JE, Ehrlich GK, Pan W, Xu ZX, Modi MW (2001) Rational design of a potent, long-lasting form of interferon: a 40 kDa branched polyethylene glycol-conjugated interferon -2a for the treatment of hepatitis C. *Bioconj. Chem* 12: 195–202

56 Wang YS, Youngster S, Grace M, Bausch J, Bordens R, Wyss DF (2002) Structural and biological characterization of pegylated recombinant interferon alpha-2b and its therapeutic implications. *Adv. Drug Delivery Rev* 54: 547–570

57 Cox GN, Rosendahl MS, Chlipala EA, Smith DJ, Carlson SJ, Doherty DHA (2007) Long-acting, mono-PEGylated human growth hormone analog is a potent stimulator of weight gain and bone growth in hypophysectomized rats. *Endocrinology* 4: 1590–1597

58 Clark R, Olson K, Fuh G, Marian M, Mortensen D, Teshima G, Chang S, Chu H, Mukku V, Canova-Davis E (1996) Long-acting growth hormones produced by conjugation with polyethylene glycol. *J. Biol. Chem* 271: 21969–21977

59 Long DL, Doherty DH, Eisenberg SP, Smith DJ, Rosendahl MS, Christensen KR, Edwards DP, Chlipala EA, Cox GN (2006) Design of homogeneous, monopegylated erythropoietin analogs with preserved *in vitro* bioactivity. *Experimental Hematology* 34: 697–704

60 Basu A, Yang K, Wang M, Liu S, Chintala R, Palm T, Zhao H, Peng P, Wu D, Zhang Z et al. (2006) Structure-function engineering of interferon-beta-1b for improving stability, solubility, potency, immunogenicity, and pharmacokinetic properties by site-selective mono-PEGylation. *Bioconjug. Chem* 17: 618–630

61 Bowen S, Tare N, Yamasaki T, Okabe M, Horii I, Eliason JF (1999) Relationship between molecular mass and duration of activity of polyethylene glycol conjugated granulocyte colony-stimulating factor mutein. *Experimental Hematology* 27: 425–432

62 Gaertner HF, Puigserver AJ (1992) Increased activity and stability of poly(ethylene glycol)-modified trypsin. *Enzyme Micro. Technol* 14: 150–155

63 Federico R, Cona A, Caliceti P, Veronese FM (2006) Histaminase PEGylation: Preparation and characterization of a new bioconjugate for therapeutic application. *J. Contr. Rel* 115: 168–174

64 Veronese FM, Caliceti P, Pastorino A, Schiavon O, Sartore L, Banci L, Scolaro LM (1989) Preparation, physico-chemical and pharmacokinetic characterization of monomethoxypoly(ethylene glycol)-derivatized superoxide dismutase. *Journal of Contr. Rel* 10: 145–154

65 Lee SH, Lee S, Youn YS, Na DH, Chae SY, Byun Y, Lee KC (2005) Synthesis, characterization, and pharmacokinetic studies of PEGylated glucagon-like peptide-1. *Bioconjug. Chem* 16: 377–382

66 Fuertges F, Abuchowski A (1990) The clinical efficacy of poly(ethylene glycol)-modified proteins. *J. Control Rel* 11: 139–148

67 Youn YS, Jeon JE, Chae SY, Lee S, Lee KC (2008) PEGylation improves the hypoglycaemic efficacy of intranasally administered glucagons-like peptide-1 in type 2 diabetic db/db mice. *Diabetes Obes. MeTab.* 10: 343–346

68 He H, Murby S, Warhurst G, Gifford L, Walker D, Ayrton J, Eastmond R, Rowland M (1998) Species differences in size discrimination in the paracellular pathway reflected by oral bioavailability of Poly(ethylene glycol) and D-peptides. *J. Pharm. Sci* 87: 626–633

69 Fishburn CS (2008) The pharmacology of PEGylation: balancing PD with PK to generate novel therapeutics. *J. Pharm. Sci* 97: 4167–4183

70 Filpula D, Zhao H (2008) Releasable PEGylation of proteins with customized linkers. *Adv. Drug Deliv. Rev* 60: 29–49
71 De Groot AS, Scott D (2007) Immunogenicity of protein therapeutics. *Trends in Immunology* 28: 482–490
72 Hermeling S, Crommelin DJ, Schellekens H, Jiskoot W (2004) Structure-immunogenicity relationships of therapeutic proteins. *Pharmaceut. Res* 21: 897–903
73 Hermeling S, Schellekens H, Maas C, Gebbink MF, Crommelin DJ, Jiskoot W (2006) Antibody response to aggregated human interferon alpha2b in wild-type and transgenic immune tolerant mice depends on type and level of aggregation. *J. Pharm. Sci* 95: 1084–1096
74 Schellekens H (2005) Factors influencing the immunogenicity of therapeutic proteins. *Nephrol. Dial. Transplant* 20: 3–9
75 Wang QC, Pai LH, Debinski W, FitzGerald DJ, Pastan I (1993) Polyethylene glycol-modified chimeric toxin composed of transforming growth factor alpha and Pseudomonas exotoxin. *Cancer Res* 53: 4588–4594
76 Filpula D, Zhao H (2008) Releasable PEGylation of proteins with customized linkers. *Adv. Drug Deliv. Rev* 60: 29–49
77 Tsutsumi Y, Onda M, Nagata S, Lee B, Kreitman RJ, Pastan I (2000) Site-specific chemical modification with polyethylene glycol of recombinant immunotoxin anti-Tac(Fv)-PE38 (LMB-2) improves antitumor activity and reduces animal toxicity and immunogenicity. *Proc. Natl. Acad. Sci* 97: 8548–8553
78 Bouvier M, Wiley DC (1996) Antigenic peptides containing large PEG loops designed to extend out of the HLA-A2 binding site form stable complexes with class I major histocompatibility complex molecules. *Proc. Natl. Acad. Sci. USA* 93: 4583–4588
79 Roseng L, Tolleshaug H, Berg T (1992) Uptake, intracellular transport, and degradation of polyethylene glycol-modified asialofetuin in hepatocytes. *J. Biol. Chem* 267: 22987–22993
80 Rajan RS, Li T, Aras M, Sloey C, Sutherland W, Arai H, Briddell R, Kinstler O, Lueras AM, Zhang Y et al. (2006) Modulation of protein aggregation by polyethylene glycol conjugation: GCSF as a case study. *Protein Sci* 15: 1063–1075
81 Veronese FM, Mero A, Caboi F, Sergi M, Marongiu C, Pasut G (2007) Site-specific pegylation of G-CSF by reversible denaturation. *Bioconjug. Chem* 18: 1824–1830
82 Schellekens H (2002) Immunogenicity of therapeutic proteins: clinical implications and future prospects. *Clin. Ther* 24: 1720–1740
83 Kamisaki Y, Wada H, Yagura T, Matsushima A, Inada Y (1981) Reduction in immunogenicity and clearance rate of Escherichia coli L-asparaginase by modification with monomethoxypolyethylene glycol. *J. Pharmacol. Exp. Ther* 216: 410–414
84 Yang Z, Wang J, Lu Q, Xu J, Kobayashi Y, Takakura T, Takimoto A, Yoshioka T, Lian C, Chen C et al. (2004) PEGylation confers greatly extended half-life and attenuated immunogenicity to recombinant methioninase in primates. *Cancer Res* 64: 6673–6678
85 Walsh S, Shah A, Mond J (2003) Improved pharmacokinetics and reduced antibody reactivity of lysostaphin conjugated to polyethylene glycol. *Antimicrobial Agents And Chemotherapy* 47: 554–558
86 An Q, Lei Y, Jia N, Zhang X, Bai Y, Yi J, Chen R, Xia A, Yang J, Wei S (2007) Effect of site-directed PEGylation of trichosanthin on its biological activity, immunogenicity, and pharmacokinetics. *Biomolec. Engineer* 24: 643–649
87 Tillmann HC, Kuhn B, Kränzlin B, Sadick M, Gross J, Gretz N, Pill J (2006) Efficacy and immunogenicity of novel erythropoietic agents and conventional rhEPO in rats with renal insufficiency. *Kidney International* 69: 60–67
88 Inada Y, Takahashi K, Yoshimoto T, Ajima A, Matsushima A, Saito Y (1986) Application of polyethylene glycol-modified enzymes in biotechnological processes: Organic solvent-soluble enzymes. *Trends in Biotechnol* 4: 190–194
89 Secundo G, Ottlina G, Carrea G (2008) Preparation and properties in organic solvents of noncovalent PEG-enzyme complexes. *Methods in Biotechnology* 15: 77–81
90 Carrea G, Riva S (2000) Properties and synthetic applications of enzymes in organic solvents. *Angewand. Chem. Intern. Ed* 39: 2226–2254
91 Yamamoto Y, Kise H (1993) Catalysis of enzyme aggregates in organic solvents: An attempt at evaluation of intrinsic activity of proteases in ethanol. *Biotechnol. Lett* 15: 647–652
92 Hernáiz MJ, Sánchez-Montero JM, Sinisterra JV (1997) Influence of the nature of modifier in the enzymatic activity of chemical modified semipurified lipase from *Candida rugosa.*

Biotechnol. Bioeng 55: 252–260

93 Jene Q, Pearson JC, Lowe CR (1997) Surfactant modified enzymes: solubility and activity of surfactant-modified catalase in organic solvents. *Enzyme Microb. Technol* 20: 69–74

94 DeSantis G, Jones JB (1999) Chemical modification of enzymes for enhanced functionality. *Current Opinion in Biotechnology* 10: 324–330

95 Castillo B, Sola R, Ferrer A, Barletta G, Griebenow K (2008) Effect of PEG modification on subtilisin carlsberg activity, enantioselectivity, and structural dynamics in 1,4-dioxane. *Biotechnology and Bioeng* 99: 9–17

96 Veronese FM, Mammuccari C, Schiavon F, Schiavon O, Lora S, Secundo F, Chilin A, Guiotto A (2000) PEGylated enzyme entrapped in poly(vinyl alcohol) hydrogel for biocatalytic application. *Il Farmaco* 56: 541–547

97 Wang M, Basu A, Palm T, Hua J, Youngster S, Hwang L, Liu HC, Li X, Peng P, Zhang Y et al. (2006) Engineering an arginine catabolizing bioconjugate: Biochemical and pharmacological characterization of PEGylated derivatives of arginine deiminase from Mycoplasma arthritidis. *Bioconjug. Chem* 17: 1447–1459

98 Sundy JS, Ganson NJ, Kelly SJ, Scarlett EL, Rehing CD, Huang W, Hershfield MS (2007) Pharmacokinetics and pharmacodynamics of intravenous PEGylated recombinant mammalian urate oxidase in patients with refractory gout. *Arthritis Rheum* 56: 1021–1028

99 Tanaka H, Satake-Ishikawa R, Ishikawa M, Matsuki S, Asano K (1991) Pharmacokinetics of recombinant human granulocyte colony-stimulating factor conjugated to polyethylene glycol in rats. *Cancer Res* 51: 3710–3714

100 Baker DP, Lin EY, Lin K, Pellegrini M, Petter RC, Chen LL, Arduini RM, Brickelmaier M, Wen D, Hess DM et al. (2006) N-terminally PEGylated human interferon-beta-1a with improved pharmacokinetic properties and *in vivo* efficacy in a melanoma angiogenesis model. *Bioconjug. Chem* 17: 179–188

101 Tsutsumi Y, Kihira T, Tsunoda S, Okada N, Kaneda Y, Ohsugi Y, Miyake M, Nakagawa S, Mayumi T (1995) Polyethylene glycol modification of interleukin-6 enhances its thrombopoietic activity. *J. Control Release* 33: 447–451

102 Yamamoto Y, Tsutsumi Y, Yoshioka Y, Nishibata T, Kobayashi K, Okamoto T, Mukai Y, Shimizu T, Nakagawa S, Nagata S (2003) Site-specific PEGylation of a lysine-deficient TNF-alpha with full bioactivity. *Nat. Biotechnol* 31: 31

103 Youn YS, Jung JY, Oh SH, Yoo SD, Lee KC (2006) Improved intestinal delivery of salmon calcitonin by Lys18-amine specific PEGylation: stability, permeability, pharmacokinetic behavior and *in vivo* hypocalcemic efficacy. *J. Control Release* 114: 334–342

104 Shin BS, Jung JH, Lee KC, Yoo SD (2004) Nasal absorption and pharmacokinetic disposition of salmon calcitonin modified with low molecular weight polyethylene glycol. *Chem. Pharm. Bull* 52: 957–960

PEGylated Protein Drugs: Basic Science and Clinical Applications
Edited by F.M. Veronese
© 2009 Birkhäuser Verlag/Switzerland

Reactive PEGs for protein conjugation

Gian Maria Bonora and Sara Drioli

Department of Chemical Sciences, Via Giorgieri 1, University of Trieste, 34127 Trieste, Italy

Abstract

Poly(ethylene glycol) (PEG) derivatives are the first choice of the water soluble, biocompatible polymers on hand for conjugation to proteins and polypeptides. This chapter deals with the PEG reagents that are available for the preparation of bioconjugates. The opportunities of different reactive groups on PEG are described and their different activities against the functional moieties of the amino acids are illustrated. Some attention is also given to the modification of the PEG backbone to increase its loading capacity and to eventually modify the stability of the conjugating bonds.

Introduction

The unique chemical and biomedical features of poly(ethylene glycol) (PEG) have been successfully exploited during these years for a variety of practical purposes, since it is possible to introduce the polymer on a desired molecules without negatively interfering with their fundamental properties. Thus, PEG itself can be easily chemically modified and joined to other units with little, if any, effect on their chemistry, but with great modification of solubility, size and stability. This polymer, at first glance, looks like a very simple molecule [1]. In fact, it is a linear neutral polyether of general structure:

$$RO-(CH_2-CH_2-O)_n-CH_2-CH_2-OH \text{ where } R = CH_3, H$$

One of its most striking features is the wide solubility, up to 50% w/w in the majority of solvents adopted in the organic reactions, as well as in water. On the other hand, by addition of ether it is easily precipitated as a powder from organic solution. When applied to organic syntheses this precipitation/filtration process allows for a fast purification of the molecules eventually bound to the polymer.

PEG as chemical reagent

Regarding its chemical modification, PEG offers the terminal primary hydroxyl groups for modification since the polyether backbone is quite chemically inert. A PEG chain offers no more than two anchoring point, namely the two OH groups at its extremities. Monomethylether of PEG (mPEG) is often used when it is desirable to have only one molecule of product bound to each polymeric chain, or to link multiple PEG chains to a substrate and avoid any possible cross-linking reaction. The covalent bond between PEG and the new molecule is achieved directly by using the original OH groups of the polymer whenever possible. Alternatively, new functionalized PEGs can be obtained or by direct transformation of hydroxyls to the wanted target functionality, or by reaction of the polymer with a bifunctional molecule where one function is used for conjugation to the polymer, while the other remains available for linking the organic moiety of interest [2]. Thus, to make PEG useful as a reagent for peptide or protein modification, it must be activated with a functional group that is reactive toward some group on the protein [3]. Modification of amino acid moieties in peptides and proteins is a well known field. Most PEG reagents use the same chemistry embodied in simpler reagents used to modify proteins. Through a series of chemical steps, one can convert the PEG hydroxyl group into a molecule containing a reactive end group (functional group) [4]. Despite the clinical successes of many PEG conjugates, conjugation of a PEG reagent to a biomolecule and providing the conjugate in a form that has the purity and structural definition required for clinical applications remains challenging. In particular, the conjugation reaction may give a relatively disperse mixture of products if there is more than one functional group that is reactive toward the PEG reagent. For example, with a typical protein, even with a monofunctional PEG, the conjugation reaction often results in an isomeric mixture of singly substituted, as well as isomeric disubstituted, and polysubstituted conjugate forms. Because of the above considerations, providing PEG reagents that can provide 'site-directed' conjugation is one of the challenges for the PEG reagent chemist. Unfortunately, reagents that are site specific are rare. More often, a specific site may be engineered into a biomolecule to allow for site specific attachment. Regardless, the selection of the appropriate reagent becomes the challenge for the bioconjugation research.

The PEG binding to proteins generally involves the amino residue of lysine, because these groups, always present in proteins, are very reactive and exposed to solvent. Furthermore a great deal of studies are reported in literature regarding methods for specific amino residue modification in proteins. The thiol modification is also of particular interest since cysteine is seldom present in proteins, but may be a very specific binding site when present. Other amino acid residues were also considered for the conjugations, as the guanidino group of arginine, the carboxylic group and the carboxamido group of glutamine, as well the sugars in case of glycoproteins. Carboxylic groups were used for the conjugation but in very few cases only, because the needed coupling agent yields the desired reaction with an amino PEG, but contemporaneously cross linking with the amino group of the protein takes place also. Prerequisite for all of the methods is that they employ chemical procedures that are compatible with the conditions of protein stability for what pH, temperature, salt concentration is concerned.

Eventually, it is well known that use of a stable bond in forming the conjugate begs for substantial loss in therapeutic activity of the protein or polypeptide. However, even if ca. 5% or more bioactivity remains, the advantages of using a PEG conjugate having a stable bond may make this option most attractive. With the use of stable PEG-conjugate bonds one does not have to worry about the rate for PEG release of the drug. In some cases, it may make a difference which reagent type is chosen because the protein charge may be important. There are other complications with certain reagents that make it prudent to consult with an experienced PEG chemist prior to taking on a major new PEGylation program [5].

The conjugate purification from excess reagent, from the non-modified protein and from the leaving groups are usually carried out by gel filtration, ultrafiltration, ion exchange or reverse-phase procedure. The use of ionic exchange chromatography is sometime hampered by the presence of PEG that could mask the protein charges; however an important advantage is given by the possible purification of large amounts of crude PEGylated derivatives in a single process. On the other hand, other methods such as based on affinity are rarely useful for the rejecting property of PEG that prevents a productive binding. Usually the analysis of the conjugation to the protein reactive groups and the estimation of the degree of modification is carried out by colorimetric method or by amino acid analysis provided that PEG was previously labeled with an unnatural amino acid as norleucine or β-alanine [6]. Moreover the ionic exchange can be also useful in analyzing the presence of different isomers due to the multiple attack sides offered by the protein. An example of such a mixture is given in Figure 1. Valuable data on protein conjugate heterogeneity are also obtained by mass spectroscopy methods, MALDI in particular.

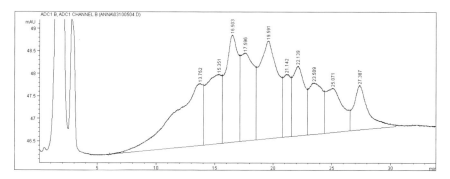

Figure 1. Ion exchange column fractionation of the reaction mixture obtained by reaction of human grow hormone with PEG-5,000-OSu, in borate buffer at pH 8.5 (Anna Mero, personal communication)

Specific PEG reagents

Modification of amino groups

Since most applications of PEG conjugation involve labile molecules, the coupling reactions require mild chemical conditions suitable also in water solutions. In the case of polypeptides, the most common reactive groups addressed in coupling are the alpha N-terminus or epsilon amino groups of lysine respectively. First-generation chemistries were generally plagued by PEG impurities, restriction to low molecular weights, unstable linkages, and lack of selectivity in modification. Examples of first-generation PEG derivatives include: (a) PEG dichlorotriazine, (b) PEG tresylate, (c) PEG succinimidyl carbonate, (d) PEG benzotriazole carbonate, (e) PEG *p*-nitrophenyl carbonate, (f) PEG trichlorophenyl carbonate, (g) PEG carbonylimidazole and (h) PEG succinimidyl succinate. In Tables 1 and 2 a survey of the different PEG-derivatives reacting with the amino group is reported: the alkylating ones maintain the charge of the native protein in the final conjugate, while the acylating reagents lead to a loss of positive charge in the final conjugate with respect to starting protein. The PEG dichlorotriazine derivative can react with multiple nucleophilic functional groups such as lysine, serine, tyrosine, cysteine, and histidine, which results in displacement of one of the chlorides and produces a conjugate with retained charge in the form of a secondary amine linkage [7]. The remaining chloride is less susceptible to reactions with nucleophilic residues. Unfortunately, the reactivity is sufficient to allow crosslinking of protein molecules containing additional nucleophilic residues. Another alkylating reagent used to nonspecifically modify multiple amino groups, forming secondary

Table 1. PEG alkylating derivatives reacting with amino groups to yield secondary amines, this preserving the protein charge

PEG—O—CH₂ (epoxide structure)	PEG-epoxide	Slowly reactive, rarely used.
PEG—O (chlorotriazine structure with N, N, N ring and Cl) PEG—O	PEG- chlorotriazine	Abandoned for therapeutic application due to toxicity.
PEG—O—SO₂—CH₂CF₃	PEG-tresyl	Not much used: it leads to a mixture of products.
PEG (aldehyde structure) H	PEG-aldehyde	A two steps reaction: it follows a reductions with NaCNBH₃. If the coupling reaction is carried out at low pH (4.5–5) it reacts only with the α-amino group.

Table 2. PEG acylating derivatives reacting with amino groups to yield amide or urethane bonds, this decreasing the protein charge by a number equal to the number of couple PEG chains

PEG-carboxilates		
PEG—O—(CH$_2$)$_n$—C(=O)—OSu	Several derivatives with one or more –CH$_2$– groups between	The rate of conjugation depends on the numbers and ramification of –CH$_2$– groups.
PEG-O—C(=O)—CH$_2$CH$_2$—C(=O)—OSu	PEG-succinimidyl-succinate	Easy hydrolysis of the ester between succinic acid and PEG
PEG–X—C(=O)—OSu	PEG-amino acid-succinimidyl ester	Easy quantification of the number of linked PEG chains by analysis of added unnatural Nle or βAla units.
PEG—(X)$_n$—C(=O)—OSu	PEG-peptide-succinimidyl ester	Met-Nle or Met- βAla allows an easy localization of PEGylation site by removing PEG with CNBr. Lysosomal cleavable sequences (i.e., H-Gly-Phe-Lue-Gly-OH) allow the release of drug inside the cell.
PEG-carbonates		
PEG–O—C(=O)—OSu	PEG-succinimidyl-carbonate	Slowly reactive, final linker as urethane
PEG–O—C(=O)—O—(2,3,5-trichlorophenyl)	PEG-2,3,5 trichloro-phenyl carbonate	Slowly reactive, final linker as urethane
PEG–O—C(=O)—O—N(benzotriazolyl)	PEG-benzotriazolyl carbonate	Slowly reactive, final linker as urethane
PEG–O—C(=O)—O—(C$_6$H$_4$)—NO$_2$	PEG-pnitrophenyl carbonate	Slowly reactive; final linker as urethane

amine linkages to proteins, viruses and liposomes, is PEG tresylate or tosylated [8]. Although more specific to amino groups than PEG dichlorotriazine, the chemistry of conjugation and the conjugation products are not unique and well defined. Therefore, a heterogeneous mixture that results from attaching PEG-tresylate to proteins may contain a population of conjugates with degradable linkages. Second-generation PEGylation chemistry has been designed to avoid the above noted problems of diol contamination, restriction to low molecular weight mPEG, unstable linkages, side reactions and lack of selectivity in substitution. One of the first examples of second-generation chemistry is mPEG-propionaldehyde [9]. mPEG-propionaldehyde is easier to prepare and use than PEG-acetaldehyde because the acetaldehyde is very susceptible to dimerization via aldol condensation. A key property is that under acidic conditions (approximately pH 5), aldehyde is largely selective for the N-terminal α-amine because of the lower pK_a of the α-amine compared to other nucleophiles [10–12]. The conjugation of electrophilic PEGs to amino acid residues on proteins is highly dependent on their nucleophilicity; nucleophilic attack will only take place when the pH of the protein solution is near or above the residue's pK_a. Although complete selectivity is not observed, the extensive heterogeneity frequently seen with lysine chemistry is greatly reduced. Coupling of aldehydes to primary amines proceeds through a Schiff base, which is reduced *in situ* to give a stable secondary amine linkage as shown in Figure 2. This is also a convenient way for conjugation when the amino positive charge is critical for the biological activity retention.

As an alternative alkylating agent, the epoxy PEG can also be used, but it is low reactive and furthermore the specificity in not sure since hydroxyl groups may also react.

Most first-generation PEG chemistries are those that produce conjugates through acylation. Two widely used first-generation activated mPEGs are succinimidyl carbonate (SC-PEG) [13, 14] and benzotriazole carbonate (BTC-PEG) [15]. SC-PEG and BTC-PEG react preferentially with lysine residues to form a carbamate linkage, but are also known to react with histidine and tyrosine residues; SC-PEG is slightly more stable to hydrolysis than BTC-PEG. Other PEG acylating reagents which produce urethane linked proteins include *p*-nitrophenyl carbonate (pNPC-PEG), trichorophenyl carbonate (TCP-PEG) and carbonylimidazole (CDI-PEG) [16, 17]. These reagents are prepared by reacting chloroformates or carbonylimidazole with the terminal hydroxyl group on mPEG, and these have much lower reactivity than either the SC-PEG or BTC-PEG. Generally, the slower the reaction the more specific the reagent is to certain amino acid groups of the protein. The formation of a carbamate bond is shown in Figure 3.

The remaining first-generation PEG reagent is succinimidyl succinate (SS-PEG) [18]. SS-PEG is prepared by reaction of mPEG with succinic anhydride,

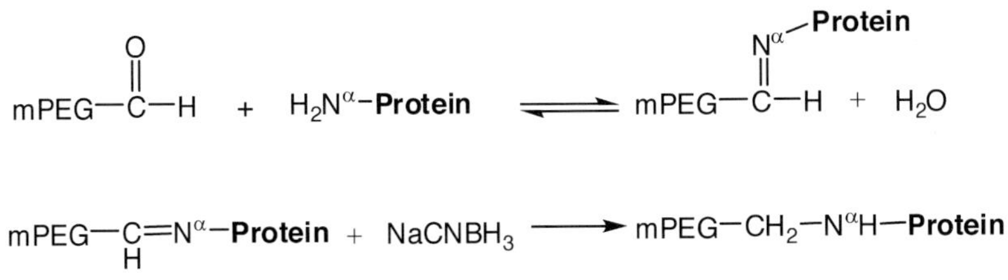

$$mPEG-\overset{\overset{\displaystyle O}{\|}}{C}-H \ + \ H_2N^{\alpha}\text{-Protein} \ \rightleftharpoons \ mPEG-\overset{\overset{\displaystyle N^{\alpha}-\textbf{Protein}}{\|}}{C}-H \ + \ H_2O$$

$$mPEG-\underset{H}{\overset{\displaystyle C}{=}}N^{\alpha}\text{-Protein} \ + \ NaCNBH_3 \ \longrightarrow \ mPEG-CH_2-N^{\alpha}H\text{-Protein}$$

Figure 2. Scheme of the formation of a stable amino linkage between mPEG and a protein.

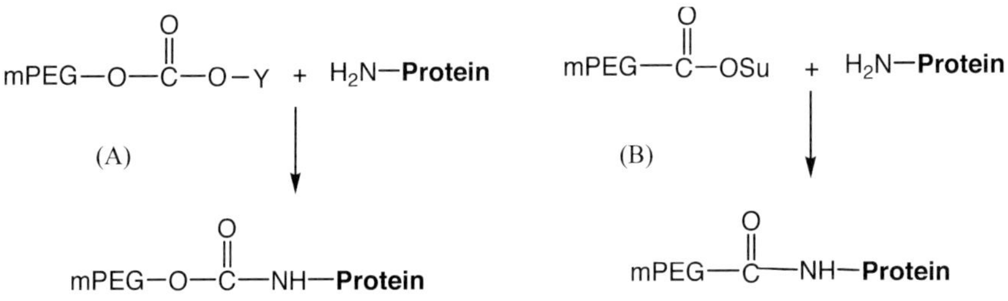

Figure 3. Scheme of the formation of a carbamate (A) and of an amide (B) bond between mPEG and a protein (Y = p-nitrophenyl, trichlorophenyl, benzotriazole)

followed by activation of the carboxylic acid to the succinimidyl ester. The polymer backbone contains a second ester linkage that remains after the conjugation reaction with a protein. This linkage is highly susceptible to hydrolysis after the polymer has been attached to the protein. Not only does this hydrolysis lead to loss of the benefits of PEG attachment, but the succinate tag that remains on the protein after hydrolysis can act as a hapten and lead to immunogenicity of the remaining protein [19]. In general, active esters of PEG carboxylic acids are the most used acylating agents for protein modification. Active esters react with primary amines near physiological conditions to form stable amides as shown in Figure 3.

The first carboxylic acid derivative of PEG not containing a degradable linkage to the PEG backbone, as in SS-PEG, was carboxymethylated PEG (CM-PEG) [20]. The succinimidyl ester of this compound (SCM-PEG) is extremely reactive and is therefore difficult to use. To have an active ester that had more favorable kinetics for protein modification, propionic acid (PEG–O–CH$_2$CH$_2$–COOH) and butanoic acid (PEG–O–CH$_2$CH$_2$CH$_2$–COOH) derivatives of PEG have been prepared [21]. Changing the distance between the active ester and the PEG backbone by the addition of methylene units had a profound influence on the reactivity towards amines and water. For example SBA-PEG, which has two additional methylene groups, has a longer hydrolysis half-life than SPA-PEG, which has one additional methylene group. Additionally, the presence of an amino acid or peptide arm between PEG and the attached macromolecule gives several advantages due to the variability of properties that may be introduced using a suitable amino acid or peptide [22]. Among these arm, norleucine is of interest for the advantage in analysis of PEGylated proteins and methionine for the identification of PEGylation site [6, 23–24].

Modification of thiol groups

PEGylation of free cysteine residues in proteins is the main approach for site-specific modification because reagents that specifically react with cysteines have been synthesized, and the number of free cysteines on the surface of a protein is much less than that of lysine residues. This rare residue may be introduced at the desired position of the sequence by genetic engineering. The advantage of this approach is that it makes possible site-specific PEGylation at areas on the protein that will minimize a loss in biological activity but decrease immunogenicity. PEG derivatives such as PEG-maleimide vinylsulfone, iodoacetamide, and orthopyridyl disulfide have been developed for PEGylation of cysteine residues, with each derivative having its own advantages and disadvantages [25–28] as resumed in Table 3. There is an activated disulphide, namely PEG-orthopyridyldisulfide (PEG-OPSS) that reacts specifically with sulfhydryl groups under both acidic and basic conditions (pH 3–10) to form a disulfide bond with the protein. Disulfide linkages are also stable, except in a reducing environment when the linkage is converted to thiols. PEG-vinylsulfone (PEG-VS) reacts slowly with thiols to form a stable thioether linkage to the protein at slightly basic conditions (pH 7–8) but will proceed faster if the pH is increased. Although PEG-VS is stable in aqueous solutions, it may react with lysine residues at elevated pH. Unlike PEG-VS, PEG-maleimide (PEG-MAL) is more reactive to thiols even under acidic conditions (pH 6–7), but it is not stable in water and can undergo ring opening or addition of water across the double bond.

PEG-iodoacetamide (PEG-IA) was also employed as a well-known reaction in protein chemistry; it should be done in slight molar excess of the PEG derivative in a dark container to limit the generation of free iodine that may react with other amino acids. This modification procedure presents the advantage that, by strong acid hydrolysis, the PEGylated cysteine gives rise to carboxymethylcysteine, a stable cysteine derivative that can be identified and quantified by standard amino acid analysis [29].

Table 3. PEG derivatives reacting with thiol groups

	PEG-iodo acetamide	Less reactive.
	PEG-vinylsulfone	Give stable linkage; may react with amines at high pH values.
	PEG-maleimide	Give stable linkage; may react with amines. It presents instability problems due to ring opening
	PEG-pyridildisulphide	Most specific towards thiol; yields a cleavable linkage by a reducing agent.

Modification of other groups

Hydroxyl groups
PEG-isocyanate is useful for hydroxyl group conjugation yielding a stable urethane linkage [30]. However, its reactivity may be best exploited for non-peptide moieties such as drugs or hydroxyl-containing matrices for chromatography and also to yield biocompatible surfaces. In fact, PEG-isocyanate is also
highly reactive with amino moieties. A possible solution, not yet fully exploited, could take advantage by the use of a PEG-phosphorylated unit playing with
the higher stability of the final phosphodiester bond in comparison with any
concurrent phosphoroamidite derivative [31]. An effective, alternative procedure for the selective modification of hydroxyl groups is given by the use of
specific enzymes as a GalNAc glycosylation at specific serine and threonine
residues in proteins [32]. This enzymatic method, termed GlycoPEGylation,
was applied to three clinically important proteins and it was demonstrated that
these proteins were exactly modified at the O-glycosylation sites of the native
proteins (see chapter by Fontana et al. in this book).

Carboxyl groups
An original strategy was devised to specifically PEGylate carboxylic groups in
proteins without cross-linking formation with amino groups: it takes place
only by the use of PEG-hydrazide [33]. The very low pK of the hydrazide
group allows its binding to carboxyl groups in acid media (pH 4.5–5) where
all the protein amino groups are protonated and not reactive. This condition
assured that intra or inter-molecular protein cross linking between carboxy-
lates and a PEG amine does not take place.

Guanido groups
The few examples of PEGylation at the level of arginine were all based on the
use of PEG-1,3-dioxocompounds. The disadvantages of this method are the
long reaction time needed for complete coupling and the non-specificity of its
chemistry, since other amino acids, histidine and lysine in particular, may also
react [34, 35].

Carboxamido groups

As for the hydroxyl groups, the specific conjugation of PEG to the amido groups of glutamines is possible under mild conditions using enzymes. Recently, has been discovered that glutamine in proteins can be a substrate for the PEGylation mediated by transglutaminase enzymes [36]. This process has been extensively investigated and the site-specific post-translational modification of proteins analyzed for an eventual prediction of the possible site of conjugation [37].

Modification of backbone

Multifunctional PEGs

As mentioned, PEG derivatives show good chemical and mechanical properties, terminal reactive functional groups, as well as advantageous solubilising properties. However, a reduced size homogeneity and loading capacity could hamper their pharmacological application, and these adverse characteristics increase with higher molecular weight. With a view to improve their overall features, many modifications of linear bifunctional PEGs have been proposed as the formation of the so-called branched PEGs, where two linear polymeric chains are linked together trough a tri-functional spacer [38]; alternatively, the introduction of dendrimeric structures has been also proposed [39, 40]. A different approach has been recently described where a simple procedure provides linear PEGs bearing multiple pendant hydroxyl groups along the final chain [41]. Recently, the production of new branched, high-molecular weight multimeric PEG-based systems (MultiPEGs), starting from inexpensive commercial PEG moieties assembled through a divergent dendrimeric fashion, have also been planned [42, 43]. A scheme of some of these PEG derivatives is drawn in Figure 4.

Reversible linkages

Most first- and second-generation PEG reagents were designed to provide a stable conjugate linkage between the PEG and protein moieties [44]. Unfortunately conjugation often results in the polymer being attached at or near a site on the active agent that is necessary for pharmacologic activity. Releasable PEGylation technologies have been developed as a means to address the typical activity loss observed with stable-bond PEGylation of many biomolecules [45]. Releasable PEG reagents provide conjugates having a degradable linkage between the PEG moiety and the therapeutic biomolecule. One of the simplest releasable linkers used in PEGylation is the PEG double-ester approach that provides an ester moiety in the PEG linker as a cleavage point [46]. Ester linkages are susceptible to both acid and base catalyzed hydrolysis, but the primary cleavage mechanism *in vivo* is expected to be enzymatic [47]. Other releasable PEG approaches are based essentially on a chemical process, as, for example, a benzyl elimination procedure. In this case

Branched PEG   **Multifunctional PEG**

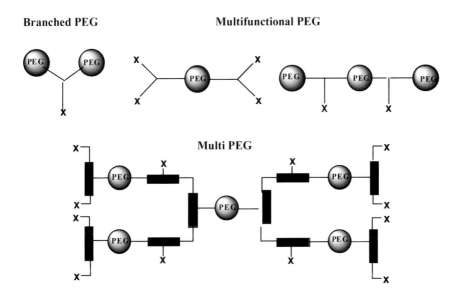

Figure 4. Scheme of the general structure of some backbone modified PEGs

a disulfide-based system allows release of the native protein by a reduction process to release the PEG polymer; this is followed by an elimination mechanism that detaches the residual linker from the protein [48].

Final remarks

Dependent on the final product desired each coupling method has been reported to have its own advantages and disadvantages. The choice of coupling agent, molecular weight of the PEG, the percentage of groups modified and the specific properties of the protein being modified can alter the characteristics of the final conjugated derivative. It is possible to change any of these variables in order to 'custom-design' the final product, though it is the inherent characteristics of the protein and peptides that dominates the final biological activity of the adduct. Proteins are not the only compounds that can benefit, since chemical drugs and other chemical compounds can be modified with PEG. The only limitations to the use of PEG are in the minds of those who develop the coupling schemes. Site-directed attachment and non covalent attachment of PEG open up a whole new category of possibilities. Changes in the molecular weight and charge of the PEG attached also provide new possibilities. Labile bonds between the PEG and the compound of interest provide the potential for pro-drug development and targeting agents. Importantly, in this era of ever increasing costs, PEG-conjugation can be used to control expenditures to replace labile compounds due to short use life times *in vivo*. By understanding the technology and having developed a database of knowledge, it is possible to produce an end-product that behaves precisely as wanted, has a long-acting life *in vivo* and is cost-effective.

References

1. Properties and Use of Polyethylene Glycol (1977), *Hoechst Polyglycols Katalog*
2. Zalipsky S (1995) Functionalized Poly(ethylene glycol) for Preparation of Biologically Relevant Conjugates. *Bioconjugate Chem.*, 6:150–165
3. Harris JM (1985) Laboratory Synthesis of Polyethylene Glycol Derivatives. *JMS-Rev. Macromol. Chem. Phys.*, C25:325–373
4. Lundblad RL (2005) *Chemical Reagents for Protein Modification* 3rd Edit. CRC Press
5. Pedder S (2001) PEGASYS®: A true once-a-week antiviral. *Annual Meeting, American Association of the Study of Liver Diseases (AASLD)*, Dallas, TX
6. Sartore L, Caliceti P, Schiavon O, Monfardini C, Veronese FM (1991) Accurate evaluation method of the polymer content in monomethoxy poly(ethylene glycol)modified proteins based on amino acid analysis. *Appl. Biochem. Biotechnol.*, 31:213–222.
7. Zalipsky S, Lee C (1992) Use of functionalized poly(ethylene glycol) s for modification of polypeptides. In: JM Harris, S Zalipsky (Eds.), *Polyethylene Glycol Chemistry, Biotechnical and Biomedical Applications*, Plenum, New York, pp. 347–370
8. Francis GE, Fisher D, Delgado C, Malik F, Gardiner A, Neale D (1998) PEGylation of cytokines and other therapeutic proteins and peptides: the importance of biological optimization of coupling techniques. *Int. J. Hematol.*, 68:1–18
9. Harris JM, Herati RM (1993) Preparation and use of poly-ethylene glycol propionaldehyde. US Patent 5,252,714
10. Kinstler OB, Brems DN, Lauren SL (1996) Characterization and stability of N-terminally PEGylated rhG-CSF. *Pharm Res.*, 13:996–1002
11. Kinstler OB, Gabriel NE, Farrar CE, DePrince RB (1999) N-terminally chemically modified protein compositions and methods, US Patent 5,985,265
12. Edwards CK (1993) PEGylated recombinant human soluble tumor necrosis factor receptor type I (rHu-sTNF-RI): A novel high-affinity TNF receptor designed for chronic inflammatory diseases. *Ann. Rheum. Dis.*, 58:173–181
13. Zalipsky S, Seltzer R, Menon-Rudolph S (1992) Evaluation of a new reagent for covalent attachment of polyethylene glicol to proteins. *Biotechnol. Appl. Biochem.*, 15:100–114
14. Miron T, Wilchek M (1993) A simplified method for the preparation of succinimidyl carbonate polyethylene glycol for coupling to proteins. *Bioconjug. Chem.*, 4:568–569
15. Dolence EK, Hu C, Tsang R, Sanders CG, Osaki S (1997) Electrophilic polyethylene oxides for the modification of polysaccharides, polypeptides (proteins) and surfaces. US Patent 5,650,234
16. Veronese FM, Largajolli R, Boccu E, Benassi CA, Schiavon O (1985) Activation of monomethoxy poly(ethylene glycol) by phenylchloroformate and modification of ribonuclease and superoxide dismutase. *Appl. Biochem. Biotechnol.*, 11:141–152
17. Beauchamp CO, Gonias SL, Menapace DP, Pizzo SV (1983) A new procedure for the synthesis of polyethylene glycolglycol protein adducts, effects on function, receptor recognition and clearance of superoxide dismutase, lactoferrin and a2-macro-globulin. *Anal. Biochem.*, 131:25–33
18. Carter MC, Meyerhoff ME (1985) Instability of succinyl ester linkages in O29-monosuccinyl cyclic AMP-protein conjugates at neutral pH. *J. Immunol. Methods*, 81:245–257
19. Abuchowski J, Kazo GM, Verhoest CR (1984) Cancer therapy with chemically modified enzymes. I. Antitumor properties of polyethylene glycol-asparaginase conjugates. *Cancer Biochem. Biophys.*, 7:175–186
20. Zalipsky S, Barany G (1986) Preparation of polyethylene glycol derivatives with two different functional groups at the termini. *Polym. Preprints*, 27:1–2
21. Zalipsky S, Barany G (1990) Facile synthesis of α-hydroxy-ω-carboxymethylpolyethylene oxide. *J. Bioact. Compat. Polym.*, 5:227–231
22. Harris JM, Kozlowski A (1997) Polyethylene glycol and related polymers monosubstituted with propionic or butanoic acids and functional derivatives thereof for biotechnical applications. US Patent 5,672,662

23. Sartore L, Caliceti P, Schiavon O, Veronese FM (1991) Enzyme modification by MPEG with amino acid or peptide as spacer arm. *Appl. Biochem. Biotechnol.*, 27:55–63
24. Veronese FM, Saccà B, Polverino de Laureto P, Sergi M, Caliceti P, Schiavon O (2001) New PEGs for peptide and protein modification, suitable for identification of the PEGylation site. *Bioconjugate Chem.*, 12:62–70
25. Goodson RJ, Katre NV (1990) Site-directed pegylation of recomproteins binant interleukin-2 at its glycosylation site. *Biotechnology*, 8:343–346
26. Kogan TP (1992) The synthesis of substituted methoxy-polymer (ethylene glycol) derivatives suitable for selective protein modification. *Synth. Commun.*, 22:2417–2424
27. Morpurgo M, Veronese FM, Kachensky D, Harris JM (1976) Preparation and characterization of poly(ethylene glycol) vinyl sulfone. *Bioconjug. Chem.*, 7:363–368
28. Woghiren C, Sharma B, Stein S (1993) Protected thiol-polycoupling ethylene glycol: a new activated polymer for reversibile protein modification. *Bioconjug. Chem.*, 4:314–318
29. Gard FRN (1972) Carboxymethylation. *Methods Enzymol.*, B25:424–449
30. Greenwald RB, Pendri A, Bolikal D (1995) Highly soluble taxol derivatives: 7-polyethylene glycol carbamates and carbonates. *J. Org. Chem.*, 60:331–336
31. Pace G, Veronese FM, Bonora GM (1999) Synthesis and reactivity of high-molecular mass phosphorylated poly(ethylene glycol). *Reactive & Functional Polimers*, 41:141–148
32. De Frees S, Wang ZG, Xing R (2006) GlycoPEGylation of recombinant therapeutic proteins produced in Escherichia coli. *Glycobiology* 16:833–843
33. Zalipski S, Menon-Rudolph S (1997) Hydrazide derivatives of poly(ethylene glycol) and their bioconjugates. Poly(ethylene glycol) chemistry and biological applications. *ACS Symp Ser*, 680:318–341
34. Yankeelov Jr AJA (1972) Modification of arginine by diketones. *Methods Enzymol.*, B25:566–584
35. Maeda H, Kai Y, Ono K (1989) Polyethylene glycol derivatives, modified, peptides and production thereof. E.P. 0.340.741
36. Sato H (2002) Enzymatic procedure for site-specific PEGylation of proteins, *Adv. Drug Deliv. Rev.*, 54:487–504
37. Fontana A, Spolaore B, Mero A, Veronese FM (2008) Site-specific modification and PEGylation of pharmaceutical proteins mediated by transglutaminase, *Adv. Drug Deliv. Rev.* 60:13–28
38. Monfardini C, Schiavon O, Caliceti P, Morpurgo M, Harris JM, Veronese FM (1995) A Branched Monomethoxypoly(ethylene glycol) for Protein Modification, *Bioconjugate Chem.*, 6:62–69
39. Schiavon O, Pasut G, Moro S, Orsolini P, Guiotto A, Veronese FM (2004) PEG- Ara C conjugation for controlled release. *Eur. J. Med. Chem.*, 39(2):123–133
40. Pasut G, Scaramuzza S, Schiavon O, Mendichi R, Veronese FM (2005) PEG-epirubicin conjugates with high drug loading. *J. of Bioactive and Compatible Polymers* 20:213–230
41. Liu X-M, Thakur A, Wang D (2007) Efficient Synthesis of Linear Multifunctional Poly(ethylene glycol) by Copper(I)-Catalyzed Huisgen 1,3-Dipolar Cycloaddition. *Biomacromolecules*, 8:2653–2658
42. Ballico M, Drioli S, Bonora GM (2005) MultiPEG: high molecular weight multifunctional poly(ethylen glycol)s assembled by a dendrimer-like approach. *EJOC*, 2064–2067
43. Drioli S, Bonora GM, Ballico M (2008) Synthesis and characterization of new multifunctional high-molecular weight PEG derivatives (MultiPEG)s. *The Open Organic Chemistry Journal* 2:17–25
44. Harris JM, Chess RB (2003) Effect of PEGylation on Pharmaceuticals. *Nature Reviews Drug Discovery* 2:214–221
45. Roberts MJ, Bentley MD, Harris JM (2002) Chemistry for peptide and protein PEGylation. *Adv. Drug Del. Rev.* 54:459–476
46. Zhao X, Harris JM (1997) Novel degradable poly(ethylene glycol) esters for drug delivery. In: JM Harris, S Zalipsky (eds): *Poly(ethylene glycol) chemistry and biological applications*. American Chemical Society, Washington, DC, 458–472
47. Testa B, Mayer JM (eds): (2003) *Hydrolysis in drug and prodrug metabolism: Chemistry, biochemistry, and enzymology*. Verlag Helvetica Chimica Acta, Wiley-VCH, Switzerland
48. Zalipsky S, Qazen M, Walker II JA, Mullah N, Quinn YP, Huang SK (1999) New detachable poly(ethylene glycol) conjugates: Cysteine-cleavable lipopolymers regenerating natural phospholipids, diacyl phosphatidylethanolamine. *Bioconjug. Chem.* 10:703–707

Rebridging disulphides: site-specific PEGylation by sequential bis-alkylation

Ji-Won Choi[1], Antony Godwin[1], Sibu Balan[1], Penny Bryant[1], Yuehua Cong[1], Estera Pawlisz[1,] Manuchehr Porssa[1], Norbert Rumpf[1], Ruchi Singh[1], Keith Powell[1] and Steve Brocchini [1,2]

[1] *PolyTherics Ltd., London Bioscience Innovation Centre, 2 Royal College Street, London, NW1 0TU, UK*
[2] *Department of Pharmaceutics, The School of Pharmacy, University of London, 29/39 Brunswick Square, London WC1N 1AX, UK*

Abstract

Site-specific PEGylation reagents have been developed that undergo thiol-specific bis-alkylation with the two cysteine sulphur atoms from a native accessible disulphide in proteins. The process for this approach of site-specific PEGylation involves two steps: (1) disulphide reduction to release the two thiols and (2) bis-alkylation of the PEG reagent to the two sulphur atoms to give a three-carbon bridge to which PEG is covalently attached. Mechanistically, the conjugation is thought to occur by a sequential, interactive bis-alkylation that requires functionalised PEG reagents that have a α,β-unsaturated β'-mono-sulphone moiety. Competitive reactions can be effectively suppressed to achieve high yield PEGylation with a stoichiometric equivalence of the reagent. The reagents are easily prepared and precursor forms of our PEG reagents can be used to control the rate of formation of the reactive PEG mono-sulphone *in situ* that undergoes conjugation with the protein. Purification is often a simple process where unPEGylated protein can be easily recycled to further increase yields. Many classes of therapeutically relevant proteins possess accessible native disulphides. Our studies have shown that peptides, proteins, enzymes and antibody fragments can be site-specifically PEGylated by bis-alkylation using a native, accessible disulphide.

Introduction: the need for site-specific PEGylation

Protein PEGylation is a clinically proven strategy to increase the therapeutic efficacy of protein-based medicines. With the discovery of PEGylation by Davis et al. [1–3] in the 1970s, many proteins and small molecules have been examined using a wide range of strategies and reactive poly(ethylene glycol) (PEG) reagents for conjugation [4–14]. A primary reason to PEGylate proteins is to increase their half-life in the vascular circulation while also maintaining a large therapeutic index [15, 16]. Pioneering work has led to important clinical products that have been developed [4, 17–22]. The clinical use of PEGylated proteins has been shown to be safe [15] while indicating that PEGylation can provide significant unexpected, and as yet unexplained, therapeutic benefits [16, 23–26]. Certainly by just reducing the frequency of dosing, PEGylation can reduce the immuno-toxicities that can be associated with frequent dosing regimes that are often necessary for protein-based medicines. PEGylation is now considered an enabling technology that can be used in the development of protein-based medicines.

It is widely recognised that there has been little toxicity observed for PEG in the wide range of consumer and healthcare products in which it is used [15]. This lack of toxicity is also generally observed when a PEGylated product is administered parenterally [15, 27]. Although some concerns have been noted due to accumulation in animal models [28], recent regulatory approvals suggest that parenteral doses of up to 200 mg or more weekly of PEG as part of a PEGylated product are possible for extended periods [15, 27, 29–33]. In terms of accumulation, there is a trend for parenteral products that are given long-term that the molecular weights for each PEG chain should be less than about 40 kDa, although this is dependent on the medical indication, total dose given, mode of administration and the particular protein being used.

While PEG is considered safe, there is much effort that is focused to make PEGylated proteins that have increased purity [11, 14, 34–42]. The vast majority of PEGylated products are mixtures with PEG conjugated at different sites on the protein, leading to positional isomerism. In recent years site-specific PEGylation has become a key goal because the issue of positional isomers has caused some questions to be asked about the nature of heterogeneous product mixtures. In particular such mixed products can have a variety of biological activities and the physical and pharmacological properties of proteins are changed depending on the site and the number of PEG molecules conjugated to the protein [43, 44]. When PEG reagents are employed that allow reactions at multiple attachment sites along the protein backbone, a heterogeneous mixture of PEG-protein conjugates is produced. This mixture is not economic to purify and results in a complex final product that is difficult to standardise, reproduce and to scale for manufacture. Site-specific PEGylation would solve these problems and will allow for a more simple development of analytical methods and an easier regulatory path for new products.

The need for more uniform protein-based products is also driving research for other approaches that are designed to favourably alter the pharmacokinetics and efficacy of protein-based medicines. These include the development of fusion proteins [45–55] and re-engineered proteins where immunological or clearance sequences have been modified [56–58]. Other synthetic polymers [59], for protein conjugation particularly polysaccharide derivatives such as hydroxyethyl starch [60] and poly(sialic acid) [61, 62] ought to be mentioned. These polysaccharides tend to be polydisperse and variable in structure and they lack the now long clinical history as protein adducts compared to PEGylation. It will be important to confirm if novel sugar containing proteins show changed immunogenicity compared to the wild-type protein or to its PEGylated forms. The more rigid structure of polysaccharides compared to PEG may result in a greater propensity for immunogenicity. Other, more recent technologies include the use of short peptide sequences, e.g., the beta sub unit of chorionic gonadotropin [63] and poly(amino acids) [64]. Again there is the need to ensure these new sequences do not create new epitopes when combined with existing protein molecules. While these other technologies, including releasable linkers for PEGylation [65–67], continue to develop, PEGylation is already a clinically proven post-translational chemical process that can make proteins into better medicines, so PEGylation will also continue to be developed and utilised [68].

Considering the need for site-specific PEGylation, positional isomers result because the conjugation of PEG to a protein is usually performed with PEGylation reagents that generally undergo alkylation or acylation reactions with amine nucleophiles on the protein [4, 6, 7, 12, 69]. The terminal amine on a protein [34, 35, 39, 70] and the amine on lysine residues along the protein backbone [4, 7–9] have been the most used for conjugation. Other residues including histidine, serine and tyrosine have also been utilised. There has also been effort to conjugate PEG to non-nucleophilic residues such as side chain and terminal carboxylic acids [12, 36, 71–74].

A considerable number of PEGylation strategies have been developed in attempts to achieve site-specific PEG conjugation [34–40, 71, 75–78]. Approaches to site-specificity have included (1) a focus on the terminal amino acid in the protein of interest [34, 35, 38, 39, 79], (2) insertion of a non-paired cysteine as a PEGylation site [40, 76, 80–84], (3) structure based approaches using prediction of the most reactive groups on the protein surface [85], (4) oxidation or enzymatic reaction of a glycosylated moiety on the protein followed by PEGylation [86], (5) use of enzyme catalysed chemistry on a specific glycosylation site [87], or a transglutaminase to conjugate to a glutamine residue [36, 72] and (6) insertion of un-natural amino acids into the protein sequence to enable mono-PEGylation at that specific site [41]. There is also considerable current attention being paid to methods to post-translationally modify proteins generally [88–92] and some of these may translate into PEGylation strategies. Of the site-specific conjugation strategies listed above, only the first two approaches have to date resulted in clinical products [21, 29, 30, 93].

Terminal amine conjugation is typically conducted by reductive amination [6, 34, 35] and is considered to be a substantially site-specific PEGylation strategy [70]. Reductive amination is a two-step procedure that first relies on the site-specific formation of an imine in water followed by reduction of the imine to give a secondary amine that links the PEG to the protein. Efforts to determine the optimal pH to achieve site-specific reaction at the amine terminus are complicated because small differences in reaction conditions result in competitive imine formation along the protein main chain. The specific conditions needed to achieve substantially terminal amine specific PEGylation are usually very narrowly defined for each protein of interest.

Conjugation reactions can be conducted efficiently and site-specifically at free thiols in the presence of amine and hydroxyl groups in mild conditions

[7]. Most amine moieties on the protein are protonated and are unable to compete with a free thiol for conjugation at pH values below approximately 7.5. Mild reaction conditions that are most optimal for proteins can usually be found where thiol conjugation is efficient in the presence of the nucleophilic groups on the protein and with water. There is a general need to ensure that the PEGylation process is as efficient as possible so that yield and purification processes are economically viable [94]. Thiol based conjugations are generally much more efficient than amine/hydroxyl based conjugations in the reaction conditions that are suitable for most proteins.

The limitation of mono-thiol selective PEGylations is that most proteins do not have a free, unpaired cysteine as a source of a free thiol. Cysteine residues in extracellular proteins tend to be paired together as disulphides [95]. Few proteins display an unpaired cysteine [95–98]. When proteins have a naturally occurring unpaired cysteine it is invariably buried within the protein to avoid aggregation and oxidation [99], and it is not generally accessible for a conjugation reaction with a PEG reagent [100]. In order to utilise thiol-specific PEG reagents, it is usually necessary to recombinantly engineer a new and free cysteine into the protein [80]. Incorporating a free cysteine is technically demanding to ensure that biological activity can be maintained. The presence of a free unpaired cysteine increases the potential for irreversible protein aggregation during its purification. Since many therapeutically relevant proteins have native disulphide bonds, the presence of a recombinantly added unpaired cysteine also increases the propensity for disulphide scrambling and protein misfolding, further complicating protein purification.

In spite of these inherent limitations at the protein level, thiol-specific PEG reagents have been developed and are being intensely evaluated [6, 7]. These reagents include those that can undergo thiol exchange reactions [75] or alkylation reactions [6, 78]. Thiol exchange reagents covalently link the PEG to the protein via a non-native disulphide bond. A non-native disulphide bond linking PEG to the surface of a protein is very different to the native disulphides in a protein. The stability of native disulphides is complex [101, 102] and is a function of the non-covalent interactions that help maintain the tertiary structure of a protein [95, 102–104]. Such interactions would not be expected to contribute to the stability of a non-native disulphide linking PEG to the protein. The inherent accessibility of disulphides that link a protein molecule to a PEG molecule also contributes their chemical lability.

PEGylation reagents that undergo a thiol alkylation reaction include those that are capable of (i) substitution (e.g., iodoacetamide derivatives) and (ii) those reagents that can undergo an addition reaction. More examples of the second set of reagents have been examined and they include vinyl sulphone, vinyl pyridine, and maleimide end-group functionalised PEGs [40, 105, 106]. Maleimide-based conjugation is by far the most common of the thiol specific reagents that are currently being utilised. Although many maleimide derivatives have been examined, as a class of compounds they are hydrolytically labile [107].

Table 1. Classes of protein-based medicines that can be improved by PEGylation

Protein type	PEGylation need	Comment
Replacement proteins	Half-life extension required for proteins that require frequent administration.	Site-specific PEGylation would improve product homogeneity. There is a need for improved efficiency in production. Proteins include cytokines, growth factors, blood factors.
Peptides	Half-life extension required Mask immunogenicity for non-endogenous proteins.	Many peptides are in development. Site-specificPEGylation is required since the relative steric-shielding effects of PEG will be much greater than with larger molecular weight polypeptides.
Full antibodies	Half-life extension is not generally required. PEGylation could mask effector function.	Full antibodies are expensive to produce at the doses required for clinical efficacy. Effector function is not required for many applications.
Antibody fragments	Half-life extension required. Mask immunogenicity for non-endogenous proteins.	Site specific PEGylation is key to maintain binding while extending half-life and minimising immunogenicity. There is considerable potential for bacterial and yeast production.
Enzymes	Half-life extension may be required.	Multi-site, hyper-PEGylation of non-human proteins is preferred to avoid immunogencity.
Other proteins	Half-life extension required for proteins that require frequent administration.	Novel protein scaffolds including ankyrins, 10FN3 domain of human fibronectin and lipocalins. Mask immunogenicity for non-endogenous proteins.

There continues to be a need to develop efficient strategies for the site-specific PEGylation of proteins (Tab. 1). The development of protein-based medicines is at a crossroads. While there are many opportunities for PEGylation to be utilised in what is essentially life-cycle management with follow-on biologics [108], it is the advent of new classes of proteins [109, 110] and peptides [111–117] where site-specific PEGylation will be most useful. In this context, site-specific PEGylation should not be treated as a 'bolt-on' platform technology merely designed to optimise pharmacokinetics. It is our view that PEGylation should become an integral part early in the design and development of protein-based medicines. Preclinical studies of PEGylated proteins are performed with model systems and these are not adequately representative of the biological activity for PEGylated biopharmaceuticals in man. There is a general lack of clinically validated animal models for many of the diseases in which PEGylated biopharmaceuticals can be evaluated systematically. Hence there is a real need for being able to evaluate and optimise PEGylated proteins of defined structure during preclinical studies. As new classes of proteins are developed we further feel that it is imperative that site-specific PEGylation be developed that utilises *only* natural amino acids.

Disulphide bridging PEGylation

The vast majority of therapeutically relevant proteins do not have free unpaired cysteines that can be used for site-specific conjugation [96, 118]. They do however tend to have an even number of cysteines that pair up to form disulphides [95, 97, 98]. Our approach to site-specific PEGylation relies on the thiol selective chemistry of the two cysteine sulphur atoms that are derived from a native accessible disulphide. Site-specific PEGylation occurs by sequential, interactive bis-alkylation using α,β-unsaturated β'-mono-sulphone functionalised PEG reagents **1** (Scheme 1). This approach to PEGylation involves using PEG reagents that are capable of the specific type of bis-alkylation shown in Scheme 1. The process for PEGylation follows two general steps: (1) mild disulphide reduction to release the two sulphur atoms as thiols from the disulphide, and (2) conjugation of both thiols by sequential, interactive bis-alkylation to yield a three-carbon bridge to which PEG is covalently attached (Scheme 1) [119–121]. The advantage of this approach is that selective and efficient PEGylation can be accomplished without the need to recombinantly engineer the protein [80] to add a free cysteine or a non-native amino acid.

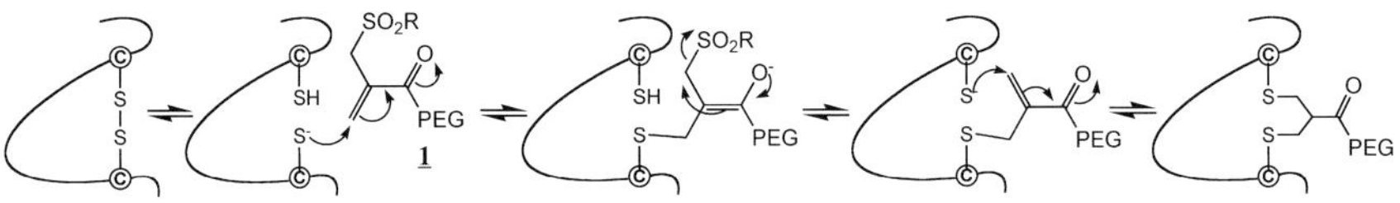

Scheme 1. Site-specific, disulphide bridging PEGylation of an accessible protein disulphide is achieved by disulphide reduction followed by reaction with the functionalised PEG **1**. These latently crossed functionalised PEGylation reagents are capable of sequential and interactive addition-elimination reactions leading to bis-thiol alkylation. Mechanistically, PEGylation involves (i) a first thiol addition to the PEG mono-sulphone, (ii) sulphinic acid elimination to generate a second double bond, and (iii) a second thiol addition. All the reagents we prepare undergo PEGylation by the same mechanism.

The PEGylation reagents that we use have a substituted propenyl group as the conjugating moiety on the terminus of PEG (e.g., PEG structure **1**, Scheme 1). This conjugation moiety comprises an electron withdrawing group (e.g., carbonyl), an α,β-unsaturated double bond, and an α,β' sulphonyl group that is prone to elimination as sulphinic acid. The electron-withdrawing group is required to promote thiol addition and to lower the pKa of the α-proton so that the elimination reaction can proceed. This juxtaposition of chemical functionality results in a latently cross-conjugated system. The conjugated double bond in the PEG mono-sulphone **1** is required to initiate a sequence of interactive and sequential addition-elimination reactions (Scheme 1). The addition of the first thiolate allows the elimination of a sulphinic acid derivative. This generates another conjugated double bond at the α, β'-position for the addition of a second thiolate. If the two thiols are derived from a protein disulphide bond, a three-carbon bridge is formed between the cysteine sulphur atoms of the original disulphide.

A key feature for site-selectivity in this approach to PEGylation is the need to first reduce a disulphide. Many relevant proteins only have one to two disulphides and thus pose no problems in terms of site-selectivity. Other proteins have more disulphides, and accessible disulphides can be selectively reduced in the presence of buried disulphides. While there are examples of small disulphide rich proteins of less than 10 kDa that can easily undergo reduction of all of their disulphides [122], selective reduction is also possible in many proteins using mild conditions and without the use of denaturants. Small proteins tend to have more disulphides than large proteins because they need to compensate for their relatively low number of hydrophobic interactions. Avoiding denaturants during reduction can ensure that inaccessible disulphides are not reduced. Buried disulphides [95, 123] usually contribute to a protein's tight packing and are therefore more difficult to reduce chemically. As a result, denaturants are often required to disrupt the protein's tertiary structure and thereby allow access of the reductant to the buried disulphide. The use of mild conditions enables the reduction of an accessible disulphide bond while still maintaining its tertiary structure. Once an accessible disulphide is reduced, the two free cysteine sulphur atoms become available for PEGylation. The chemical modification of an accessible disulphide with our bis-alkylation PEG reagent **1** (Scheme 1) completely differs from other forms of chemical modification. Instead of chemically functionalising each cysteine sulphur atom with a separate molecule to block the reformation of the disulphide bond – as many modifications do – a covalent link through the three-carbon bridge is maintained between the two cysteines of the original disulphide bond.

The therapeutic targets of protein-based medicines are in the extracellular environment. Unlike the cell cytoplasm, the extracellular environment is non-reducing, so proteins in the extracellular environment usually have intact disulphides [99, 101]. Currently, most classes of proteins that are used therapeutically are related to endogenous extracellular proteins, so it is not surprising that most [124] therapeutic proteins have disulphides. In the context of the fully folded protein, disulphide bonds influence the physicochemical and biological properties of the protein in subtle and complex ways [101, 102]. The tertiary structure of a protein is more stable than the unfolded forms of the protein [125–127]. While the folded protein displays a net gain in energy due to decreased entropy, the decrease in energy of the folded form of the protein occurs from the loss of enthalpy from many non-covalent interactions within the folded states [103, 104, 125–127]. While our method of PEGylation exploits the effective conjugation chemistry of thiols, it relies on re-bridging disulphides such that protein structure is not unduly perturbed.

There are many protein folding pathways [128] that have been studied *ex vivo* and protein folding while complicated, is influenced by varying degree by non-covalent interactions [103, 104]. As folding proceeds, there is usually an increase in these cooperative interactions. Disulphide bonds stabilise these intermediate folded structures as well as the final folded tertiary structure [129–132]. The controlled reduction of disulphides to study the unfolding of proteins has been useful to examine the mechanisms underlying the correct folding of a protein and to study those non-covalent interactions that stabilise its tertiary structure [133–140]. These studies have led to an understanding of how disulphide bonds function to stabilise a protein's structure and one concept that has emerged is that solvent accessible disulphides can contribute primarily to the stability of the protein rather than primarily to its biological function [95]. This essentially means such disulphides contribute to maintaining the non-covalent interactions within the tertiary structure of the protein.

In this context, there are several examples [141] where disulphides are derived from cysteines that are well separated in the protein's primary structure. For example the Cys6–Cys120 bond in α-lactalbumin is important for the stability of the protein [142]. There are many therapeutically relevant proteins with solvent accessible disulphides that are derived from cysteines that are well separated along the protein main chain. The cysteines in these disulphides are particularly relevant as targets for PEGylation. Computational studies can aid the process of selecting the target protein by determining the potential effects of inserting a three-carbon bridge into the accessible disulphides of a protein. Our search of publicly accessible protein databases suggests that there are many therapeutically relevant proteins that have at least one disulphide bond that is close to the surface of the protein which can be chemically modified without resulting in a loss of the protein's structure or function.

It is often quite straight forward using publicly accessible databases to determine which disulphides are solvent accessible, and we have used computational approaches to predict whether the insertion of a three-carbon bridge will lead to the loss of the protein's tertiary structure [143–145]. Modelling studies examined the insertion of a three-carbon bridge into each of the two disulphides in interferon-α2 (IFN). They predicted only minimal disturbance of the overall tertiary structure of the protein would occur [143, 144]. This conclusion was subsequently confirmed experimentally by (1) circular dichroism – which indicated that the disulphide reduced IFN and the bridged PEG-IFN maintained their native structure, and (2) biological studies – which showed that the mono PEG-IFN had antiviral activity that was similar to amine-PEGylated versions of IFN [119]. Such computationally generated modelling information is useful to gain some insight about the protein being studied. It can also be applied to proteins that have to be recombinantly engineered to incorporate an optimised disulphide that will be amenable to PEGylation using a three carbon bridge [146, 147].

Our disulphide bridging PEGylation reagents are fundamentally different from other PEGylation reagents because they undergo bis-alkylation as shown in Scheme 1. Other PEGylation reagents have been described that have two, or more, separate thiol selective moieties; e.g., two maleimides [76, 148] or two vinyl sulphones [149]. The reactive moieties in these reagents while separated by many more atoms than in our reagents, are not capable of undergoing bis-alkylation as shown in Scheme 1. In contrast, the sequential nature of the mechanism for our reagents ensures that only one alkylation reaction can occur at any point in time. The sequential, interactive nature of the addition reactions with our reagents **1** is important to ensure that efficient rebridging of the original disulphide bond can occur. The combination of (a) maintaining the protein's tertiary structure after reduction of a disulphide, (b) bis-thiol selectivity of the PEG reagent, and (c) PEG associated steric shielding generally ensures that only one PEG molecule is conjugated to bridge the sulphur atoms derived from a disulphide. Using the bis-alkylation reagents **1**, we have PEGylated peptides, proteins, enzymes and antibody fragments without abolishing their biological activity (Tab. 2). The PEGylation is efficient and often occurs with just one equivalent or a slight excess of reagent being required. Un-PEGylated protein can be isolated and recycled, or the reduction step can be repeated and followed by a second addition of the PEGylation reagent **1**. Our process for PEGylation can be accomplished with a range of reagents.

Table 2. Examples of proteins that have been PEGylated by disulphide bridging

Protein class	Example proteins	Activity
Cytokines, hormones, haemapoietic proteins (helical barrel proteins)	Interferon α-2a	✓
	Interferon α-2b	✓
	Erythropoietin	✓
	Leptin	✓
Blood proteins	Coagulation factor	✓
Antibody fragments	Anti-TNFα	✓
	Anti-CD4	✓
Enzymes	Asparaginase	✓
	Lipase	✓
Peptides	Imaging ligand	✓

Reagent synthesis

The first reagent that we examined that is analogous to the generalised reagent **1** (Scheme 1) was the PEG mono-sulphone **8** (Scheme 2). This PEG reagent **8** is typically prepared from the known carboxylic acid bis-sulphone **5** [150–152] (Scheme 2). Coupling of mono-amine terminated PEG is routinely accomplished using the active ester bis-sulphone **6** or by direct carbodiimide mediated coupling with the acid bis-sulphone **5**.

Analogous coupling reactions to append PEG can also be conducted with the carboxylic acid bis-sulphide **4** (or its NHS ester) to give the corresponding PEG bis-sulphide. These can then be gently oxidised to give the corresponding PEG bis-sulphone without any changes to the structural characteristics of the

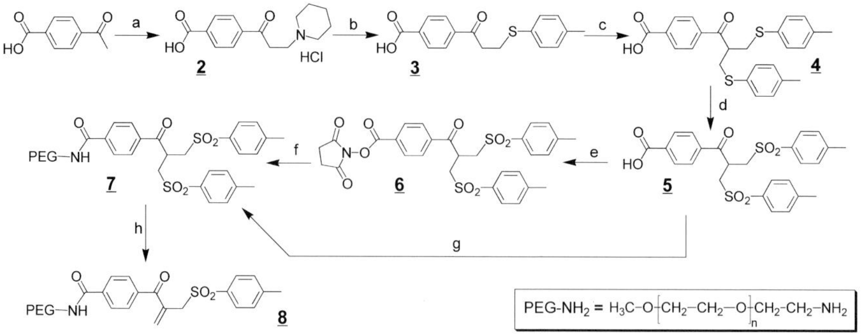

Scheme 2. Preparation of PEG mono-sulphone **8**: (a) paraformaldehyde, piperdine hydrochloride, ethanol, reflux; (b) toluene thiol, ethanol, reflux; (c) formalin, ethanol, toluene thiol, reflux; (d) OXONE; (e) DIPC, NHS; (f) methoxy-PEG-NH$_2$; (g) DCC, CH$_3$O-PEG-NH$_2$; (h) phosphate buffer, pH 7.8.

PEG. We used this strategy to prepare the PEG bis-sulphone **11** (Scheme 3) and with other polymers such as the water-soluble, biocompatible polymer, poly(2-methacryloyloxyethyl phosphorylcholine) (PMPC) [153]. In the case of PMPC, we found the most effective way to prepare the PEGylation reagents from this polymer was to prepare a bis-sulphide initiator that was used to give the analogous PMPC-bisulphide. This was then oxidised to give the desired PMPC-bis-sulphone conjugation reagent that was used to PCylate interferon [154].

Although the reaction pathway for PEG conjugation requires the PEG mono-sulphone **8** to be present for the first addition reaction, the precursor PEG bis-sulphones (e.g., **7** and **11**) tend to be used for PEGylation reaction. This is because the PEG bis-sulphone can be used to generate the mono-sulphone (e.g., **7** → **8**) *in situ* (Scheme 4) using mild conditions that are generally applicable for solubilising proteins. The PEG mono-sulphone can be gener-

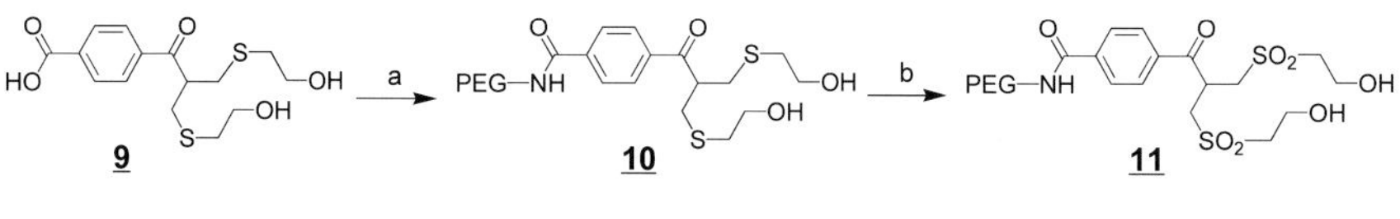

Scheme 3. Synthesis of PEG bis-sulphone **11**: (a) DIPC and (b) OXONE.

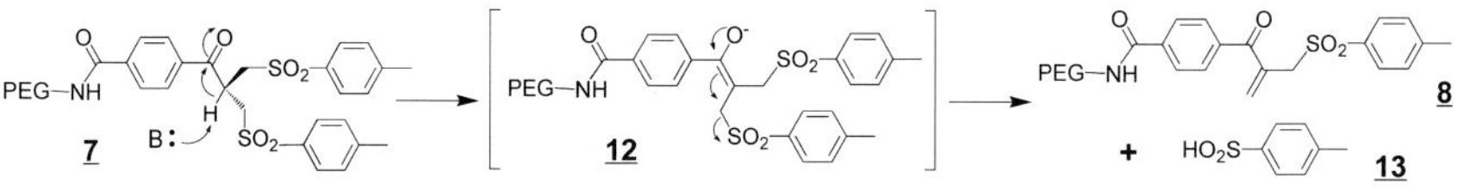

Scheme 4. Elimination of toluene sulphinic acid **13** from PEG *bis*-sulphone **7** generates PEG mono-sulphone **8**. It is required for the first thiol addition reaction to occur.

ated in a controlled fashion while in the presence of the protein. Specifically, we have observed that the PEG bis-sulphone **7** undergoes elimination to give the corresponding PEG mono-sulphone **8** at pH values of 6.0 or higher. The side product from this reaction is toluene sulphinic acid **13**. We have characterised this elimination reaction by ^{1}H-NMR and RP-HPLC [120]. The rate at which this elimination reaction occurs is dependent upon the structure of the PEGylation reagent, pH, concentration and temperature. Preliminary computational studies of the elimination reaction from the PEG bis-sulphone have been published [143].

There are two advantages for using the PEG bis-sulphone reagents. Firstly, the PEG bis-sulphones are stable compounds that are not readily susceptible to hydrolytically based degradation reaction and can be easily stored. Secondly, a single reagent can be used to site-specifically PEGylate a wide range of proteins. The facility to control the generation of the reactive conjugating molecule (i.e., PEG mono-sulphone) provides a means by which the rate of the first addition reaction to the protein can be controlled. Hence PEG bis-sulphone reagents are important because the rate at which the corresponding PEG mono-sulphone is generated *in situ* can be matched to the thiol reactivity of the protein after the reduction of its disulphide(s). This means that the conjugation conditions can be tailored by simple means to optimise the PEGylation for a wide range of proteins. When it is necessary to conduct a PEGylation reaction at an acidic pH value because of, for example, issues related to the protein's solubility or its stability, it is best to isolate and use a PEG mono-sulphone reagent directly for the conjugation reaction [120, 121]. Some peptides and proteins are less prone to aggregation or are more soluble in slightly acidic conditions. Also a range of pH conditions can be useful for optimising the reduction step; e.g., reduction of somatostatin at pH 6.2 followed by conjugation with the PEG mono-sulphone **8** [120]. The synthesis and isolation of pure PEG mono-sulphone (e.g., **8**) is accomplished by incubation of the corresponding PEG bis-sulphone in 50 mM phosphate buffer (pH 7.8) followed by purification on reverse-phase (RP) HPLC.

For the majority of protein PEGylation reactions, it is more practical to use the PEG bis-sulphone to generate the PEG mono-sulphone *in situ*. We have prepared the PEG bis-sulphone reagents using PEGs of molecular weights up to 40 kDa. The reagents are easily prepared on a multigram scale. It is important with polymer-based reagents to ensure that there is adequate availability of the reactive moiety in a range of conditions that are appropriate for proteins. We have certainly observed that with some water-soluble polymers [155] the reactivity of the functionality at the terminus can be reduced [154]. For example, we were not able to observe protein conjugation with aldehyde or N-hydroxysuccinimide ester terminal functionalities on PMPC polymers. It was not possible to prepare efficiently the PMPC bis-sulphone directly from the corresponding PMPC amine because the amine reactivity was low.

It is also known that thiol reactivity in a protein is a function of neighbouring amino acid residues [156]. To ensure our reagents can be optimised for use with any given protein while also following the same mechanism for conjugation (Scheme 1), we have prepared PEGylation reagents with different leaving groups (e.g., PEG bis-sulphone **11**, Scheme 2). The overall hydrophilicity of the reactive moiety can be varied by simple changes in the leaving group. Such changes in the hydrophilicity of the polymer terminus could be important for presenting the reactive moiety optimally to a protein of interest. Also, since the overall conjugation reaction is thought to be thermodynamically driven, it is anticipated that differences in leaving group solubility will influence the reaction.

It is possible to make a wide range of PEGylation reagents simply by using the easily prepared PEG bis-sulphone **7** (shown in Scheme 2). Since the PEG bis-sulphone **7** can undergo reaction with thiols via the corresponding mono-sulphone **8**, it is a simple matter to allow the PEG bis-sulphone **7** to react with functionalised thiols of interest as a means to make PEGylation reagents with different leaving groups. The bis-sulphide products from these reactions can then be oxidised to give the corresponding bis-sulphones.

It is also possible to utilise the PEG bis-sulphone reagent **7** to prepare reagents with completely different types of leaving group. For example amines, which can be quarternised or are protonated during protein conjugation to form a charged PEGylation reagent, can be utilised. In this case, a charged PEGylation reagent could more optimally interact with the protein of interest. For example (Scheme 5), the PEG bis-sulphone **7** was used to give the PEG mono-amine **14** which was found to readily undergo protein PEGylation.

In addition to varying the leaving groups, we are also examining reagents that link the PEG to the protein entirely by an aliphatic linkage. An example aliphatic PEG bis-sulphone **18** is shown in Scheme 6. The electron-withdrawing group in this reagent **18** is an amide rather than an aryl carbonyl that is present in the other PEG bis-sulphones that have been described. While the reactivity of amide **18** is reduced, these reagents are still observed to undergo

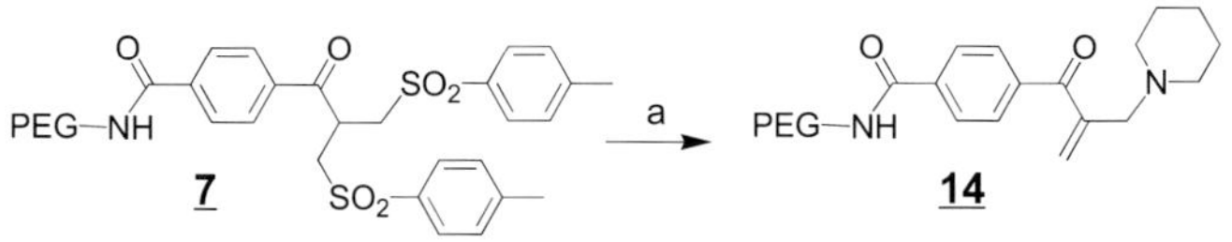

Scheme 5. Synthesis of PEG mono-amine **14**: (a) piperidine, ambient temperature, THF, 3.5 h.

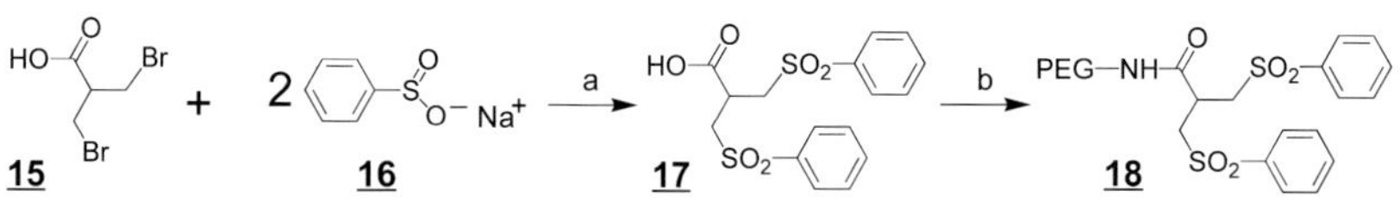

Scheme 6. Synthesis of aliphatic amide PEG bis-sulphone **18**: (a) NaOH, methanol, reflux, 4 h and (b) DCC.

protein PEGylation. The practical synthesis of these aliphatic reagents uses sulphinic acid salts [157] as nucleophiles and offers the opportunity to prepare many types of reagent that could be used to conjugate proteins by the mechanism shown in Scheme 1.

The process for disulphide PEGylation

The process that we follow to PEGylate proteins comprises two steps: (1) reduce an accessible disulphide and (2) rebridge the disulphide by sequential bis-alkylation PEGylation. We generally observe that an equivalent or a small excess of the PEGylation reagent is required for each disulphide to be rebridged with PEG. The reagents do not typically undergo reaction with water, salts, buffer or common denaturants. We always conduct control PEGylation reactions to confirm that no conjugation has occurred unless a disulphide in the protein is first reduced. Purification is usually straightforward because little excess PEG is present and the number of products formed is related to the number of disulphides that have been reduced.

Considering the mechanism for the sequential bis-alkylation reactions (Scheme 1), there can only be one double bond available in the reagent for thiol addition at any point in time. Both β positions are activated by the same carbonyl electron-withdrawing group and thus are not chemically independent of each other. The reversible nature of thiol addition reactions to conjugated double bonds [158, 159] contribute to the efficient formation of the three-carbon bridge for a given disulphide. When thiols are present, then amines display less relative reactivity at the reaction conditions that are conducive for thiol addition reactions. Thiols are nucleophilic at neutral pH and therefore most amine residues will be protonated. Amine alkylation and acylation reactions often require more basic conditions and often the reagents undergo competitive reactions with water. While thiol specificity and the equilibrium nature of conjugate addition chemistry are key, there are two additional physicochemical characteristics that help to underpin the success of our approach for site-specific PEGylation: (1) maintain the propinquity of the two cysteines after the reduction of a disulphide bond and (2) steric shielding by the PEG after the first addition reaction has occurred.

Throughout the disulphide reduction and PEGylation steps for most proteins, it is generally possible to preserve a varying degree of the protein tertiary structure. This helps to keep the free thiols close to each other and to minimise protein aggregation. Maintenance of the protein's tertiary structure after the reduction of the disulphide ensures that the second cysteine sulphur atom is nearby so that it favours the second addition reaction that is required to connect the three-carbon bridge. Lawton et al. [150–152, 160] first demonstrated that the β,β'-bis-sulphone functionality could undergo bis-thiol alkylation via the alkenyl β-mono-sulphone to crosslink proteins by the mechanism shown in Scheme 1 [151, 152, 160]. The bis-sulphone functionality was also examined with antibody fragments [161], and others have studied the chemical reactivity of related propenyl reactive moieties in organic synthesis [150, 162–166]. In the protein cross-link studies multiple intermolecular and competitive reactions with other amino acid residues were observed [150–152, 160, 161]. Without PEG attached, the early compounds were not water-soluble. Exposure of therapeutic proteins to organic solvents generally results in a greater propensity for denaturation due to the disruption of the hydrophobic interactions within the protein structure. Denaturation will thus increase when protein disulphides are reduced in such conditions.

After the first addition reaction (Scheme 1), the steric shielding of the PEG will inhibit the approach of a second protein molecule with a free thiol to compete for the second conjugation reaction. The steric shielding of PEG after the first conjugation reaction should also inhibit the approach of a second PEG, and thereby prevent one molecule of PEG becoming conjugated to each of the cysteine sulphur atoms from a disulphide. The combination of PEG steric shielding and broadly maintaining protein tertiary structure are thought important for disulphide-bridging PEGylation.

Different proteins have been examined to determine the scope for disulphide bridging PEGylation (Tab. 2). Digestion, NMR and MALDI studies are all consistent with PEGylation having occurred by disulphide bridging [119, 120]. Control reactions are often conducted. A typical example is asparaginase [120]. This tetrameric protein has one disulphide and 22 lysines per monomeric unit. There was no observed reaction when asparaginase was incubated with 1.3 equivalents of the PEG mono-sulphone **8** at pH values ranging from 6.0–8.6 overnight at ambient temperature. Once the disulphide was reduced with DTT, quantitative PEGylation was observed in these conditions. This example illustrates the efficiency and the selectivity for PEGylating proteins with the PEG mono-sulphone **8**.

For proteins with one accessible disulphide we have found that quantitative PEGylation occurs with one equivalent of the PEG bis-sulphone **7**. This is observed for asparaginase and also for leptin, which is α-helical bundle protein. Leptin shares many structural features with therapeutic proteins such as interferons α and β, granulocyte-colony stimulating factor, growth hormone and erythropoietin. Computational studies showed that the disulphide bond in leptin was accessible and stochastic modelling indicated that leptin would maintain its tertiary structure if the disulphide was modified with a three-carbon bridge to which PEG was attached [167]. Compared to many other α-helical bundle proteins where disulphides are formed by two cysteines each located in a helix, one cysteine in the disulphide bond in leptin (Fig. 1) is located in a very flexible loop of the protein. The other cysteine in leptin is located in a helix. It is therefore possible that reduction of the disulphide in leptin could lead to the separation of the two free cysteine sulphur atoms making rebridging of the disulphide difficult. However, PEGylation using the PEG bis-sulphone **7** (1.3 equivalents) in phosphate buffer at pH 7.8 with 2 M L-arginine gave a high yield of the mono-PEGylated leptin (~67% by SEC). In a cell pro-

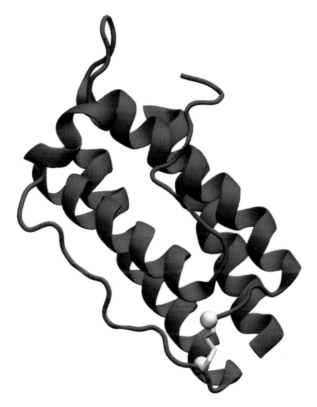

Figure 1. Computer generated image of leptin showing the single disulphide bond that links a cysteine located on one helical barrel and a cysteine located on the protein with little secondary structure.

liferation assay, the mono-PEGylated leptin showed optimum activity at a concentration range of 0.5–1 µg/ml. This is equivalent to ~10% of the proliferative activity of leptin which has an optimum proliferative activity between 0.05–0.1 µg/ml [168].

If a protein has two accessible disulphides, such as interferon-α2, and if both are reduced (e.g., interferon-α2) followed by the addition of one equivalent of the PEG bis-sulphone **7**, we observe yields of 60–70% of the mono-PEGylated IFN. The remainder of the product mixture was unPEGylated protein and diPEGylated IFN (~5%). If two equivalents of the PEG bis-sulphone **7** were added, then diPEGylated-IFN was the exclusive product. Each disulphide in interferon holds two helical barrels together. While the disulphides in interferon are easily reduced, the hydrophobic interactions between the α-helices aid in maintaining the tertiary structure of the reduced protein as observed by circular dichorism. With a single PEG attached, the biological activity of the PEG-IFN was similar to the published *in vitro* activity of the amine linked PEGylated IFN-α2a in clinical use [43, 44, 169–172]. The di-PEGylated-IFN also displayed *in vitro* biological activity [120]. As expected, the *in vivo* half-life of the PEG-IFN directly correlated with the size of the PEG [119, 120].

Interestingly, the biological activity of the disulphide bridged PEG-IFNs was independent of the molecular weight of the PEG attached. This is in contrast to the experience with protein amine PEGylation where biological activity varies with the PEG [43]. We have also observed this independence of biological activity as a function of the size of the PEG with asparaginase. We believe that this is due to the site-specific nature of the disulphide bridging PEGylation, which is independent of the size of the PEG reagent used. Often with amine selective reagents, a large excess of reagent is used and different positional isomers are isolated because of the kinetic nature of the reaction. Different product mixtures are observed by just varying the size of the PEG reagent. Product mixtures also depend on the protein and the specific PEGylation process that is used. For example no conjugation of PMPC to IFN was observed with MPC reagents that were prepared for amine alkylation (aldehyde) or amine acylation (NHS ester). Using a PMPC bis-sulphone reagent the corresponding PMPC-IFN molecule was made, and both *in vitro* activity and an extended half-life for IFN were observed [154].

It is important to ensure that any disulphides that have been reduced, but not PEGylated are then reoxidised [120]. This will ensure the retention of the protein's biological activity. For example, when both of the disulphides of IFN were reduced, and one equivalent of the PEG bis-sulphone **7** was added, the unPEGylated and reoxidised protein retained its biological activity. Therefore any reduced disulphides that have not been PEGylated, can be reoxidised to enable recycling of any unPEGylated protein for another disulphide bridging PEGylation reaction.

Other proteins with multiple numbers of disulphides have also been evaluated. Examples include ribonuclease and lipase. Lipase B has three disulphide bonds (Fig. 2). Using a 20 kDa disulphide-bridging PEG reagent, the major product was monoPEGylated lipase (40%), and as with interferon, the reaction could shift towards diPEGylated lipase formation with the addition of increased amounts of the PEGylation reagent. Without optimising this reaction, we determined the *in vitro* activity of the PEGylated lipase by measuring the amount of *p*-nitrophenol released from the substrate *p*-nitrophenyl palmitate. Compared to the native lipase, the mono-PEGylated lipase retained 25–50% activity. Also in a similar fashion to interferon, no significant difference was observed when different molecular weights of PEG were attached, corroborating our hypothesis of the importance of site specificity on the independence of bioactivity with PEG size.

Antibody fragments do not have the effector functionalities that are associated with the constant region and they can be manufactured more economically from bacterial and yeast sources. There is significant effort to develop fragments, novel protein scaffolds [109, 110] and peptides [173] that can display antibody or protein-like binding properties. In antibodies generally, there are several accessible disulphides that can be PEGylated. For Fabs, there is an

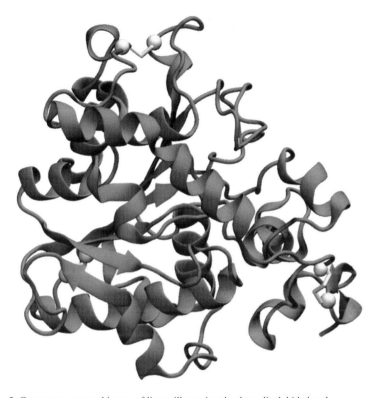

Figure 2. Computer generated image of lipase illustrating the three disulphide bonds.

accessible disulphide in the vicinity of the hinge region. It is distal to the Fab's receptor binding site. There are multiple non-covalent interactions between the light and heavy chains that maintain the Fab's tertiary structure when the hinge disulphide is reduced. It has been reported that this disulphide in the Fab can be removed with little loss of Fab stability [174], hence this disulphide is a particularly good target for our PEGylation strategy. The conditions required to reduce the accessible disulphide are much milder than those required to reduce the other more buried disulphides located within a Fab. This enables practical site-specific PEGylation of Fabs [119]. We have shown with disulphide bridging PEGylation that the anti-HIV-1 activity of an anti-CD4 Fab is retained. As expected its affinity for the CD4 cell surface receptor on lymphocytes is reduced. These results are consistent with the observations of Kubetzko et al. [175] where the attachment of a 20 kDa PEG near the hinge region of an antibody fragment resulted in a five-fold reduction in the apparent *in vitro* affinity of a PEGylated antibody fragment to its receptor. This loss of affinity was due to a reduction in the association rate of the PEGylated molecule for its receptor. There was no change in the dissociation rate. Although PEGylation reduces the observed association rate, all of the PEGylated molecules can eventually bind to the receptor. Receptor clustering may also affect binding affinity because the steric shielding of PEG would be expected to influence the approach of neighbouring PEGylated molecules. The deconvolution of avidity and PEG steric effects is complex [176] and much still remains to be elucidated. Related observations have been made for PEG-IFN [43, 172]. PEGylated molecules may intermolecularly inhibit association events of the protein to which it is covalently bound, thus resulting in an apparent reduction in the number of binding sites, minimising protein aggregation and increasing protein stability [177].

One general paradigm in medicinal chemistry during analogue optimisation is to rigidify the structure of candidate compounds to decrease their conformational mobility in order to better match a profile of interactions with the target of interest. The same general strategy may apply with the development of peptides because they can readily be accessed by phage libraries and optimised as cyclic peptides designed to mimic the relevant properties of a protein of interest [173, 178]. Our PEGylation strategy works well with peptides, particularly cyclic peptides that have a disulphide. For example, somatostatin is comprised of 14 amino acid residues (molecular mass – 1,637.9 Da) and has a disulphide bond between Cys3 and Cys14. This disulphide when reduced, readily underwent PEGylation using the PEG mono-sulphone **8**. The disulphide reduction of somatostatin was optimally accomplished with TCEP-HCl at slightly acidic values because at higher pH values peptide precipitated out of solution. TCEP-HCl [179] was ideal for disulphide reduction at acidic pH values and it could be used at stoichiometric equivalence. Once oxidised, TCEP was not nucleophilic and it did not react with the PEG mono-sulphone **8**; therefore, its removal [180] before conjugation was unnecessary.

All our data show that the PEG-to-protein linkage is stable in the proteins and peptides that we have examined. The thiol-ether bonds formed during PEGylation are not susceptible to reduction. Although in principle it is possible that a reverse Michael reaction can occur in the PEGylated protein, we have not observed this to happen, even in the presence of excess dithiothritol (DTT) in water at 95 °C. Since the electron-withdrawing group in our reagents is a carbonyl group, it is always possible to reduce it to ensure there is no possibility for thiol elimination. It is also reassuring from the published literature that the consensus is that PEGylated biologicals have a low exposure-toxicity relationship in humans [15]. Therefore, if the toxicological evaluation of the biological is complete, and efficacy is confirmed, the PEGylation of a protein is not expected to present an additional or unexpected risk in human studies. Nonetheless, as for any therapeutic protein that is being developed for use in humans, a toxicology package in relevant species will be required prior to starting clinical trials. Since our strategy for PEGylation is site specific with both the PEGylation process and activity broadly independent of PEG size, preclinical optimisation essentially only requires the determination of the disulphide reducing conditions and the size of PEG that is required for optimal pharmacokinetic and pharmacodynamic properties.

Conclusions

PEGylation is a clinically proven strategy to increase the efficacy of protein-based medicines. Our approach to site-specific PEGylation exploits the thiol selective chemistry of the two cysteine sulphur atoms from an accessible, naturally occurring disulphide. The efficiency of the process is exemplified by the near stoichiometric use of our reagents during PEGylation. The PEGylation reagents are designed to undergo bis-thiol alkylation by an interactive mechanism that involves addition-elimination reactions. The result is that PEG is attached to a three-carbon bridge that is linked to the two sulphur atoms from an accessible protein disulphide. Our site-selective approach to protein PEGylation is chemically efficient and provides a unique PEGylation technology where the PEG size can be tailored to a desired *in vivo* half-life without influencing the *in vitro* activity. Since no recombinant engineering is required to incorporate a cysteine or a non-native amino acid more homogenous PEG-protein products may be achievable more easily. Disulphide bridging PEGylation is therefore a technology for the rapid development of PEGylated biopharmaceuticals that may become more cost-effective medicines.

Acknowledgements
SB is grateful to the Wellcome Trust (068309), BBSRC-UK (BB/D003636/1) and PolyTherics for funding.

References

1. A. Abuchowski, T. Es, N. Palczuk, and F. Davis, *Alteration of immunological properties of bovine serum albumin by covalent attachment of polyethylene glycol.* J. Biol. Chem., 1977. 252(11): p. 3578–3581.
2. A. Abuchowski, J. McCoy, N. Palczuk, T. Es, and F. Davis, *Effect of covalent attachment of poly-ethylene glycol on immunogenicity and circulating life of bovine liver catalase.* J. Biol. Chem., 1977. 252(11): p. 3582–3586.
3. F. Davis, *The origin of pegnology.* Adv. Drug Del. Rev., 2002. 54: p. 457–458.
4. S. Zalipsky, *Chemistry of poly(ethylene glycol) conjugates with biologically active molecules.* Adv. Drug Del. Rev., 1995. 16: p. 157–182.
5. R. Greenwald, Y. Choe, J. McGuire, and C. Conover, *Effective drug delivery by PEGylated drug conjugates.* Adv. Drug Del. Rev., 2003. 55: p. 217–250.
6. M. Roberts, M. Bentley, and J. Harris, *Chemistry for peptide and protein PEGylation.* Adv. Drug Del. Rev., 2002. 54: p. 459–476.
7. J. Harris and R. Chess, *Effect of Pegylation on pharmaceuticals.* Nat. Rev. Drug Dis., 2003. 2(3): p. 214–221.
8. G. Pasut, A. Guiotto, and F. Veronese, *Protein, peptide and non-peptide drug PEGylation for therapeutic application.* Expert Opin. Ther. Patents, 2004. 14(5): p. 1–36.
9. F. Veronese and G. Pasut, *PEGylation, successful approach to drug delivery.* Drug Discovery Today, 2005. 10(21): p. 1451–1458.
10. G. Pasut and F. Veronese, *Polymer-drug conjugation, recent achievements and general strategies.* Prog. Polym. Sci., 2007. 32: p. 933–961.
11. G. Kochendoerfer, *Site-specific polymer modification of therapeutic proteins.* Curr. Opin. Chem. Biol., 2005. 9: p. 555–560.
12. M. Gauthier and H. Klok, *Peptide/protein-polymer conjugates: synthetic strategies and design concepts.* Chem. Comm., 2008: p. 2591–2611.

13. J. Lutz and H. Borner, *Modern trends in polymer bioconjugates design.* Prog. Polym. Sci., 2008. 33: p. 1–39.
14. K. Heredia and M. Maynard, *Synthesis of protein-polymer conjugates.* Org. Biomol. Chem., 2007. 5(1): p. 45–53.
15. R. Webster, E. Didier, P. Harris, N. Siegel, J. Stadler, L. Tilbury, and D. Smith, *PEGylated proteins: evaluation of their safety in the absence of definitive metabolism studies.* Durg Metabol. Dispos., 2007. 35(1): p. 9–16.
16. P. Caliceti and F. Veronese, *Pharmacokinetic and biodistribution properties of poly(ethylene glycol) – protein conugates.* Adv. Drug Del. Rev., 2003. 55: p. 1261–1277.
17. F. Fuertges and A. Abuchowski, *The clinical efficacy of poly(ethylene glycol) modified proteins.* J. Cont. Rel, 1990. 11: p. 139–148.
18. Y. Wang, S. Youngster, M. Grace, J Bausch, and DFWyss, *Structural and biological characterization of pegylated recombinant interferon alpha-2b and its therapeutic implications.* Adv. Drug Del. Rev., 2002. 54: p. 547–570.
19. K. Reddy, M. Modi, and S. Pedder, *Use of peginterferon alfa-2a (40 KD) (Pegasys) for the treatment of hepatitis C.* Adv. Drug Del. Rev., 2002. 54: p. 571–586.
20. G. Pasut, M. Sergi, and F. Veronese, *Anti-cancer PEG-enzymes: 30 years old, but still a current approach.* Adv. Drug Del. Rev., 2008. 60: p. 69–78.
21. D. Piedmonte and M. Treuheit, *Formulation of Neulasta (pegfilgrastim).* Adv. Drug Del. Rev., 2008. 60: p. 50–58.
22. M. Sherman, M. Saifer, and F. Ruiz, *PEG-uricase in the management of treatment-resistant gout and hyperuricemia.* Adv. Drug Del. Rev., 2008. 60: p. 59–68.
23. M. Fried, M. Shiffman, K. Reddy, C. Smith, G. Marinos, F. Goncales, D. Haussinger, M. Diago, G. Carosi, D. Dhumeaux, A. Craxi, A. Lin, J. Hoffman, and J. Yu, *Peginterferon alfa-2a plus ribavirin for chronic hepatitis C virus infection.* N. Engl. J. Med., 2002. 34: p. 975–982.
24. S. Hadziyannis, *Peginterferon-alpha2a and ribavirin combination therapy in chronic hepatitis C: a randomized study of treatment duration and ribavirin dose.* Ann. Intern. Med., 2004. 140: p. 346–355.
25. A. Talal and e. al., *Pharmacodynamics of PEG-IFN alpha differentiate HIV/HCV coinfected sustained virological responders from nonresponders.* Hepatology, 2006. 43: p. 943–953.
26. C. Fishburn, *The pharmacology of PEGylation: Balancing PD with PK to generate novel therapeutics.* J. Pharm. Sci., 2008. In press: p. DOI 10.1002/jps.
27. M. Young, A. Malavalli, N. Winslow, K. Candegriff, and R. Winslow, *Toxicity and hemodynamic effects after single dose administration of MalPEG-hemoglobin (MP4) in rhesus monkeys.* Translational Res., 2007. 149(6): p. 333–342.
28. A. Bendele, J. Seely, C. Richey, G. Sennello, and G. Shopp, *Reneal tubular vacuolation in animals treated with poly(ethylene glycol) conjugated proteins.* Tox. Sci., 1998. 42: p. 152–157.
29. W. Sandborn, B. Feagan, S. Stoinov, P. Honiball, P. Rutgeerts, D. Mason, R. Bloomfield, and S. Schreiber, *Certolizumab Pegol for the treatment of Crohn's disease.* N. Engl. J. Med., 2007. 357(3): p. 228–238.
30. S. Schreiber, M. Kareemi, I. Lawranc, O. Thomsen, S. Hanauer, J. McColm, R. Bloomfield, and W. Sandborn, *Maintenance therapy with certolizumab Pegol for Crohn's disease.* N. Engl. J. Med., 2007. 357(3): p. 239–250.
31. C.p. label., *Cimzia package label.* 2008.
32. C. Parkinson, J. Scarlett, and P. Trainer, *Pegvisomant in the treatment of acromegaly.* Adv. Drug Del. Rev., 2003: p. 1303–1314.
33. D. Yin, F. Vreeland, L. Schaaf, R. Millham, B. Duncan, and A. Sharma, *Clinical pharmacodynamic effects of the growth hormone receptor antagonist Pegvisomant: implications for cancer therapy.* Clin. Cancer Res., 2007. 13(3): p. 1000–1009.
34. O. Kinstler, D. Brems, S. Lauren, A. Paige, J. Hamburger, and M. Treuheit, *Characterization and stability of N-terminally PEGylated rhG-CSF.* Pharm. Res., 1996. 13(7): p. 996–1002.
35. O. Kinstler, G. Molineux, M. Treuheit, D. Ladd, and C. Gegg, *Mono-N-terminal poly(ethylene glycol)-protein conjugates.* Adv. Drug Del. Rev., 2002. 54(4): p. 477–485.
36. H. Sato, *Enzymatic procedure for site-specific pegylation of proteins.* Adv. Drug Del. Rev., 2002. 54: p. 487–504.
37. C. Edwards, S. Martin, J. Seely, O. Kinstler, S. Buckel, A. Bendele, M. Cosenza, U. Feige, and T. Kohno, *Design of PEGylated soluble tumor necrosis factor receptor type I (PEG sTNF-RI) for chronic inflammatory diseases.* Adv. Drug Del. Rev., 2003. 55(10): p. 1315–1336.

38. H. Lee, H. Jang, S. Ryu, and T. Park, *N-terminal site-specific mono-PEGylation of epidermal growth factor.* Pharm. Res., 2003. 20(5): p. 818–825.
39. Y. Yamamoto, Y. Tsutsumi, Y. Yoshioka, T. Nishibata, K. Kobayashi, T. Okamoto, Y. Mukai, T. Shimizu, S. Nakagawa, S. Nagata, and T. Mayumi, *Site-specific PEGylation of a lysine-deficient TNF-alpha with full bioactivity.* Nature Biotechnology, 2003. 21: p. 546–552.
40. I. Gentle, D. DeSouza, and M. Baca, *Direct production of proteins with N-Terminal cysteine for site-specific conjugation.* Bioconjugate Chem., 2004. 15: p. 658–663.
41. A. Deiters, T. Cropp, D. Summerer, M. Mukherji, and P. Schultz, *Site-specific PEGylation of proteins containing unnatural amino acids.* Bioorg. Med. Chem. Lett., 2004. 14: p. 5743–5745.
42. P. Thordarson, B. Droumaguet, and K. Velonia, *Well-defined protein-polymer conjugates – synthesis and potential applications.* Appl. Microbiol. Biotechnol., 2006. 73: p. 243–254.
43. M. Grace, S. Lee, S. Bradshaw, J. Chapman, J. Spond, S. Cox, M. DeLorenzo, D. Brassard, D. Wylie, S. Cannon-Carlson, CCullen, S. Indelicato, M. Voloch, and R. Bordens, *Site of pegylation and polyethylene glycol molecule size attenuate interferon-α antiviral and antiproliferative activities through the JAK/STAT signaling pathway.* J. Biol. Chem., 2005. 280(8): p. 6327–6336.
44. M. Grace, S. Youngster, G. Gitlin, W. Sydor, L. Xie, L. Westreich, S. Jacobs, D. Brassard, J. Bausch, and R. Bordens, *Structural and biologic characterization of pegylated recombinant IFN-α2b.* J. Interferon Cytokine Res., 2001. 21: p. 1103–1115.
45. M. Goldenberg, *Etanercept, a novel drug for the treatment of patients with severe, active rheumatoid arthritis.* Clinical Thera., 1999. 21(1): p. 75–87.
46. A. Li and Z. Qian, *Transferrin/transferrin receptor-mediated drug delivery.* Med. Res. Rev, 2002. 22(3): p. 225–250.
47. N. Kavimandan, E. Losi, J. Wilson, J. Brodbelt, and N. Peppas, *Synthesis and characterization of insulin-transferrin conjugates.* Bioconjugate Chem., 2006. 17: p. 1376–1384.
48. K. Schriebl, E. Trummer, C. Lattenmayer, R. Weik, R. Kunert, D. Muller, H. Katinger, and K. Uhl, *Biochemical characterization of rhEpo_Fc fusion protein expressed in CHO cells.* Protein Exp. Purif., 2006. 49: p. 265–275.
49. G. Cox, D. Smith, S. Carlson, A. Bendele, E. Chilpala, and D. Doherty, *Enhanced circulating half-life and hematopeietic properties of a human granulocyte colony-stimulating factor/immunoglobulin fusion protein.* Exp. Hematol., 2004. 32: p. 441–449.
50. I. Wilkinson, E. Ferrandis, P. Artymiuk, M. Teillot, C. Soulard, C. Touvay, S. Pradhananga, S. Justice, Z. Wu, K. Leung, C. Strasburger, J. Sayers, and R. Ross, *A ligand-receptor fusion of growth hormone forms a dimer and is a potent long-actin agonist.* Nat. Med., 2007. 13(9): p. 1108–1113.
51. Y. Jiang, C. Liu, M. Hong, S. Zhu, and Y. Pei, *Tumor cell targeting of transferrin-PEG-TNFa conjugate via a receptor-mediated delivery system: design, synthesis and biological evaluation.* Bioconjugate Chem., 2007. 18: p. 41–49.
52. G. Subramanian, M. Fiscella, ALSmith, S. Zeuzem, and J. McHutchison, *Albinterferon a-2b: a genetic fusion protein for the treatment of chronic hepatitis C.* Nat. Biotechnol., 2007. 25(12): p. 1411–1419.
53. D. Muller, A. Karle, B. Meissburger, I. Hofig, R. Stork, and R. Kontermann, *Improved pharmacokinetics of recombinant bispecific antibody molecules by fusion to human serum albumin.* J. Biol. Chem., 2007. 282(17): p. 12650–12660.
54. R. Stork, D. Muller, and R. Kontermann, *A novel tri-functional antibody fusion protein with improved pharmacokinetic properties generated by fusing a bispecific single-chain diabody with an albumin-binding domain from streptococcal protein G.* Protein Eng. Des. Sel., 2007. 20(11): p. 569–576.
55. Y. Haung, P. Lundy, A. Lazaris, Y. Huang, H. Baldassare, B. Wing, C. Turcotte, M. Cote, A. Bellemare, A. Bilodeau, S. Brouillard, M. Ouati, P. Herskovits, I. Begin, N. Neveu, E. Brochu, J. Pierson, D. Hockley, D. Cerasoli, D. Lenz, H. Wilgus, C. Karatzas, and S. Langerman, *Substantially improved pharmacokinetics of recombinant human butyrylcholinesterase by fusion to human serum albumin.* BMC Biotech., 2008. 8: p. 49.
56. S. Marshall, G. Lazar, A. Chirino, and J. Desjarlais, *Rational design and engineering of therapeutic proteins.* Drug Discovery Today, 2003. 8(5): p. 212–221.
57. A. DeGroot and D. Scott, *Immunogenicity of protein therapeutics.* Trends Immun., 2007. 28(11): p. 482–490.
58. H. Schellekens, *Immunogenicity of therapeutic proteins: clinical implications and future prospects.* Clinical Thera., 2002. 24(11): p. 1720–1740.

59. P. Caliceti, O. Schiavon, and F. Veronese, *Biopharmaceutical properties of uricase conjugated to neutral and amphiphilic polymers.* Bioconjugate Chem., 1999. 10: p. 638–646.
60. M. Orlando, *HESylation-a new technology for polymer conjugation to biologically active molecules. Modification of proteins and low molecular weight substances with hydroxyethyl starch (HES).* 2003, Justus-Liebig Universitat Giessen. p. 191.
61. A. Constantinou, A. Epenetos, D. Hreczuk-Hirst, S. Jain, and M. Deonarain, *Modulation of antibody pharmacokinetics by chemical polysialylation.* Bioconjugate Chem., 2008. 19: p. 643–650.
62. G. Gregoriadis, S. Jain, I. Papaioannou, and P. Laing, *Improving the therapeutic efficacy of peptides and proteins: A role for polysialic acids.* Int. J. Pharm., 2005. 300: p. 125–130.
63. F. Fares, N. Suganuma, K. Nishimori, P. LaPolt, A. Hsueh, and I. Boime, *Design of a long-acting follitropin agonist by fusing the C-terminal sequence of the chorionic gonadotropin beta subunit to the follitropin beta subunit.* Proc. Natl. Acad. Sci. U. S. A., 1992. 89: p. 4304–4308.
64. M. Schlapschy, I. Theobald, H. Mack, M. Shottelius, H. Wester, and A. Skerra, *Fusion of a recombinant antibody fragment with a homo-amino-acid polymer: effects on biophysical properties and prolonged plasma half-life.* Protein Eng. Des. Sel., 2007. 20(6): p. 273–284.
65. R. Greenwald, J. Yang, H. Zhao, C. Conover, S. Lee, and D. Filpula, *Controlled release of proteins from their poly(ethylene glycol) conjugates: Drug delivery systems employing 1,6-elimination.* Bioconjugate Chem., 2003. 14(2): p. 395–403.
66. D. Filpula and H. Zhao, *Releasable PEGylation of proteins with customized linkers.* Adv. Drug Del. Rev., 2008. 60: p. 29–49.
67. S. Zalipsky, N. Mullah, C. Engbers, M. Hutchins, and R. Kiwan, *Thiolytically cleavable dithiobenzyl urethane-linked polymer-protein conjugates as macromolecular prodrugs: reversible PEGylation of proteins.* Bioconjugate Chem., 2007. 18: p. 1869–1878.
68. V. Gaber-Porekar, I. Zore, B. Podobnik, and V. Menart, *Obstacles and pitfalls in the PEGylation of therapeutic proteins.* Curr. Opin. Drug Dis. Devel., 2008. 11(2): p. 242–250.
69. Y. Kodera, A. Matsushima, M. Hiroto, H. Nishmura, A. Ishii, T. Ueno, and Y. Inada, *PEGylation of proteins and bioactive substances for medical and technical applications.* Prog. Polym. Sci., 1998. 23: p. 1233–1271.
70. D. Lee, I. Sharif, S. Kodihalli, D. Stewart, and V. Tsvetnitsky, *Preparation and characterization of monopegylated human granulocyte-macrophage colony-stimulating factor.* J. Interferon Cytokine Res., 2008. 28: p. 101–112.
71. H. Sato, K. Yamamoto, E. Hayashi, and Y. Takahara, *Transflutaminase-mediated dual and site-specific incorporation of poly(ethylene glycol) derivatives into a chimeric interleukin-2.* Bioconjugate Chem., 2000. 11: p. 502–509.
72. A. Fontana, B. Spolaore, A. Mero, and F. Veronese, *Site-specific modification and PEGylation of pharmaceutical proteins mediated by transglutaminase.* Adv. Drug Del. Rev., 2008. 60: p. 13–28.
73. T. Sakane and W. Pardridge, *Carboxyl-directed Pegylation of brain-derived neurotrophic factor markedly reduces systemic clearance with minimal loss of biologic activity.* Pharm. Res., 1997. 14(8): p. 1086–1091.
74. F. Veronese, *Peptide and protein PEGylation: a review of problems and solutions.* Biomaterials, 2001. 22: p. 405–417.
75. C. Woghiren, B. Sharma, and S. Stein, *Protected thio-poly(ethylene glycol): A new activated polymer for reversible protein modification.* Bioconjugate Chem., 1993. 4(5): p. 314–318.
76. A. Chapman, *PEGylated antibodies and antibody fragments for improved therapy: a review.* Adv. Drug Del. Rev., 2002. 54: p. 531–545.
77. D. Wylie, M. Voloch, S. Lee, Y. Liu, S. Cannon-Carlson, C. Cutler, and B. Pramanik, *Carboxyalkylated histidine Is a pH-dependent product of pegylation with SC-PEG.* Pharm. Res., 2001. 18: p. 1354–1360.
78. B. Manjula, A. Tsai, R. Upadhya, K. Perumalsamy, P. Smith, A. Malavalli, I. Vandegriff, R. Winslow, I. Intaglietta, M. Prabhakaran, J. Friedman, and A. Acharya, *Site-specific PEGylation of hemoglobin at cys-93: Correlation between the colligative properties of the PEGylated protein and the length of the conjugated PEG chain.* Bioconjugate Chem., 2003. 14: p. 464–472.
79. D. Baker, E. Lin, K. Lin, M. Pellegrini, R. Petter, L. Chen, R. Arduini, M. Brickelmaier, D. Wen, D. Hess, L. Chen, D. Grant, A. Whitty, A. Gill, D. Lindner, and R. Pepinsky, *N-terminally PEGylated human interferon-beta-1a with improved pharmacokinetic properties and in vivo efficacy in a melanoma angiogenesis model.* Bioconjugate Chem., 2006. 17: p. 179–188.
80. M. Rosendahl, D. Doherty, S. Smith, S. Carlson, E. Chlipala, and G. Cox, *A long-acting, highly potent interferon a-2 conjugate created using site-specific PEGylation.* Bioconjugate Chem.,

2005. 16: p. 200–207.
81. D. Doherty, M. Rosendahl, D. Smith, J. Hughes, E. Chlipala, and G. Cox, *Site-specific PEGylation of engineered cysteine analogues of recombinant human granulocyte-macrophage colony-stimulating factor.* Bioconjugate Chem., 2005. 16: p. 1291–1298.
82. A. Basu, K. Yang, M. Wang, S. Liu, R. Chintala, T. Palm, H. Zhao, P. Peng, D. Wu, Z. Zhang, J. Hua, M. Hsieh, J. Zhou, G. Petti, Z. Li, A. Janjua, M. Mendez, J. iu, C. Longley, Z. Zhang, M. Mehlig, V. Borowski, M. Viswanathan, and D. Filpula, *Structure-function engineering of interferon-beta-1b for improving stability, solubility, potency, immunogenicity, and pharmacokinetic properties by site-selective mono-PEGylation.* Bioconjugate Chem., 2006. 17: p. 618–630.
83. S. Bell, C. Fam, E. Chlipala, S. Carlson, J. Lee, M. Rosendahl, D. Doherty, and G. Cox, *Enhanced circulating half-life and antitumor activity of a site-specific Pegylated interferon-a protein therapeutic.* Bioconjugate Chem., 2008. 19(299–305).
84. X. Wu, X. Liu, Y. Xiao, Z. Huang, J. Xiao, S. Lin, L. Cai, W. Feng, and X. Li, *Purification and modification by polyethylene glycol of a new human basic fibroblast growth factor mutant-hb-FGF.* J. Chromat. A, 2007. 1161: p. 51–55.
85. K. Hinds and S. Kim, *Effects of PEG conjugation on insulin properties.* Adv. Drug Del. Rev., 2002. 54(4): p. 505–530.
86. S. DeFrees, Z. Wang, R. Xing, A. Scott, J. Wang, D. Zopf, D. Gouty, R. Sjoberg, K. Panneerselvam, E. Brinkman, R. Bayer, M. Tarp, and H. Clausen, *GlycoPEGylation of recombinant therapeutic proteins produced in Escherichia coli.* Glycobio., 2006. 16(9): p. 833–843.
87. R. Bayer, H. Ostergaard, M. Kalo, P. Holm, K. Kinealy, B. Sorensen, S. Bjorn, D. Zopf, and H. Stennicke, *Development of long-acting FVIIA derivatives by glycopegylation.* J Thromb. Haem., 2007. 5(Supplement 1): p. Abstract number P-T-016.
88. I. Carrico, B. Carlson, and C. Bertozzi, *Introducing genetically encoded aldehydes into proteins.* Nat. Chem. Biol., 2007. 3(6): p. 321–322.
89. A. Wang, N. Nairn, R. Johnson, D. Tirrell, and K. Grabstein, *Processing of N-terminal unnatural amino acids in recombinant human interferon-b in Escherichia coli.* Chem. Bio. Chem., 2008. 9: p. 324–330.
90. G. Bernardes, J. Chalker, J. ERREY, and B. Davis, *Facile conversion of cystein and alkyl cysteines to dehydroalanine on protein surfaces: versatile and switchable access to functionalized proteins.* J. Am. Chem. Soc., 2008. 130: p. 5052–5053.
91. D. Romanini and M. Francis, *Attachment of peptide building blocks to proteins through tyrosine bioconjugation.* Bioconjugate Chem., 2008. 19: p. 15–157.
92. G. Walsh and R. Jefferis, *Post-translational modifications in the context of therapeutic proteins.* Nat. Biotechnol., 2006. 24: p. 1241–1252.
93. G. Molinuex, *The design and development of pegfilgrastim.* Curr. Pharm. Des., 2004. 10(11): p. 1235–1244.
94. C. Fee and J. Alstine, *PEG-proteins: Reaction engineering and separation issues.* Chem. Eng. Sci., 2006. 61: p. 924–939.
95. J. Thorton, *Disulphide bridges in globular proteins.* J. Mol. Biol., 1981. 151: p. 261–287.
96. A. Pavlou and J. Reichert, *Recombinant protein therapeutics – success rates, market trends and values to 2010.* Nat. Biotechnol., 2004. 22: p. 1513–1519.
97. M. Petersen, P. Jonson, and S. Petersen, *Amino acid neighbours and detailed conformational analysis of cysteines in proteins.* Protein Eng., 1999. 12(7): p. 535–548.
98. H. Leung, G. Xu, M. Narayan, and H. Scheraga, *Impact of an easily reducible disulfide bond on the oxidative folding rate of multi-disulfide-containing proteins.* J. Peptide Res., 2005. 65: p. 47–54.
99. E. Fahey, J. Chaudhuri, and P. Binding, *Refolding of low molecular weight urokinase plasminogen activator by dilution and size exlusion chromatography. A comparative study.* Sep. Sci. Tech., 2000. 35(11): p. 1743–1760.
100. F. Veronese, A. Mero, F. Caboi, M. Sergi, C. Maronjiuio, and G. Pasut, *Site-specific Pegylation of G-CSF by reversible denaturation.* Bioconjugate Chem., 2007. 18: p. 1824–1830.
101. Z. Gugolya, A. Dosztanyi, and I. Simon, *Interresidue interactions in protein classes.* Proteins, 1997. 27: p. 360–366.
102. S. Betz, *Disulfide bonds and the stability of globular proteins.* Protein Sci., 1993. 2: p. 1551–1558.
103. C. Levinthal, *Are there pathways for protein folding?* J. Chim. Phys., 1968. 85: p. 44–45.
104. V. Ittah and E. Haas, *Nonlocal interactions stabilize long range loops in the initial folding inter-*

mediates of reduced bovine pancreatic trypsin inhibitor. Biochemistry, 1995. 34: p. 4493–4506.

105. B. Manjula, A. Tsai, R. Upadhya, K. Perumalsamy, P. Smith, A. Malavalli, I. Vandegriff, R. Winslow, I. Intaglietta, M. Prabhakaran, J. Friedman, and A. Acharya, *Site-Specific PEGylation of Hemoglobin at Cys-93(â): Correlation between the Colligative Properties of the PEGylated Protein and the Length of the Conjugated PEG Chain.* Bioconjugate Chem., 2003. 14: p. 464–472.

106. A. Chapman, *PEGylated antibodies and antibody fragments for improved therapy: a review.* Adv. Drug Del. Rev., 2002. 54: p. 531–545.

107. A. Kozlowski, R. Gross, and S. McManus, *Hydrolytically stable maleimide-terminated polymers.* 2004. WO 2004/060965: 118 pages.

108. Editorial, *The other path for follow-ons.* Nat. Biotechnol., 2008. 7: p. 715.

109. D. Gill and N. Damle, *Biopharmaceutical drug discovery using novel protein scaffolds.* Curr. Opin. Biotech., 2006. 17: p. 653–658.

110. T. Hey, E. Fiedler, R. Rudolph, and M. Fiedler, *Artificial, non-antibody binding proteins for pharmaceutical and industrial applications.* Trends Biotech., 2005. 23(10): p. 514–522.

111. Y. Youn, M. Kwon, D. Na, S. Chae, S. Lee, and K. Lee, *Improved intrapulmonary delivery of site-specific PEGylated salmon calcitonin: optimization by PEG size selection.* J. Cont. Rel., 2008. 125: p. 68–75.

112. Y. Youn, D. Na, and K. Lee, *High-yield production of biologically active mono-PEGylated salmon calcitonin by site-specific PEGylation.* J. Cont. Rel., 2007. 117: p. 371–379.

113. Y. Imura, M. Nishida, and K. Matsuzaki, *Action mechanism of PEGylated magainin 2 analogue peptide.* Biochim. et Biophys. Acta, 2007. 1768: p. 2578–2585.

114. Y. Imura, M. Nishida, Y. Ogawa, Y. Takakura, and K. Matsuzaki, *Action mechanism of tachyplesin I and effects of PEGylation.* Biochim. et Biophys. Acta, 2007. 1768: p. 1160–1169.

115. S. Chae, C. Jin, H. Shin, Y. Youn, S. Le, and K. Lee, *Preparation, characterization, and application of biotinylated and biotin-PEGylated glucagon-like peptide-1 analogues for enhanced oral delivery.* Bioconjugate Chem., 2008. 19: p. 334–341.

116. I. Verbaeys, F. Tamariz, J. Buyse, M. DeCuyper, H. Pottel, M. VanBoven, and M. Cokelaere, *PEGylated cholecystoknin prolongs satiation in rats: dose dependency and receptor involvement.* Brit. J. Pharmcol., 2007. 152: p. 396–403.

117. F. Tamariz, I. Verbaeys, M. VanBoven, M. DeCuyper, J. Buyse, E. Clynen, and M. Cokelaere, *PEGylation of cholecystokinin prolongs it anorectic effect in rats.* Peptides, 2007. 28: p. 1003–1011.

118. R. Fahey, J. Hunt, and G. Windham, *On the cysteine and cystine content of proteins-differences between intracellular and estracellular proteins.* J. Mol. Evol., 1977. 10: p. 155–160.

119. S. Shaunak, A. Godwin, J. Choi, S. Balan, E. Pedone, D. Vijayarangam, S. Heidelberger, I. Teo, M. Zloh, and S. Brocchini, *Site-specific PEGylation of native disulfide bonds in therapeutic proteins.* Nat. Chem. Bio., 2006: p. 312–313.

120. S. Balan, J. Choi, A. Godwin, I. Teo, C. Laborde, S. Heidelberger, M. Zloh, S. Shaunak, and S. Brocchini, *Site-specific PEGylation of protein disulfide bonds using a three-carbon bridge.* Bioconjugate Chem., 2007. 18(1): p. 61–76.

121. S. Brocchini, S. Balan, A. Godwin, J. Choi, M. Zloh, and S. Shaunak, *PEGylation of native disulfide bonds in proteins.* Nat. Protocols, 2006. 1(5): p. 2241–2252.

122. R. Singh and A. Rao, *Reductive unfolding and oxidative refolding of a Bowman-Birk inhibitor from horsegram seeds (Dolichos biflorus): evidence for 'hyperreactive' disulfide bonds and rate-limiting nature of disulfide isomerisation in folding.* Biochim. et Biophys. Acta, 2002. 1597: p. 280–291.

123. A. Saunders, G. Young, and G. Pielak, *Polarity of disulfide bonds.* Protein Sci., 1993. 2: p. 1183–1184.

124. M. Parker, Y. Chen, F. Danehy, K. Dufu, J. Ekstrom, E. Getmanova, J. Gokemeijer, L. Xu, and D. Lipovsek, *Antibody mimics based on human fibronectin type three domain engineered for thermostability and high-affinity binding to vascular endothelial growth factor receptor two.* Protein Eng. Des. Sel., 2005. 18(9): p. 435–444.

125. W. Guo, J. Shea, and R. Berry, *The physics of the interactions governing folding and association of proteins.* Ann. N.Y. Acad. Sci., 2005: p. 34–53.

126. A. Jungbauer and W. Kaar, *Current status of technical protein refolding.* J. Biotechnol., 2007. 128: p. 587–596.

127. J. Yon, *Protein folding: concepts and perspectives.* CMLS, 1997. 53: p. 557–567.

128. A. Fernandez and A. Colubri, *Pathway heterogeneity in protein folding.* Proteins, 2002. 48: p. 293–310.
129. T. Creighton, *Disulphide-coupled protein folding pathways.* Phil. Trans. R. Soc. Lond. B, 1995. 348: p. 5–10.
130. A. Robinson and J. King, *Disulphide-bonded intermediate on the folding and assembly pathway of a non-disulfphide bonded protein.* Nat. Struct. Biol., 1997. 4(6): p. 450–455.
131. E. Collins, J. Wirmer, K. Hirai, H. Tachibana, S. Segawa, C. Dobson, and H. Schwalbe, *Characterisation of disulfide-bond dynamics in non-native states of lysozyme and its disulfide deletion mutants by NMR.* ChemBiochem, 2005. 6: p. 1619–1627.
132. T. Creighton, *Toward a better understanding of protein folding pathways.* Proc. Natl. Acad. Sci. U. S. A., 1988. 85: p. 5082–5086.
133. J. Chang, L. Li, and A. Bulychev, *The underlying mechanism for the diversity of disulfide folding pathways.* J. Biol. Chem., 2000. 275(12): p. 8287–8289.
134. J. Chang, *Evidence for the underlying cause of diversity of the disulfide folding pathway.* Biochemistry, 2004. 43: p. 4522–4529.
135. W. Wedemeyer, E. Welker, M. Narayan, and H. Scheraga, *Disulfide bonds and protein folding.* Biochemistry, 2000. 39: p. 4207–4216.
136. M. Narayan, G. Xu, D. Ripoll, H. Zhal, K. Breuker, C. Wanjalla, H. Leung, A. Navon, E. Welker, F. McLafferty, and H. Scheraga, *Dissimilarity in the reductive unfolding pathways of two ribonuclease homologues.* J. Mol. Biol., 2004. 338: p. 795–809.
137. G. Xu, M. Narayan, I. Kurinov, D. Ripoll, E. Welker, M. Khalili, S. Ealick, and H. Scheraga, *A localised specific interaction alters the unfolding pathways of structural homologues.* J. Am. Chem. Soc., 2006. 128: p. 1204–1213.
138. S. Jiang-Ning, L. Wei-Jiang, and X. Wen-Bo, *Cooperativity of the oxidation of cysteines in globular proteins.* J. Theor. Biol., 2004. 231: p. 85–95.
139. K. Takeda, K. Ogawa, M. Ohara, S. Hamada, and Y. Moriyama, *Conformational changes of a-lactalbumin induced by the stepwise reduction of its disulfide bridges: The effect of the disulfide bridges on the structural stability of the protein in sodium dodecyl sulfate solution.* J. Protein Chem., 1995. 14(8): p. 679–684.
140. G. Graziano, F. Catanzano, and E. Notomista, *Enthalpic and entropic consequences of the removal of disulfide bridges in ribonuclease A.* Thermochimica Acta, 2000. 364: p. 165–172.
141. M. Denton and H. Scheraga, *Spectroscopic, immunochemical, and thermodynamic properties of carboxymethyl (cys6, cys127)-hen egg white lysozyme.* J. Protein Chem., 1991. 10(2): p. 213–232.
142. K. Kuwajima, M. Ikeguchi, T. Sugawara, Y. Hiraoka, and S. Sugai, *Kinetics of disulfide bond reduction in a-lactalbumin by dithiothreitol and molecular basis of superreactivity of the cys6–cys120 disulfide bond.* Biochemistry, 1990. 29: p. 8240–8249.
143. M. Zloh, S. Balan, S. Shaunak, and S. Brocchini. *Modeling study of disulfide bridged PEGylated proteins.* In: *6th European Conference on Computational Chemistry.* 2006. Slovakia.
144. A. Godwin, J. Choi, E. Pedone, S. Balan, R. Jumnah, S. Shaunak, S. Brocchini, and M. Zloh, *Molecular dynamics simulations of proteins with chemically modified disulfide bonds.* Theoretical Chem. Acc., 2007. 117(2): p. 259–265.
145. M. Zloh, S. Balan, S. Shaunak, and S. Brocchini, *Identifying protein disulfides for the insertion of a 3-carbon bridge.* Nat. Protocols, 2007. 2: p. 1070–1083
146. J. Bae, L. Yang, C. Manithody, and A. Rezaie, *Engineering a disulfide bond to stabilize the calcium-binding loop of activated Protein C eliminates its anticoagulant but not its protective signaling properties.* J. Biol. Chem., 2007. 282(12): p. 9251–9259.
147. D. Humphreys, Modified antibody fragments. 2007. WO/2007/010231.
148. J. Casey, R. Pedley, D. King, R. Boden, A. Chapman, G. Yarranton, and R. Begent, *Improved tumour targeting of di-Fab' fragments modified with polyethylene glyco.* Tumor Target, 2000. 4: p. 235–244.
149. M. Harris and A. Kozlowski, Poly(ethylene glycol) derivative with proximal reactive groups. 1999. WO 1999/45964: 30 pages.
150. S. Brocchini, M. Eberle, and R. Lawton, *Molecular yardsticks. Synthesis of extended equilibrium transfer alkylating cross-link reagents and their use in the formation of macrocycles.* J. Am. Chem. Soc., 1988. 110: p. 5211.
151. F. Liberatore, R. Comeau, J. McKearin, D. Pearson, B. Belonga, S. Brocchini, J. Kath, T. Phillips, K. Oswell, and R. Lawton, *Site directed modification and cross-linking of a monoclonal antibody with equilibrium transfer alkylating cross-link reagents.* Bioconjugate Chem., 1990. 1: p. 36–50.

152. R. Rosario, S. Brocchini, R. Lawton, R. Wahl, and R. Smith, *Sulfydral site-specific cross-linking of a monoclonal antibody by a fluorescent equilibrium transfer alkylating cross-link reagent.* Bioconjugate Chem., 1990. 1: p. 51–65.

153. A. Lewis, Y. Tang, S. Brocchini, J.W. Choi, and A. Godwin. *Comparative study of MPC and PEG for protein conjugation.* In: *International Symposium on Polymer Therapeutics ISPT-07.* 2007. Berlin.

154. A. Lewis, Y. Tang, S. Brocchini, J. Choi, and A. Godwin, *Poly(2-methacryloyloxyethyl phosphorylcholine) for protein conjugation.* Bioconjugate Chem. 2008. 19(11): p. 2144–2155.

155. D. Miyamoto, J. Watanabe, and K. Ishihara, *Effect of water-soluble phospholipid polymers conjugated with papain on the enzymatic stability.* Biomaterials, 2004. 25: p. 71–76.

156. M. Lutolf, N. Tirelli, S. Cerritelli, L. Cavalli, and J. Hubbel, *Systematic modulation of Michael-type reactivity of thiols through the use of charged amino acids.* Bioconjugate Chem., 2001. 12: p. 1051–1056.

157. Z. Guan, W. Zuo, L. Zhao, Z. Ren, and Y. Liang, *An economical and convenient synthesis of vinyl sulfones.* Synthesis, 2007: p. 1465–1470.

158. B. Shi and M. Greaney, *Reversible Michael addition of thiols as a new tool for dynamic combinatorial chemistry.* Chem. Comm., 2005: p. 886–888.

159. P. Corbett, J. Leclaire, L. Vial, K. West, J. Wietor, J. Sanders, and S. Otto, *Dynamic combinatorial chemistry.* Chem. Rev., 2006. 106: p. 3652–3711.

160. S. Mitra and R. Lawton, *Reagents for the cross-linking of proteins by equilibrium transfer alkylation.* J. Am. Chem. Soc., 1979. 101(11): p. 3097–3110.

161. D. Wilbur, J. Stray, D. Hamlin, D. Curtis, and R. Vessella, *Monoclonal antibody Fab' fragment cross-linking using equilibrium transfer alkylation reagents. A strategy for site-specific conjugation of diagnostic and therapeutic agents with F(ab')2 fragments.* Bioconjugate Chem., 1994. 5: p. 220–235.

162. S. Kim and C. Lim, *Tin-free radical-mediated C-C bond formations with alkyl allyl sulfones as radical precursors.* Angew. Chem. Int. Ed., 2002. 41(17): p. 3265–3267.

163. Y. Pan, D. Hutchinson, M. Nantz, and P. Fuchs, *Synthesis via vinyl sulfones. 34. Sn2' additions of cuprates to sulfone and ester-polarized cyclopentenylic systems.* Tetrahedron, 1989. 45(2): p. 467–478.

164. D. Seebach and P. Knochel, *2'-nitro-2'-propen-1-yl 2,2-dimethylpropanoate (NPP), a multiple coupling reagent.* Helv. Chim. Acta, 1984. 67(1): p. 261–283.

165. R. Nelson and R. Lawton, *On the a,a' annelation of cyclic ketones.* J. Am. Chem. Soc., 1966. 88(16): p. 3884–3885.

166. A. Padwa, D. Kline, S. Murphree, and P. Yeske, *Use of 2,3-bis(phenylsulfonyl)-l-propene as a multicoupling reagent.* J. Org. Chem., 1992. 57: p. 298–306.

167. S. Balan, M. Zloh, S. Shaunak, and S. Brocchini. *Disulfide bridged PEGylated leptin.* in *Pharmaceutical Sciences World Congress.* 2007. Amsterdam.

168. T. Aparicio, L. Kotelevets, A. Tsocas, J. Laigneau, I. Sobhani, E. Chastre, and T. Lehy, *Leptin stimulates the proliferation of human colon cancer cells in vitro but does not promote the growth of colon cancer xenografts in nude mice or intestinal tumorigenesis in Apc (Min/+) mice.* Gut, 2005. 54(8): p. 1136–1145.

169. P. Bailon, A. Palleroni, C. Schaffer, C. Spence, W. Fung, JE Porter, G. Ehrlich, W. Pan, Z. Xu, M. Modi, A. Farid, and W. Berthold, *Rational design of a potent, long-lasting form of interferon: A 40 kDa branched polyethylene glycol-conjugated interferon a-2a for the treatment of Hepatitis C.* Bioconjugate Chem., 2001. 12: p. 195–202.

170. R. Rajender, M. Modi, and S. Pedder, *Use of peginterferon alfa-2a (40 KD) (Pegasys) for the treatment of Hepatitis C.* Adv. Drug Del. Rev., 2002. 54: p. 571–586.

171. S. Youngster, Y. Wang, M. Grace, J. Bausch, R. Bordens, and D. Wyss, *Structure, biology, and therapeutic implications of pegylated interferon alpha-2b.* Curr. Pharm. Des., 2002. 8: p. 2139–2157.

172. C. Dhalluin, A. Ross, W. Huber, P. Gerber, D. Brugger, B. Gsell, and H. Senn, *Structural, kinetic, and thermodynamic analysis of the binding of the 40 kDa PEG-interferon-alpha 2a and its individual positional isomers to the extracellular domain of the receptor IFNAR2.* Bioconjugate Chem., 2005. 16: p. 518–527.

173. K. Shibata, K. Maruyama-Takahashi, M. Yamasaki, and N. Hirayama, *G-CSF receptor-binding cyclic peptides designed with artificial amino-acid linkers.* Biochem. Biophys. Res. Comm., 2006. 341: p. 483–488.

174. D. Humphreys, S. Heywood, A. Henry, L. Ait-Lhadj, P. Antoniw, R. Palframan, K. Greenslade, B. Carrington, D. Reeks, L. Bowering, S. West, and H. Brand, *Alternative antibody Fab' fragment PEGylation strategies: combination of strong reducing agents, disruption of the interchain disulphide bond and disulphide engineering.* Protein Eng. Des. Sel., 2007. 20(5): p. 227–234.
175. S. Kubetzko, C. Sarkar, and A. Pluckthun, *Protein PEGylation decreases observed target association rates via a dual blocking mechanism.* Mol. Pharmacol., 2005. 68: p. 1439–1454.
176. S. Kubetzko, E. Balic, R. Waibel, U. Wittke, and A. Pluckthun, *PEGylation and multimerization of the anti-p185 HER-2 single chain Fv fragment 4D5.* J. Biol. Chem., 2006. 281: p. 35186–35201.
177. C. Dhalluin, A. Ross, L. Leuthold, S. Foser, B. Gsell, F. Muller, and H. Senn, *Structural and biophysical characterization of the 40 kDa PEG-interferon-a2a and its individual positional isomers.* Bioconjugate Chem., 2005. 16: p. 504–517.
178. N. Wrighton, F. Farrell, R. Chang, A. Kashyap, F. Barbone, L. Mulcahy, D. Johnson, R. Barrett, L. Jolliffe, and W. Dower, *Small peptides as potent mimetics of the protein hormone erythropoietin.* Science, 1996. 273(5274): p. 458–463.
179. J. Burns, J. Butler, J. Moran, and G. Whitesides, *Selective reduction of disulfides by tris(2-carboxyethyl)phosphine.* J. Org. Chem., 1991. 56: p. 2648–2650.
180. D. Shafer, J. Inman, and A. Lees, *Reaction of Tris(2-carboxyethyl)phosphine (TCEP) with Maleimide and a-Haloacyl Groups: Anomalous Elution of TCEP by Gel Filtration.* Anal. Biochem., 2000. 282: p. 161–164.

Enzymatic techniques for PEGylation of biopharmaceuticals

Mauro Sergi, Francesca Caboi, Carlo Maullu, Gaetano Orsini and Giancarlo Tonon

Bio-Ker S.r.l, Parco Scientifico e Tecnologico della Sardegna, 09010 Pula, Cagliari, Italy

Abstract

Modification of therapeutic proteins and peptides by polyethylene glycol conjugation is a well known method to improve the pharmacological properties of such drugs.

Here we describe an alternative way of PEGylation from classic chemical methods, taking advantage of enzymes able to specifically modify some amino acid side chains, in particular glycosyltransferases and transglutaminases.

A few examples are here described, in particular granulocyte-colony stimulating factor, which has been successfully PEGylated by enzymatic methods leading to a new long-lasting compound presently under evaluation in clinical studies.

Introduction

Pharmacological therapy for human diseases has involved a growing number of proteins and peptides as therapeutic agents. Although great effort has been dedicated to finding new biomolecules and to improve their therapeutic properties, most of them still suffer drawbacks such as short half-life, low therapeutic index, immunogenicity and other side effects. To overcome these limitations several approaches have been proposed, the most successful one to date being the covalent conjugation of polyethylene glycol (PEG) to the protein surfaces. For an approach to PEGylation strategy, among the numerous papers and reviews published so far, the reader may be referred to two dedicated books [1, 2] or a recent collections of reviews [3–5].

The predominant method used to link PEG to proteins or peptides takes advantage of reactive electrophiles at the terminal end of methoxy-PEG suitable for reaction with nucleophiles in proteins, usually amino residues but also thiol ones. This is the case of some marketed PEGylated drugs, such as Adagen® (PEG-adenosine deaminase) [6], Oncaspar® (PEG-asparaginase) [7] and PEG-Intron® (PEG-interferon alpha-2b) [8]. These products have been obtained using the so-called first generation PEGylation chemistry, characterised by a non-specific protein or peptide modification with low molecular weight polyethylene glycol chains (≤12 kDa). An improvement in PEGylation relies on the use of higher molecular weight polymer (≥20 kDa), linear or

branched, possibly attached in a site-specific fashion as in the case of Pegasys®
[9], a form of interferon α-2a having a branched polyethylene glycol chain of
40 kDa bound to primary amine groups of lysine. However, also in this case,
the coupling reaction is not site specific, the products containing a mixture of
mono-PEGylated positional isomers and as a consequence the biological activity is the combined result of the single different species. To overcome the issue
of isomer mixtures, site selective PEGylations were devised taking advantage
of a polyethylene glycol functionalised with an aldehyde group which, at low
pH values, reacts specifically with the N-terminal amino group of proteins to
yield a labile Schiff base, which upon reduction yields a stable secondary
amine. The greater the difference in pK between the N-terminals and the ε-side
chains of lysines, the greater selectivity of this approach. The most relevant
example of this conjugate chemistry is represented by Neulasta® (PEG-filgrastim) [10], where a 20 kDa linear PEG chain is covalently bound to the
N-terminal Met1 residue through an alkyl-amino bond.

A different approach to site-specific PEGylation is based on the introduction into the protein or peptide of an engineered cysteine residue that specifically binds to a thiol reactive polyethylene glycol chain [11, 12]. This method
allows for a high yield and selectivity of conjugation but, involving a change
in the protein sequence, it may eventually lead to structural changes in the protein, decreased activity or enhanced immunogenicity. Introduction of an additional free thiol into the protein may also create issues with unwanted dimeri-
sation during protein production.

Emerging techniques for an increased selectivity of PEGylation take advantage of the properties of enzymes which recognise and specifically modify
only a single or a few amino acid residues in the native proteins as it is the case
of glycosyl-transferases and transglutaminases.

Glycosylation and glyco-PEGylation

Carbohydrates are the most diverse biopolymers in nature and their specific
interactions with physiological receptors make their role crucial in many biological processes. When bound to a protein they can modulate or mediate a
wide variety of events such as folding, secretion, serum lifetime, molecular
recognition, etc. They are enzymatically attached to proteins in a process
called glycosylation that represents the most extensive and complicated co- or
post-translational modification into eukaryotic cells [13]. Carbohydrate moieties are naturally attached to a selected amino acid residue by linkages catalysed by a family of glycosyltransferases [14, 15] which are specific for both
the donor nucleotide-activated sugar and the acceptor amino acid. Since a different enzyme usually catalyses each step in a pathway leading to the final
oligosaccharide, the chain elongation is realised by a multiglycosyltransferase
system. The binding of carbohydrate moieties to the amino acid residue
involves two different types of glycosylation: N-linked and O-linked ones. The

N-linked glycosylation consists of the addition of a large lipid-linked oligosaccharide precursor followed by ordered removal of certain sugar residues and addition of others residues, each step being catalysed by different enzymes. On the other hand, O-linked sugars are added one at a time by a different glycosyltransferase enzyme. The biosynthetic pathway for O-glycosylation is a post-translational process that takes place in the Golgi apparatus and most commonly involves initially the linking of GalNAc (N-acetyl galactosamine) to the hydroxyl group of serine or threonine or hydroxylysine residues and is typically completed by the terminal addition of one or two negatively charged N-acetylneuraminic acid residues from a nucleotide-activated precursor.

Since glycosylation can result in the improvement of physical-chemical and biological properties of recombinant proteins, glycobiology and carbohydrate chemistry have become increasingly important in modern biotechnology and in manufacturing of biodrugs [16]. In one application, isolated glycosyltransferases are used *in vitro* to add desired sugars to proteins of therapeutic interest. A number of such transferases have been cloned and expressed for use in large-scale processes.

A successful example of the industrial application of glycobiology is offered by "GlycoAdvance™" technology (Neose Technologies Inc.), an important tool that can serve as a remedy for the shortcomings of the expression systems of commercial interest. By using *in vitro* isolated glycosyltransferases, Neose is able to perform either a complete sialylation that enhances proteins *in vivo* half-life (without the terminal sialic acid glycoproteins are rapidly cleared from the blood), or a fucosylation that increases bioactivity (incompletely fucosylated proteins lack the recognition site essential for interaction with selectins).

To date numerous neo-glycoproteins have been produced with applications in vaccine development [17], disease diagnosis [18], drug targeting [19] and immunosuppressive therapy [20].

As an extension of the GlycoAdvance™ method, Neose developed a novel strategy that uses glycosyltranferases for enzymatic conjugation of sialic acid covalently substituted with PEG at the 5'-amino position [21] to glycan acceptor sites on glycoproteins. This new enzymatic PEGylation approach was termed "GlycoPEGylation" (Fig. 1).

The method involves a two-step procedure, namely enzymatic GalNAc glycosylation at serine and threonine residues in proteins, followed by enzymatic transfer of PEG-linked sialic acid to the previously introduced GalNAc residues.

GlycoPEGylation has been applied to a growing number of clinically important recombinant proteins, such as granulocyte colony stimulating factor, interferon alpha-2b and granulocyte/macrophage colony stimulating factor that are naturally O-glycosylated when expressed in mammalian cells but not when expressed in *E. coli*.

The main advantage of GlycoPEGylation resides in the enhanced structural homogeneity of the oligosaccharide chains conjugated to proteins as compared to those obtained by random chemical methods.

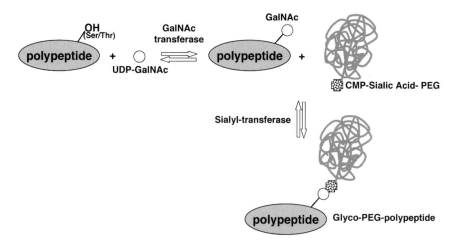

Figure 1. Schematic representation of two-step glycoPEGylation process. In the first reaction Ser or Thr side chain is glycosylated by attachment of GalNAc, in the second one PEG-sialic acid is bound to the previously added sugar.

Granulocyte Colony Stimulating Factor (G-CSF)

G-CSF is a 174 amino acids glycosylated cytokine that stimulates the proliferation, survival and differentiation of neutrophil granulocyte progenitor cells and mature neutrophils. The development of recombinant G-CSF (e.g., methionyl G-CSF produced in *E. coli* or filgrastim) allowed the development of new therapies to treat several kinds of neutropenia and related infective and febrile phenomena. In particular filgrastim is administered in high-risk patients being treated with myelotoxic antitumoral drugs or for myeloablation followed by bone marrow transplant or in late-stage HIV infection.

G-CSF is rapidly eliminated from blood ($t_{1/2}$ of 3–4 h) by a combination of non-specific and non-saturable renal clearance mechanisms and receptor-mediated endocytosis. Its therapeutic use requires multiple daily injections to obtain an extended pharmacological effect [22]. To overcome this limit, the fusion of G-CSF to carrier proteins such as immunoglobulin or albumin, the incorporation in slow-release nano- and microspheres and the conjugation to PEG have been proposed. PEGylation was initially performed by nonselective strategies leading to mixtures of positional isoforms and later by a chemical selective route linking PEG to a cysteine mutant using a thiol specific reagent [23]. A further approach takes advantage of the difference in pKa of the α and ε amino groups for N-terminal PEGylation of methionyl-G-CSF by selective acylation at the α amino group only occurring at low pH. The product, possesses a single PEG chain of 20 kDa and is marketed under the trade name of Neulasta® [10]. The drug has achieved great success for reducing the incidence of febrile neutropenia in patients with non-myeloid cancer receiving myelo-suppressive chemotherapy. Administered as a single subcutaneous injection, once per chemotherapy cycle, is effective as a daily administration of the non-conjugated filgrastim [24, 25].

GlycoPEGylation was used [26] to modify G-CSF. In this case *E. coli* expressed G-CSF was selectively GalNAc O-glycosylated at its natural glycosylation site. The (GalNAc)-G-CSF product was purified by size exclusion chromatography before the second step of enzymatic PEGylation by sialyltransferases using the substrate cytidine monophosphate (CMP)-sialic acid. CMP-Sia-PEG-20K was prepared at Neose Technologies, while human GalNAc-transferases and sialyltransferase were expressed in a soluble form in insect cells and purified to homogeneity by Gen Bank. The conjugate was purified by HPLC on SP Sepharose and Superdex 75. MALDI-TOF/MS time-course analysis (performed on a Voyager-DE MALDI time-of-flight mass spectrometer equipped with delayed extraction), revealed that glycosylation of G-CSF is essentially complete at a single site, Thr133.

A comparative pharmacokinetic study in rat, carried out with Neulasta® as reference compound, showed comparable values of elimination half-life ($t_{1/2}$), AUC and clearance. Similarities were also found in other preclinical assays, including G-CSF competition binding and cellular proliferation.

Granulocyte Macrophage Colony Stimulating Factor (GM-CSF)

GM-CSF is a cytokine active as a white blood cell growth factor which stimulates stem cells to produce granulocytes and macrophages. In its human mature form, it is O-glycosylated at multiple sites (Ser5, Ser7, Ser9, and/or Thr10) [27, 28], but can be produced as a non-glycosylated protein by recombinant expression in *E. coli*. Recombinant GM-CSF is usually administered to patients after bone marrow transplant or chemotherapy.

A glyco-PEGylation reaction [21], carried out as for G-CSF, yielded a more heterogeneous product. The combined MS/MS and peptide Edman degradation analysis demonstrated that galactosyl N-acetyl-transferase transfers GalNAc residues to the amino acid residues Ser7 and Ser9, with traces of incorporation into an unidentified site.

Interferon Alpha-2b (IFN α-2b)

IFN α-2b is a 165 amino acid protein, naturally O-glycosylated at Thr106 residue. It is extensively used as antiviral or antineoplastic agent to treat chronic myelogenous leukemia, lymph nodes spread melanoma multiple myeloma, non-Hodgkin's lymphoma, kidney cancer and Kaposi's sarcoma. Produced in bacteria by recombinant DNA technology, it still retains biological activity since glycosylation is not required for receptor binding or folding. IFN α-2b was glycoPEGylated [21] following the Neose strategy with a sialyl PEG 20 kDa and purified by HPLC on SP Sepharose followed by HPLC on a Superdex column. MALDI-TOF/MS analysis of the product showed complete conjugation at Thr106. GlycoPEGylated IFN α-2b labelled with a radioactive tracer showed a significantly slower clearance rate in rats than the unmodified protein.

Other proteins of clinical interest which have been glycoPEGylated include follicle stimulating hormone [29], erythropoietin [30] and Factor VII [31].

Transglutaminase mediated PEGylation

Transglutaminases (TG-ases) are a group of enzymes of prokaryotic or eukaryotic origin acting at the glutamine side chain, as described in Figure 2, where A represents a polypeptidic chain bearing one or more glutamine residues whose γ-carboxamide groups act as acyl donors and B can be a variety of unbranched primary amines, including the amino groups of lysine side chains or a generic alkylamine acting as acyl acceptors [32–34].

The mammalian enzymes are Ca^{2+} dependent and are coded by a family of genes that lead to several structurally and functionally correlated isoenzymes (from TG-1 to TG-7; Factor XIIIA; Band 4.2). The enzymes are expressed in various tissues yielding specific post-translational modifications of proteins under physiological or pathological conditions [35].

Microbial TG-ases isolated from different micro-organisms, in particular from *Streptoverticillium* spp, do not require Ca^{2+}, are less specific and are used for industrial applications in the food chain, in particular to improve the texture of meat, cheese and their derivatives [36].

No specific consensus sequences that are targeted by TG-ases have been identified, but it was recently found that glutamine residues should be located on flexible and solvent accessible polypeptide chain regions in order to be recognised and modified by such enzymes [37]. Furthermore, it appears that positively charged or sterically bulky side chains preceding or following a glutamine residue in the substrate can positively influence recognition by the enzymes [38, 39].

On the other hand, no selectivity exists for the acyl acceptor moiety since both microbial and mammalian TG-ases can interact with different primary aliphatic alkylamines, preferentially bearing a linear chain of at least four carbon atoms as in the case of the ε-amino group of lysine [40]. One of the most studied bacterial transglutaminases is that produced by *Streptoverticillium mobaraense,* which is a smaller protein than the mammalian enzymes.

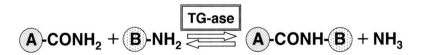

Figure 2. Conjugation reaction mediated by transglutaminase. A is a glutamine containing molecule (protein or peptide), B a primary amine donor (such as lysine side chain, PEG-amino).

Structural investigations [41] suggested that the lower specificity of microbial TG-ases are due to higher flexibility of the active site region and to the smaller size of the protein, making the interaction with the glutamine substrate residues easier. Microbial TG-ases are therefore more suitable for the modification of proteins and peptides, using an amino functionalised PEG as amine donor [42–44].

Selected examples of the successful use of TG-ases in site-specific PEGylation of biodrugs chosen from among those reported so far in several papers and patents are presented below.

Enzymatic PEGylation of granulocyte-colony stimulating factor

As an alternative to the commercial product Neulasta®, where PEG is linked by a chemical strategy, a new G-CSF-PEG 20 kDa was recently produced by BioKer s.r.l, using site specific mono-conjugation of an *E. coli* produced Met-G-CSF, catalysed by microbial transglutaminase [45]. The product, currently in clinical phase I, is named BK0026.

The PEGylation takes place under mild conditions (few hours, 4 °C, pH 7.4), by incubation of the protein with a 10-fold molar excess of 20 kDa PEG-NH_2 in the presence of 0.1–0.5 units of microbial TG-ase per ml of reaction mixture. After cationic exchange chromatography followed by size exclusion chromatography and a final ultrafiltration concentration step, a homogenous product was obtained with 90% yield. RP-HPLC, electrophoresis, SEC-HPLC and MALDI mass spectrometry were used to assess the purity and homogeneity of the conjugate and confirmed the presence of one chain of PEG per protein molecule. The conjugation site was assessed by analysis of the peptides obtained by enzymatic hydrolysis with endoproteinase Glu-C from *Staphylococcus aureus* V8.

As shown in Figure 3, the chromatographic profiles of the digested peptide mixtures of the starting and conjugated proteins are identical with the exception of two peaks that were attributed to the peptides comprising the conjugation site and originated by specific and non-specific endoproteinase Glu-C cuts. These peptides were isolated and identified, showing that they both contain the residues Gln132 and Gln135. Further characterisation by Edman sequencing demonstrated that the PEGylation only occurred at Gln135.

The assignment of the PEGylation site was confirmed by an orthogonal approach based on the enzymatic PEGylation of G-CSF with a monodisperse polymer form (556 Da Boc-PEG-NH_2, obtained from LCC, SW). The conjugation with this new PEG reagent independently validated the attribution of Gln135 as the unique site of conjugation through the direct characterisation of the endoproteinase Glu-C digestion mixture by ESI-MS and tandem MS (MS/MS) [46].

The pharmacokinetic parameters of BK0026 evaluated after subcutaneous administration in Sprague-Dawley male rats, as compared to those of native

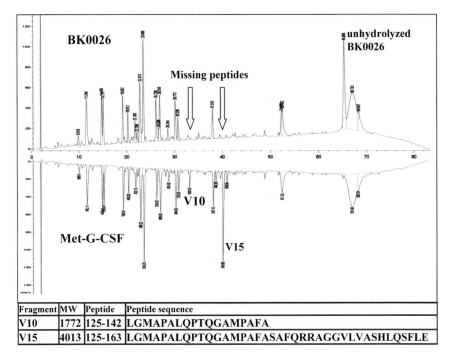

Fragment	MW	Peptide	Peptide sequence
V10	1772	125-142	LGMAPALQPTQGAMPAFA
V15	4013	125-163	LGMAPALQPTQGAMPAFASAFQRRAGGVLVASHLQSFLE

Figure 3. Reversed Phase HPLC profiles of endoproteinase Glu-C digested BK0026 and Met-G-CSF. Missing peptides correspond to V10 and V15 fragments, whose sequences are here reported.

Met-G-CSF and of Neulasta[®] showed a similar behaviour in terms of half-life, C max and AUC for both BK0026 and Neulasta[®] (Tab. 1).

The *in vitro* biological activity of BK0026, evaluated in terms of cell pro-liferation effects on the mieloblastic NFS-60 cell line, demonstrated a compa-rable stimulation of the proliferation of NFS-60 cells for BK0026 and Neulasta[®.] The EC_{50} values were 247.9 ± 0.14 and 201.5 ± 0.08 pg/ml respec-tively (unpublished observations).

To complete the BK0026 biological characterisation, its pharmacodynamic properties compared to Neulasta[®] were investigated by subcutaneous adminis-

Table 1. Pharmacokinetic parameters calculated after subcutaneous administration of Met-G-CSF, BK0026 and Neulasta[®] to Sprague-Dawley male rats

	Met-G-CSF	Neulasta[®]	BK0026
Tmax (h)	2	8	8
Cmax(ng/ml)	132	29.2	36.9
AUC (ng h/ml h)	735	920	794
$T_{1/2}$ (h)	2.1	8.1	8.2

tration of both products at three different dosages in rats. As shown in Figure 4, in all BK0026 treated animals a moderate to marked elevation of neutrophil count was evident on day 1, reaching its maximum level on days 2–3 and returning to baseline after about 6–8 days. A comparable dose-dependence trend was observed following Neulasta® treatment. The increase of absolute neutrophil count (ANC) peak time at higher doses (Fig. 4c) is likely due to the saturation of the dose-dependent neutrophil-mediated clearance mechanism, as already reported after administration of Neulasta® [47].

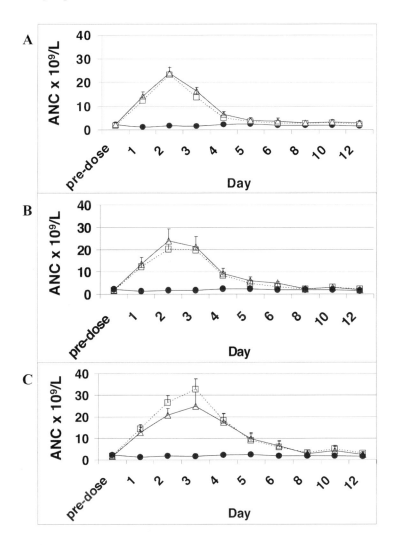

Figure 4. Absolute neutrophil count (ANC) *versus* time after subcutaneous administration of BK0026 (□), Neulasta® (△) and negative control (buffer, ●), at three different dosages (A: 40 μg/kg; B: 120 μg/kg; C: 360 μg/kg) on Sprague-Dawley male rats.

These results demonstrate that BK0026, besides being a product easier to obtain in pure and homogeneous form and simple to scale-up, may represent a convenient clinical alternative to Neulasta®.

Interleukin-2 (IL-2)

IL-2, a cytokine involved in the immune signalling system and produced during immune response, is responsible for growth and differentiation of cytotoxic T cells. A recombinant IL-2 is marketed as Proleukin® for the treatment of some forms of cancer such as metastatic melanoma.

Guinea pig liver transglutaminase (G-TG-ase), a form of mammalian TG-ase commercially available, was unsuccessful in catalysing PEG conjugation of IL-2, most probably since, as mentioned above, TG-ases from eukaryotic sources possess high substrate specificity.

To get PEGylation, it was necessary to produce a chimeric form of IL-2 [42] carrying at the N-terminus a suitable glutamine containing tag with the sequence AQQIVM to yield a fusion protein called rTG2-IL-2. This protein, incubated with 10 kDa PEG-amino at a 500-fold molar excess in the presence of G-TG-ase and calcium ion, gave mono and di-PEGylated species. The formation of a di-PEGylated product was unexpected considering the hindrance of the two adjacent glutamines, but it can be probably explained by the large excess of polymer used as well as by its small size.

Di-PEGylated rTG2-IL-2 was characterised for in vitro activity and pharmacokinetic profile in rats in comparison with a random PEGylated IL-2 obtained by chemical reaction. Both products showed prolonged half-life and similar pharmacokinetic profiles, but the adduct obtained by G-TG-ase reaction PEGylation exhibited a higher in vitro activity than the random PEGylated product, most probably due to the conjugation of the polymer at the N-terminal tag not involved in receptor interaction. However, this experiment, first described by Sato, demonstrated that a TG-ase could be successfully used to modify a protein in a site-specific fashion with the unnatural substrate PEG amine [44].

Sato also described IL-2 PEGylation using microbial transglutaminase (M-TG-ase) from a *Streptoverticillium* sp. strain under quite similar reaction conditions. In this case, one of the six glutamine residues of natural IL-2 underwent PEGylation. Peptide mapping coupled with MS analysis allowed identification of Gln74 as the only PEGylation site.

These results confirmed the higher specificity of G-TG-ase as compared to the microbial form of the enzyme, although M-TG-ase can still show some specificity towards the amino acidic sequence due to the flexibility of the protein chain or the presence of polar and charged residues surrounding the reactive glutamine.

Growth hormone (GH)

GH is a 191 amino acid single chain protein that is synthesised by the anterior pituitary gland, and plays an important role in somatic growth through its effects on the metabolism of proteins, carbohydrates and lipids. Recombinant human GH is produced by *E. coli* or mammalian cell fermentation and is marketed under different brand names. To date, no PEGylated derivatives have been approved for clinical use, although many different strategies have been tested for conjugation, mainly through chemical coupling to lysine N or C terminus, and cysteines, (obtained by disulfide bridge reduction or insertion by mutagenesis) [48].

Use of microbial transglutaminase was also studied as an alternative for growth hormone PEGylations. By incubating GH with PEG amine of different MW in the presence of M-TG-ase, the formation of two main mono PEGylation products at Gln40 or Gln141 has been observed in spite of the presence of thirteen Gln residues on the polypeptide chain. The identification of the conjugation sites was accomplished by peptide mapping followed by MS analysis. It would be interesting to make one the two reactive sites silent, to allow the formation of a single mono-PEGylated product, which would be easier to characterise and obtain in a pure, and reproducible form.

Erythropoietin (EPO)

Erythropoietin or haematopoietin, a cytokine primarily synthesized by foetal liver and adult kidneys, is responsible of regulation of red blood cell production. Recombinant EPO, currently used as treatment for anaemia due to chronic kidney diseases, cancer treatment and heart failure, is a glycoprotein produced by mammalian cell fermentation and sold under different brand names, such Epogen®, Eprex® or Neorecormon®.

Many studies on PEG modification of erythropoietin have been carried out, but the FDA only recently approved a mono-PEGylated EPO, called Mircera®, with positional isomers bearing PEG chains both at the N-terminus and at some lysine residues [49].

PEG attachment to erythropoietin through TG-ase catalysis has been proven by Centocor [50]. In some patents Centocor claims to have obtained a PEGylated form, mainly mono-PEGylated EPO, by incubating the protein with amine functionalised PEG in the presence of guinea pig liver transglutaminase. Microbial forms are also declared to be suitable biocatalysts for the enzymatic conjugation.

Of general interest is the fact that a reversed TG-ase catalysed reaction can also take place when polyethylene glycol end capped with glutamine-glycine dipeptide is used as the acyl donor while the lysine side chains of the proteins act as acyl acceptors.

Conclusions

PEG conjugation (known as PEGylation) has been established as a strategy to overcome several limitations connected to the use of proteins in therapy. So far, the chemical methods of PEGylation have received the most attention and have allowed the development of a number of PEGylated therapeutic proteins marketed for clinical use.

An emerging, but so far not sufficiently exploited technique, is represented by the use of enzymes, such as glucosyl-transferases and transglutaminases in the PEG coupling step which are able to covalently bind sugars, PEG or other chemical moieties to protein and peptides, in a fairly specific fashion with high yield and limited costs. The most noticeable example of conjugated proteins obtained by enzymatic PEGylation is represented by G-CSF which was glyco-PEGylated by a strategy developed by Neose Technologies Inc., or site-specifically mono-PEGylated by microbial TG-ase to yield a product in clinical Phase I according to a method first described by Bio-Ker S.r.l.

To determine the basis of TG-ase specificity we recently examined the PEGylation sites of proteins reported in the literature or independently studied in our laboratory in terms of their known 3D structure and dynamics. We have observed a clear-cut correlation between sites of TG-ase attack and regions of enhanced backbone flexibility, this last being detected by the crystallographic profile of the B-factor along the protein polypeptide chain [37]. A mechanism of local unfolding of the polypeptide substrate will therefore strongly favour a site-specific modification by TG-ase, provided that a Gln residue is encompassed in that region. Since chain flexibility was earlier found to dictate the sites of limited proteolysis of proteins [51], we propose that both TG-ases and proteases require an unfolded polypeptide substrate for their selective enzymatic attachment and consequently it should be possible to predict the site(s) of TG-ase mediated modification of a protein on the basis of its 3D structure and molecular dynamics.

It is also worthy to note that this correlation holds for other enzymes. A recent analysis of the structure and dynamic features of the O-glycosylation sites targeted by the enzymatic GlycoPEGylation technique [21], demonstrated the correlation between sites of chain flexibility or local disorder determined crystallographically and sites of O-conjugation.

References

1 Harris JM (ed.) (1992) *Polyethylene glycol chemistry, biotechnical and biomedical applications.* Plenum Press, New York
2 Harris JM, Zalipsky S (eds) (1997) *Poly(ethylene glycol) chemistry and biological applications.* American Chemical Society, Washington DC
3 Roberts MJ, Bentley MD, Harris JM (2002) Chemistry for peptide and protein PEGylation. *Adv Drug Deliv Rev* 54: 459–476
4 Harris JM, Veronese FM (2003) Peptide and protein pegylation II – Clinical evaluation. *Adv Drug Deliv Rev* 55: 1259–1260

5 Veronese FM, Harris JM (2008) Peptide and protein PEGylation III: Advances in chemistry and clinical applications. *Adv Drug Deliv Rev* 60: 1–2
6 Levy Y, Hershfield MS, Fernandez-Mejia C et al. (1988) Adenosine deaminase deficiency with late onset of recurrent infections: response to treatment with polyethylene glycol-modified adenosine deaminase. *J Pediatr* 113(2): 312–317
7 Graham LM (2003) Pegaspargase: a review of clinical studies. *Adv Drug Deliv Rev* 55: 1293–1302
8 Wang YS, Youngster S, Grace M et al. (2002) Structural and biological characterization of PEGylated recombinant interferon alpha-2b and its therapeutic implications. *Adv Drug Deliv Rev* 54: 547–570
9 Bailon P, Palleroni A, Schaffer CA et al. (2001) Rational design of a potent, long-lasting form of interferon: a 40 kDa branched polyethylene glycol-conjugated interferon alpha-2a for the treatment of hepatitis. *Bioconjug Chem* 12(2): 195–202
10 Kinstler OB, Brems DN, Lauren SL et al. (1996) Characterization and stability of N-terminally PEGylated rhG-CSF. *Pharm Res* 13(7): 996–1002
11 Cox G (1999) Derivatives of growth hormone and related proteins. WO 1999/003887
12 Goodson RJ, Katre NV (1990) Site-directed pegylation of recombinant interleukin-2 at its glycosylation site. *Biotechnology (NY)* 8(4): 343–346
13 Spiro RG (1973) Glycoproteins. *Adv Protein Chem* 27: 349–467
14 Roseman S (2001) Reflections on glycobiology. *J Biol Chem* 276(45): 41527–41542
15 Ernst B, Sinay P, Hart G (eds): (2000) *Oligosaccharides in Chemistry and Biology – A Comprehensive Handbook.* Wiley-VCH Verlag GmbH, Germany
16 Shriver Z, Raguram S, Sasisekharan R (2004) Glycomics: a pathway to a class of new and improved therapeutics. *Nat Rev Drug Discov* 3: 863–873
17 Verez-Bencomo V, Fernández-Santana V, Hardy E et al. (2004) A synthetic conjugate polysaccharide vaccine against Haemophilus influenzae type b. *Science* 305(5683): 522–525
18 Legendre H, Decaestecker C, Goris Gbenou M et al. (2004) Prognostic stratification of Dukes B colon cancer by a neoglycoprotein. *Int J Oncol* 25(2): 269–276
19 Davis BG, Robinson MA (2002) Drug delivery systems based on sugar-macromolecule conjugates. *Curr Opin Drug Discov Devel* 5(2): 279–288
20 Teranishi K, Gollackner B, Bühler L et al. (2002) Depletion of anti-gal antibodies in baboons by intravenous therapy with bovine serum albumin conjugated to gal oligosaccharides. *Transplantation* 73(1): 129–139
21 DeFrees S, Wang ZG, Xing R et al. (2006) GlycoPEGylation of recombinant therapeutic proteins produced in *Escherichia coli. Glycobiology* 16(9): 833–843
22 Roskos LK, Lum P, Lockbaum P et al. (2006) Pharmacokinetic/pharmacodynamic modeling of pegfilgrastim in healthy subjects. *J Clin Pharmacol* 46(7): 747–757
23 Rosendahl MS, Doherty DH, Smith DJ et al. (2005) Site-specific protein PEGylation: application to cysteine analogs of recombinant human granulocyte colony-stimulating factor. *BioProcess International* 3: 52–62
24 Molineux G (2004) The design and development of pegfilgrastim (PEG-rmetHuG-CSF, Neulasta). *Curr Pharm Des* 10(11): 1235–1244
25 Frampton JE, Keating GM (2005) Spotlight on pegfilgrastim in chemotherapy-induced neutropenia. *BioDrugs* 19(6): 405–407
26 Defrees S, Clausen M, Zopf DA et al. (2007) Glycopegylated Granulocyte Colony Stimulating Factor. US Patent Application 20070254836
27 Kaushansky K, Lopez JA, Brown CB (1992) Role of carbohydrate modification in the production and secretion of human granulocyte macrophage colony-stimulating factor in genetically engineered and normal mesenchymal cells. *Biochemistry* 31: 1881–1886
28 Forno G, Fogolin MB, Oggero M et al. (2004) N- and O-linked carbohydrates and glycosylation site occupancy in recombinant human granulocyte-macrophage colony-stimulating factor secreted by a Chinese hamster ovary cell line. *Eur J Biochem* 271: 907–919
29 Defrees S, Bayer RJ, Bowec et al. (2008) Glycopegylated Follicle Stimulating Hormone. US Patent Application 20080015142
30 Defrees S, Bayer RJ, Zopf DA et al. (2006) Glycopegylated erythropoietin formulations. US Patent Application 20060287224
31 Klausen NK, Bjorn S, Behrens C et al. (2008) Pegylated Factor VII Glycoforms. US Patent Application 20080039373

32 Folk JE, Finlayson JS (1977) The epsilon-(gamma-glutamyl)lysine crosslink and the catalytic role of trans-glutaminases. *Adv Protein Chem* 31: 1–133

33 Folk JE (1980) Transglutaminases. *Annu Rev Biochem* 49: 517–531

34 Lorand L, Conrad SM (1984) Transglutaminases. *Mol Cell Biochem* 58: 9–35

35 Griffin M, Casadio R, Bergamini CM (2002) Transglutaminases: nature's biological glues. *Biochem J* 368(Pt 2): 377–396

36 Nielsen PM (1995) Reactions and Potential Industrial Applications of Transglutaminase. Review of Literature and Patent. *Food Biotechnol* 9: 119–156

37 Fontana A, Spolaore B, Mero A, Veronese FM (2008) Site-specific modification and PEGylation of pharmaceutical proteins mediated by transglutaminase. *Adv Drug Deliv Rev* 60(1): 13–28

38 Coussons PJ, Price NC, Kelly SM, Smith B, Sawyer L (1992) Factors that govern the specificity of transglutaminase-catalysed modification of proteins and peptides. *Biochem J* 282(Pt 3): 929–930

39 Ohtsuka T, Ota M, Nio N et al. (2000) Comparison of substrate specificities of transglutaminases using synthetic peptides as acyl donors. *Biosci Biotechnol Biochem* 64(12): 2608–2613

40 Ohtsuka T, Sawa A, Kawabata R et al. (2000) Substrate specificities of microbial transglutaminase for primary amines. *J Agric Food Chem* 48(12): 6230–6233

41 Kashiwagi T, Yokoyama K, Ishikawa K et al. (2002) Crystal structure of microbial transglutaminase from *Streptoverticillium mobaraense*. *J Biol Chem* 277(46): 44252–44260

42 Sato H, Yamamoto K, Hayashi E et al. (2000) Transglutaminase-mediated dual and site-specific incorporation of poly(ethylene glycol) derivatives into a chimeric interleukin-2. *Bioconjug Chem* 11(4): 502–509

43 Sato H, Hayashi E, Yamada N et al. (2001) Further studies on the site-specific protein modification by microbial transglutaminase. *Bioconjug Chem* 12(5): 701–710

44 Sato H (2002) Enzymatic procedure for site-specific pegylation of proteins. *Adv Drug Deliv Rev* 54(4): 487–504

45 Tonon G, Orsini G, Schrepper R et al. (2008) G-CSF site-specific mono-conjugates. International Patent Application WO 2008/017603

46 Veronese FM, Mero A, Spolaore B et al. (2008) Site-specific PEGylation of pharmaceutical proteins mediated by transglutaminase (Poster presented at the 35th Annual Meeting and Exposition of the Controlled Release Society, New York)

47 Bowen S, Tare N, Inoue T et al. (1999) Relationship between molecular mass and duration of activity of polyethylene glycol conjugated granulocyte colony-stimulating factor mutein. *Exp Hematol* 27(3): 425–432

48 Zundel M, Peschke B (2006) C-terminally pegylated growth hormones. WO 2006/084888

49 Macdougall IC (2005) CERA (Continuous Erythropoietin Receptor Activator): a new erythro-poiesis-stimulating agent for the treatment of anemia. *Curr Hematol Rep* 4(6): 436–440

50 Pool CT (2004) Formation of novel erythropoietin conjugates using transglutaminase. International Patent Application WO 2004/108667

51 Fontana A, Fassina G, Vita C et al. (1986) Correlation between sites of limited proteolysis and seg-mental mobility in thermolysin. *Biochemistry* 25: 1847–1851

The site-specific TGase-mediated PEGylation of proteins occurs at flexible sites

Angelo Fontana[1], Barbara Spolaore[1], Anna Mero[2] and Francesco M. Veronese[2]

[1] *CRIBI Biotechnology Centre, University of Padua, Viale G. Colombo 3, 35121 Padua, Italy*
[2] *Department of Pharmaceutical Sciences, Via F. Marzolo 5, University of Padua, 35131 Padua, Italy*

Abstract

Transglutaminase (TGase) is able to catalyse the acyl transfer reaction between the γ-carboxamide group of a protein-bound glutamine (Gln) residue and an amino-derivative of poly(ethylene glycol) (PEG-NH$_2$), thus leading to a PEGylated protein. Several proteins of therapeutic interest have been PEGylated by means of TGase, among them interleukin-2, granulocyte colony-stimulating factor, human growth hormone and erythropoietin. Surprisingly, PEGylation occurred at specific Gln residue(s), despite the fact that these proteins contained several Gln residues. An analysis of the TGase-mediated reactions in terms of structure and dynamics of protein substrates revealed a correlation between sites of TGase attack and chain regions of enhanced backbone flexibility, as detected by the crystallographic profile of the B-factor along the protein polypeptide chain. Moreover, the TGase-mediated reactions often occurred at chain regions characterized by missing electron density, indicating that these regions are disordered. In particular, it was noted that in a number of cases the sites of TGase attack occurred at the same chain regions prone to limited proteolysis phenomena. Since chain flexibility or local unfolding was earlier found to dictate the sites of limited proteolysis of proteins, it is concluded that both TGase and a protease require an unfolded polypeptide substrate in an extended conformation for the site-specific enzymatic attack.

Introduction

Protein drugs may possess several shortcomings that can limit their usefulness in therapy, including susceptibility to degradation by proteases, rapid kidney clearance and propensity to generate neutralizing antibodies [1–4]. Among the techniques so far explored for the development of safer and more useful protein drugs, undeniably the protein surface modification by covalent attachment of poly(ethylene glycol) (PEG) became an extremely valuable technique for producing protein drugs more water-soluble, non-aggregating, non-immunogenic and more stable to proteolytic digestion [5–10]. In numerous studies it has been demonstrated that PEGylated proteins possess improved half-life in the blood serum, because the large size of the PEG-conjugated molecule slows down renal clearance [11–14]. Finally, PEGylation can mask the protein's surface and strongly reduce protein degradation by the action of proteolytic enzymes. PEG is a water-soluble, biocompatible polymer commonly utilized as an additive in protein formulations, as well as for facilitating crystallization of proteins. Many attempts have been undertaken to develop new polymers with improved properties, but none of these new polymers has been able yet to compete with PEG for protein conjugation. This can be explained by the bio-compatibility of PEG and the valuable experience with PEG as a low-cost additive for the pharmaceutical industry over the last decades [15, 16].

The most used chemical methods of PEGylation of proteins involve the covalent conjugation of PEG at the level of the ε-amino group of lysine residues by using acylating PEG derivatives [17–20]. A drawback of these pro-cedures resides in multiple sites of conjugation and thus in the substantial het-erogeneity of the PEGylated proteins. Nowadays, there is a stringent need of a very detailed chemical characterization of the proteins used for therapy and thus the heterogeneity of PEGylated proteins is clearly a serious problem [20, 21]. Since PEGylation is a permanent protein modification, the national and international authorities for drug approval make high demands on the stringent characterisation of PEG reagents and the final PEGylated product. Major requirements are the specification of the degree of PEGylation, analysis of the dispersity index of PEG and determination of the PEGylation sites on the pro-tein conjugate [21]. Therefore, in order to prepare more homogeneous protein conjugates, the site-specific PEGylation of pharmaceutical proteins will be an active area of research for the coming years.

Several chemical approaches were developed for a site-specific PEGylation of proteins, such as the selective PEGylation at the level of the thiol group of cysteine (Cys) residues [22–24] or at the N-terminal amino group [25–29] of a polypeptide chain. Recently, novel procedures of transg-lutaminase (TGase)-mediated PEGylation of pharmaceutical proteins were proposed [30–34]. The key step for the TGase-mediated catalysis involves the interaction of a γ-carboxamide group of a glutamine (Gln) residue of a polypeptide substrate with the TGase's active site, forming a reactive thioa-cyl-moiety and then reaction with an amino donor, thus leading to a new isopeptide amide bond [35–39] (Fig. 1). When the amino-donor is the ε-amino group of a protein-bound lysine (Lys) residue, the TGase-mediated reaction results in an intermolecular protein crosslinking due to the formation of ε-γ-glutamyl)lysine isopeptide bonds. This crosslinking reaction is a most relevant and physiologically important enzymatic reaction of TGase, being involved in wound healing and blood coagulation [40]. Of course, the TGase-mediated reaction between Gln and Lys residues can also lead to an intramol-ecular protein crosslinking. Other nucleophiles can react with the acyl-enzyme derivative, including primary amines, hydroxylamine and other amino-containing chemical moieties. The hydrolysis of the reactive thioester intermediate can be a competitive reaction, thus leading to the modification of the former Gln into a Glu residue [41].

The PEG polymer can be covalently linked to a protein-bound Gln residues by using an amino-derivative of PEG (PEG-NH$_2$) as an amino donor in the TGase-mediated reaction, resulting in a protein–CONH–PEG conjugate (see

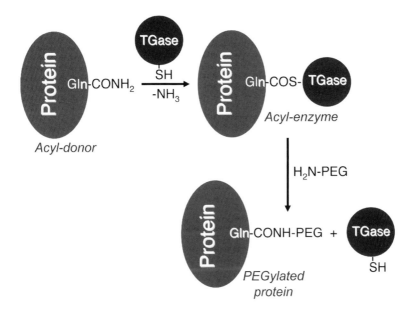

Figure 1. Schematic drawing of the TGase-mediated PEGylation of a protein. The γ-carboxamido group of protein-bound glutamine (Gln) residue serves as an acyl-donor to form the acyl-enzyme intermediate at the level of the active-site cysteine (Cys) residue located in position 64 of the 331-residue polypeptide chain of microbial TGase [46] and a concomitant release of ammonia. Then, an amino-donor reacts with the highly reactive thioester intermediate leading to the formation of a novel γ-amide derivative of glutamic acid (Glu) and the free native enzyme [36]. The acyl-enzyme reacts with nucleophilic primary amines, including the amino-derivative of poly(ethylene glycol) (PEG-NH$_2$). In this case, the resulting product of the reaction is a PEGylated protein.

Fig. 1) [42]. Thus, TGase can act as a reverse-protease, catalysing a coupling reaction instead of hydrolysis. Here, we will briefly present and discuss few cases of site-specific TGase-mediated PEGylation of proteins and we will highlight the molecular features of the protein substrate that enable the often observed site-specific PEGylation at the level of one or very few protein-bound Gln residue(s). We will show that the Gln-residues being attacked by TGase are encompassed in chain regions of the protein substrate displaying enhanced flexibility or being unfolded. These features are those earlier identified as determining the sites of limited proteolysis of globular proteins. Indeed, in a number of cases the same chain regions being sites of TGase's reaction are also sites of limited proteolysis.

Microbial transglutaminase (TGase)

Transglutaminase (TGase; protein-glutamine γ-glutamyl-transferase, EC 2.3.2.13) catalyzes an acyl transfer reaction in which the γ-carboxyamide groups of peptide- or protein-bound Gln residues act as the acyl donors [36, 40, 41]. The most common acyl acceptors of TGase are the ε-amino groups of lysine residues within proteins or the primary amino groups of some naturally occurring polyamines [37, 40]. TGases are widely distributed in various organisms, including vertebrates, invertebrates, plants and microorganisms [35, 36]. Among these TGases, the human blood coagulation factor XIII has been most studied and its molecular and enzymological properties have been analyzed in great detail [40]. A microbial TGase has been isolated from the culture medium of *Streptoverticillium* sp. S-8112, which has been identified as a variant of *Sv. mobaraense* [43–46]. The crystal structure of this microbial TGase consists of a compact domain with overall dimensions 65 × 59 × 41 Å [47]. The cysteine (Cys) residue in position 64 of the 331-residue chain of the enzyme, essential for the catalytic activity, is located at the bottom of a deep cleft, its depth being 16 Å (see Fig. 2). The crystal structure of microbial TGase revealed that the overall fold of this enzyme is different from that of factor XIII-like TGase [48]. Nevertheless, a similar cysteine protease-like catalytic mechanism for the microbial TGase has been proposed [46]. Indeed, the Cys-His-Asp triad of cysteine-proteases is conserved, being given in the microbial TGase by Cys64, Asp255 and His274 residues. Of interest, the catalytic triad

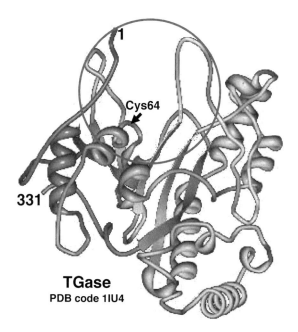

Figure 2. Three-dimensional structure of microbial TGase from *Streptomyces mobaraense*. The protein model was prepared from the X-ray structure of the enzyme [46] using the software MBT (Molecular Biology Toolkit; http://mbt.sdsc.edu) available in PDB (code 1IU4) [47]. The numbers 1 and 331 indicate the N- and C-terminus, respectively, of the 331-residue chain of TGase. The active site area of the enzyme is circled and the location of Cys64 is indicated by an arrow.

of microbial TGase almost superimposes the same triad of factor XIII-like TGase, implying a similar mechanism of catalysis. There are a number of acidic residues (Asp1, Asp3, Asp4, Glu249, Asp255 and Glu300) in the TGase's active site cleft and aromatic residues (Trp59, Tyr62, Trp69, Tyr75, Tyr278, Tyr291 and Tyr302) on the surface around the cleft. These characteristic distributions of the acidic and aromatic residues appear to have an effect on the substrate specificity of TGase [46].

The optimum pH for TGase activity is between 5 and 8, but the enzyme displays some activity also at pH 4 or 9. The optimum temperature for enzymatic activity is 55 °C (for 10 min at pH 6.0) and the enzyme maintains full activity for 10 min at 40 °C, but looses activity within a few minutes at 70 °C [49]. At variance from other mammalian TGases, including the well-characterized guinea pig liver enzyme [40], microbial TGase is totally independent of calcium ions. Heavy metals such as Cu^{2+}, Zn^{2+} and Pb^{2+} are strong inhibitors of the enzyme, since they bind to the thiol group of the active-site Cys64 [43, 46]. Several molecular and functional characteristics of microbial TGase, including calcium-independence, protein stability, higher reaction rate and small molecular size, are advantageous for its industrial applications [50–53]. Actually, microbial TGase is widely used to improve the physical properties of food proteins, including legume globulins, wheat glutens, egg yolk and albumin proteins, actins, myosins, fibrins, milk caseins, α-lactalbumin, β-lactoglobulin and other proteins. Because of the usefulness and generality of the enzymatic reaction mediated by TGase outlined in Figure 1, additional and novel applications of this enzyme in protein research and technology are expected to be further explored [50].

Examples of TGase-mediated PEGylation of proteins

The use of microbial TGase for the enzymatic PEGylation of proteins is quite recent [30–34, 42], but the examples of successful production of homogeneous protein conjugates are indeed very interesting and clearly demonstrate that the TGase methodology offers great promises [34]. A striking result of these studies is that PEGylation mediated by TGase can be site-specific, sometimes leading to the conjugation of a PEG moiety at the level of only one Gln residue among the many Gln residues of the protein substrate [34, 42]. Here, we will summarize the results obtained with few proteins, emphasizing those obtained with proteins of pharmaceutical interest.

Interleukin-2

Interleukin-2 (IL-2) was identified more than 30 years ago and shown to consist of 133 amino acid residues [54, 55]. Two cysteine residues at positions 58 and 105 form a disulfide bridge. A third residue at position 125 is not essential for biological activity and sometimes is replaced in commercially available IL-2 to avoid misfolding and aggregation and to increase shelf-life of the protein [55]. This pleiotropic cytokine is produced mainly by activated T cells and it was first described as a T cell growth factor. IL-2 has a wide spectrum of effects in the immune system. Some of the possible mechanisms by which IL-2 carries out its anticancer effects include the augmentation of cytotoxicity of immune cell, enabling delivery of immune cells and possibly serum components into tumor [54, 55].

A site-specific PEGylated-derivative of recombinant IL-2 was prepared utilizing the microbial TGase from *Streptomyces mobaraense* and an amino derivative of PEG (PEG-NH$_2$) [30–33]. It was found that TGase mediates a selective and stoichiometric incorporation of the PEG polymer chain at the level of Gln74, despite the protein contains six Gln residues in positions 11, 13, 22, 57, 74 and 126. An IL-2 derivative with a galactose-terminated triantennary glycoside Gal$_3$ covalently bound selectively to Gln74 was also prepared by the TGase method, with the goal of obtaining a therapeutic protein that can allow selective hepatic delivery [33].

In order to explain the striking site-specific reaction of TGase on IL-2 we must consider molecular aspects of the protein–protein interaction phenomenon occurring when TGase attacks its globular protein substrate and thus the structure and dynamics of IL-2. As shown in Figure 3, IL-2 is a four-helix bundle protein, in which helix B (residues 59–73) and C (residues 83–97) are connected by a long loop [56–59]. The TGase-reactive Gln74 residue is encompassed by this connecting loop. An interesting observation can emerge if one considers the *B*-factor profile of the polypeptide chain of IL-2 (Fig. 3B). The *B*-factor (or crystallographic temperature factor) represents the mean-square displacement of the polypeptide backbone and is frequently used as a measure of chain flexibility of folded globular proteins [60–63]. When plotted against residue number, *B*-factor values provide a graphic image of the degree of mobility existing along the polypeptide chain of a protein. Moreover, X-ray crystallography can define missing electron density in protein structures, which corresponds to disordered regions [64–66]. The absence of interpretable electron density for some sections of the protein structure is associated with the increased mobility of atoms in these regions, which leads to non-coherent X-ray scattering, making atoms invisible [64–66]. Upon inspection of the *B*-factor profile shown in Figure 3B, it is seen that the site of TGase attack at Gln74 occurs at a disordered chain region of the IL-2 molecule, since there is no electron density for chain segment 74–81. Therefore, the Gln74 residue is embedded in a chain region which is unfolded and thus a most suitable substrate for the TGase action, considering that the enzymatic reaction requires the binding and adaptation of a 10–12 residue polypeptide segment at the enzyme's active site ([34], see also below).

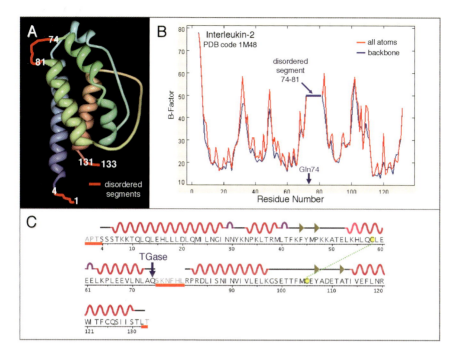

Figure 3. TGase-mediated PEGylation of human interleukin-2 (IL-2). (A) Three-dimensional structure of recombinant IL-2 derived from the X-ray structure of the protein (PDB code 1M48) [56]. The figure was prepared using the MBT (Molecular Biology Toolkit; http://mbt.sdsc.edu) software available in PDB [47]. The 133-residue chain of IL-2 is depicted with rainbow colors from the N-terminus (blue) to the C-terminus (red). The protein chain is characterized by missing electron density in segments 1–4, 74–81 and 131–133 and thus these chain segments are very flexible or disordered. In the model these segments are arbitrarily drawn by a red line. (B) Profile of the B-factor along the 133-residue chain of the protein, as determined from the X-ray structure of the protein taken from PDB (code 1M48). There are no figures of B-factor for the N- and C-terminal regions, as well as for chain region 74–81. The site of specific TGase-mediated PEGylation of IL-2 at Gln74 is indicated by an arrow. (C) Amino acid sequence and secondary structure of human IL-2. The helical segments along the protein chain are indicated by squiggled red lines above the amino acid sequence. The connectivity of Cys58 to Cys105 forming the single disulfide bridge of the protein is indicated by a dashed green line. The disordered chain segments are indicated by red lines below the amino acid sequence. The site of TGase attack at Gln74 is indicated by an arrow.

Granulocyte colony-stimulating hormone

Granulocyte-colony stimulating factor (G-CSF) is the major regulator of granulopoiesis *in vivo* and stimulates proliferation, differentiation and survival of cells of the granulocyte lineage [67–69]. Structurally, G-CSF is constituted by a single polypeptide chain of 174 amino acid residues and belongs to a group of proteins that share a four-helix bundle architecture [68–69]. While wild-type G-CSF has a glycosylation site at Thr133, recombinant G-CSF lacks glycosylation. The clinical use of G-CSF involves treatment of neutropenia, a condition that occurs in a variety of diseases, including congenital defects, bone marrow suppression following pharmacological manipulation and infection, as well as in treatments of cancer patients undergoing cytotoxic chemotherapy. Recombinant G-CSF is already in clinical practice for those patients suffering from neutropenia during or after chemotherapy, as well as for use in bone marrow transplantation [70–75]. PEGylated G-CSF has been prepared by using reactive PEG derivatives [76] and nowadays a variety of effective G-CSF drugs (Neupogen, Filgrastin) are in use [70–75].

The site-specific incorporation of PEG-NH$_2$ into recombinant G-CSF can be achieved by using microbial TGase using polydisperse PEG of high molecular weight [77] or a monodisperse PEG of low molecular weight [42]. In both cases, it was possible to achieve essentially quantitative conjugation of PEG-NH$_2$ and to isolate a mono-PEGylated derivative of G-CSF. Considering that G-CSF contains 17 Gln residues, this selectivity of the TGase reaction is clearly striking. Since the determination of the site of conjugation in the PEGylated protein by fingerprinting techniques combined with mass spectrometry using polydisperse PEG is hampered by the molecular heterogeneity of the resulting conjugate, the G-CSF derivative prepared using TGase and a monodisperse PEG-NH$_2$ of low molecular weight was used for the identification of reactive Gln residue(s). To this aim, PEGylated G-CSF was digested with proteases and the resulting peptides analyzed by electrospray and tandem mass spectrometry [42]. The results of these analyses allowed the unambiguous identification of Gln134 as the site of PEGylation. Of note, Gln134 is located nearby the Thr133 residue, which is the site of glycosylation in wild-type G-CSF. Therefore, we may anticipate that the TGase-mediated reaction likely mimics molecular aspects of the *O*-glycosylation reaction of the protein *in vivo* (see below).

Since the X-ray structure of human G-CSF is known, we can try to explain the selective reaction of TGase by considering the structural features of the protein substrate [68–69]. First of all, we can observe that Gln134 is embedded in the long chain segment connecting helices C and D and, therefore, that modification does not occur at the level of chain regions encompassing the four main helices of the protein [69] (see Fig. 4A). Moreover, a main feature of this loop is that it encompasses the chain segment 121–137, for which there is no electron density and thus is disordered by crystallographic criteria. Therefore, TGase selectively attacks the G-CSF substrate at a disordered region of the protein chain. However, there are other Gln residues located in disordered regions, but not susceptible to the TGase reaction. We explain the lack of TGase reaction at the level of Gln131 and Gln173 by considering that these two Gln residues are followed by a Pro residue (see Fig. 4C). If one accepts our view that the mechanism of reaction of TGase is similar to that of a protease ([34], see also below), we can propose that TGase acts similarly, for example, to trypsin, which does not cleave the Lys–Pro or Arg–Pro peptide bond (see the web site shttp://www.expasy.ch/tools/peptidecutter/). The lack of TGase reaction at the level of Gln67 and Gln70 can be explained by the fact that, even if these residues are embedded in the flexible/disordered chain seg-

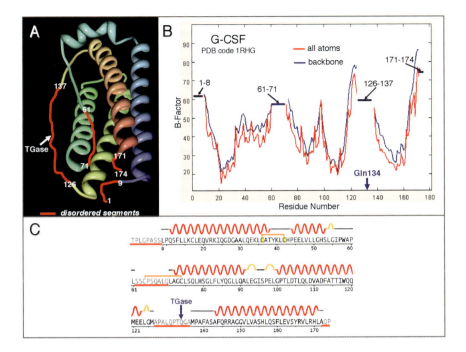

Figure 4. TGase-mediated PEGylation of human granulocyte colony-stimulating factor (G-CSF). (A) Three-dimensional structure of G-CSF derived from the X-ray structure of the protein (PDB code 1RHG) [69]. The figure was prepared using the MBT (Molecular Biology Toolkit; http://mbt.sdsc.edu) software available in PDB [47]. The 174-residue chain of G-CSF is depicted in rainbow colors from the N-terminal (blue) to the C-terminal (red). The protein chain is characterized by missing electron density in segments 1–9, 61–71, 126–137 and 171–174 and thus these segments are disordered. In the model these segments are arbitrarily drawn as red lines. (B) Profile of the *B*-factor along the 174-residue chain of G-CSF as determined from the X-ray structure of the protein and taken from PDB (code 1RHG). There are no figures for *B*-factor in chain regions 1–9, 61–71, 126–137 and 171–174 due to the fact that there is no electron density for them. The site-specific TGase-mediated PEGylation at Gln134 is indicated by an arrow. (C) Amino acid sequence and secondary structure of human G-CSF. Helical segments along the protein chain are indicated by squiggled red lines above the amino acid sequence. The connectivities of the two disulfide bonds Cys36–Cys42 and Cys64–Cys74 are indicated by yellow lines above the sequence. The disordered chain segments are indicated by red lines below the sequence. The site of TGase attack at Gln134 is indicated by an arrow.

ment 61–71 (see Fig. 4B), this chain region is crosslinked by the disulfide bridge Cys63–Cys74 and this chain segment may be too short for an effective binding and adaptation at the TGase's active site. Indeed, we remind herewith that a chain segment of 10–12 residues in an extended conformation is required for an enzymatic reaction on a polypeptide substrate ([34], see also below). Finally, there is no TGase reaction at the level of the disordered N-terminal chain segment 1–8 (Fig. 4B), since at this N-terminal region there is no Gln residue that could be attacked by TGase (see Fig. 4C). Overall, these considerations allow us to conclude that, in analogy to the IL-2 case above discussed, the selective reaction of TGase on the G-CSF substrate can be explained by the main requirement of a Gln residue to be embedded in a disordered or unfolded chain region.

Human growth hormone

Human growth hormone (hGH) is a single chain polypeptide hormone of 191 amino acid residues displaying several biological activities, including effects on growth, lactation, activation of macrophages and insulin-like and diabetogenic effects [78]. Nowadays hGH is extensively used to treat hGH deficiency, including treatment of short stature resulting from hGH inadequacy, as well as renal failure in children. The protein has a short functional half-life *in vivo* and must be administered daily by subcutaneous injection for maximum effectiveness. Improved versions of hGH were obtained by PEGylation of the hormone at the level of α- and ε-amino groups using the N-hydroxysuccinimidyl ester derivative of PEG. Eight to nine lysine residues and the N-terminal amino acid were modified to varying extents and the various hGH derivatives differed from one another for their *in vitro* bioactivity [79]. To achieve a selective protein modification, a Cys residue was introduced by site-directed mutagenesis into the hormone at position 3 in place of a threonine residue and then the mutant hGH was reacted with a Cys-reactive PEG derivative [80].

Recently, in our laboratory we have studied the TGase-mediated PEGylation of hGH using a PEG-NH_2 derivative [34, 42]. The PEG derivatives thus obtained were isolated by chromatography and analyzed for their content of PEG chains, as well as their location along the 191-residue chain of the hormone. Briefly, it was found that a major product was a hGH derivative PEGylated at Gln40 and Gln141, in analogy to similar results reported in a patent [81]. This selective or preferential PEGylation of hGH appears to be quite striking, if one considers that the 191-residue chain of the hormone contains 13 Gln residues.

An analysis of the structure and dynamics of the hGH substrate can help us to find out possible molecular features of the hGH substrate that enable the TGase-mediated reactions at Gln40 and Gln141. A four-helix bundle architecture characterizes the 3D architecture of hGH [82, 83]. Interestingly, the sites of attack by TGase and proteases lie outside the four helical segments of the four-helix bundle (residues 9–34, 72–92, 106–128 and 155–184), implying that rigid chain regions embedded in a hydrogen-bonded secondary structure do not react with TGase. The long chain loop 129–154 that connects helix C to helix D displays high *B*-factor values or no electron density, implying that this loop is highly flexible or even disordered (see the hGH crystal structure, PDB code 3HHR) [83]. Also the region approximately from residue 30 to residue 60 displays high *B*-factor values. Therefore, in analogy to the cases of IL-2 and G-CSF above presented and discussed, we can conclude that chain flexibility/disorder explains the site-specific TGase's reactions on the hGH substrate. In addition, we should here emphasize that several proteases, including plasmin, trypsin, thrombin and subtilisin, cleave hGH at region 134–149 [78]. Pepsin cleaves the peptide bond Phe44–Leu45, thus leading to the complementing fragments 1–44 and 45–191 [84]. Staphylococcal V8-protease cleaves hGH at Glu33, followed by slower reactions at Glu56 and Glu66 [85]. Therefore, several proteases selectively hydrolyse hGH at the same chain regions that are sites also of TGase's attack, implying that both TGase and proteases require chain flexibility/disorder for their site-specific reactions on the hGH substrate. Other details of the TGase reactions on hGH can be found in recent publications from our laboratory [34, 42].

Erythropoietin

Erythropoietin (EPO) is a 166-residue protein that acts as the major regulator of erythropoiesis in the body [86]. The secreted protein is heavily glycosylated and this post-translational modification is important for the stability, solubility and *in vivo* bioactivity of the protein, but not for its interaction with the receptor. Recombinant EPO in being used as an effective therapeutic agent in the treatment of various forms of anemia, including anemia associated with chronic renal failure and cancer patients on chemotherapy. Trade names of EPO presently in use include Epogen, Procrit and Aranesp. Several approaches have been used to prepare derivatives of EPO with a longer half-life in the blood serum and reduced immunogenicity, including hyperglycosylation and PEGylation [87].

The PEGylation of recombinat EPO using TGase was described in a patent [88] and evidence was provided that the TGase-mediated PEGylation allows the production of a protein with one to three PEG chains attached to the protein. In order to identify the protein-bound Gln residues that can be attacked by TGase, dansyl-cadaverine was used as an amino-donor (see Fig. 1) and then the labeled protein was analyzed by fingerprinting techniques combined with mass spectrometry. It was possible to conclude that dansyl-cadaverine was linked to two out of the seven Gln residues, one being Gln115. The second modified Gln was encompassed in chain segment 53–97, but since several Gln residues are contained in this chain region of the protein, the identity of the modified residue could not be ascertained [88].

Clearly additional experimental work is required for a detailed analysis of the TGase-mediated PEGylation of EPO. Nevertheless, an attempt can be made to explain why Gln115 is a site of the enzymatic reaction. The structure of the receptor-bound EPO has been solved by X-ray crystallography [89] and that of the free from in solution by NMR [90]. It is interesting to observe that, in the 166-residue chain of EPO, Gln115 is located at the long chain loop connecting helix C (helix 90–111) to helix D (helix 138–160). The high flexibility of this loop is evidenced by the *B*-factor profile along the polypeptide chain of the protein (see PDB file 1CN4) [89]. Moreover, Gln115 is located nearby the chain region 124–130, for which there is no electron density and thus is disordered [89]. Of interest, a noticeable difference between the NMR [90] and crystal structure [89] of EPO is found in the region from Leu112 to Thr132 of the loop connecting helix C to helix D. The NMR data show that the chain segment 112–132, encompassing Gln115, has an extended conformation and is highly flexible. Therefore, the limited data so far available for the PEGylation reaction of EPO are in line with the proposal that the TGase reaction requires substrate's flexibility at the site of enzymatic attack.

Apomyoglobin

Apomyoglobin (apoMb), myoglobin without the heme group, is a small monomeric protein of 153 amino acid residues that has been used as a model protein for numerous studies of protein structure, folding and stability. While the holo form of the protein is highly helical and consists of eight helices (named A through H) [91], apoMb shows a slightly reduced helical content due to unfolding of helix F, as demonstrated by NMR measurements [92, 93] and limited proteolysis experiments [94, 95]. Indeed, while the holo protein is resistant to proteolytic attack, several proteases with different specificities selectively cleave apoMb in a very restricted chain segment encompassing the helix F (residues 82–97) of the holo protein [94, 95]. For example, thermolysin cleaves apoMb at the level of the peptitde bond Pro88–Leu89, leading to the two complementary fragments 1–88 and 89–153 [96]. The mobility or unfolding of helix F in apoMb is deduced also from molecular dynamics simulations [97–100]. Recently, we have reacted apoMb with a monodisperse amino-derivative of PEG in the presence of microbial TGase at neutral pH [42]. A homogeneous mono-PEGylated apoMb derivative modified at the level of Gln91 was thus prepared, despite the protein contains six Gln residues along its 153-residue chain. Therefore, both TGase and proteases attack at the same and rather restricted flexible/unfolded chain region of the apoMb substrate.

Other examples of TGase reacting at flexible sites of the protein substrate

Here, we will add few examples of experimental studies that demonstrate that TGase can react with globular proteins at flexible/unfolded regions of their polypeptide chain. First of all, in a number of studies, relatively short Gln-peptides were added to globular proteins as an N- and C-terminal tails using recombinant methods [101–103]. It was found that TGase mediates enzymatic reactions selectively at the protein end(s) of these chimeric proteins. The added short peptide tags, lying outside the core of the globular protein, certainly are flexible or disordered, considering that the N- and C-terminal ends of globular proteins are usually rather flexible or even disordered. For example, the reader can observe that the ends of the polypeptide chain of IL-2 (Fig. 3) and G-CSF (Fig. 4) are indeed rather flexible or disordered, as evidenced by the high B-factor values for these chain ends.

Among the chimeric proteins prepared for their subsequent labeling using TGase, we may mention that a derivative of IL-2 fused to the TGase substrate Ala-Gln-Ile-Val-Met was prepared and used for a TGase-mediated specific PEGylation at the level of the Gln residue of the peptide tag [31]. Similarly, a chimera given by glutathione-S-transferase with the flexible peptide tag Pro-Lys-Pro-Gln-Gln-Phe-Met linked at its N- or C-terminus was labeled specifically at the level of the peptide tag using TGase and dansyl- or fluorescein-cadaverine as amino-donor [101]. Also the 20-residue Gln-containing S-peptide of bovine ribonuclease A was fused at the N-terminal region of green fluorescent protein (GFP) and shown to be the site of protein cross-linking reactions mediated by TGase [104]. In other studies, TGase has been used for the attachment of small-molecule probes to bioengineered proteins containing a Gln-peptide linked at the N- or C-terminus of the protein [101, 103]. Overall, it is clear that the appended tags at the ends of a protein chain in the chimeric recombinant proteins are flexible or unfolded, so that these sites can be easily attacked by TGase.

As an additional example, we may here mention that the 375-residue chain of actin can be selectively labeled at Gln41 by a TGase-mediated reaction using dansyl-ethylenediamine or dansyl-cadaverine used as amino-donor [105]. Moreover, actin can be selectively cleaved at peptide bond Met47–Gly48 and Gly42–Val43 by digestion with subtilisin and a protease from *E. coli*, respectively [106–109]. That chain disorder controls the selectivity of the reaction by both TGase and proteases is substantiated by the X-ray structure of actin [110]. Indeed, the B-factor profile along the polypeptide chain of actin shows a discontinuity at chain region 39–51 due to missing electron density, indicative of a disordered segment (see the crystallographic data reported in PDB-code 1QZ6).

It is relevant to add here that TGase can react on protein substrates given by partly or fully denatured proteins. The apo-form of bovine α-lactalbumin in its partly folded or molten globule state at neutral pH was shown to react with TGase, while the native calcium-loaded protein was not a substrate for TGase [111–114]. Of interest, it was shown that TGase can selectively incorporate dansyl-cadaverine at the level of Gln residues encompassed by the chain segment which was shown, by means of spectroscopic measurements, to be disordered in the molten globule state of α-lactalbumin [115, 116]. Moreover, the same chain region acting as substrate for TGase is also selectively hydrolyzed by several proteases [117, 118], thus demonstrating that both TGase and proteases act on the flexible/disordered protein regions of α-lactalbumin. Finally, mention should be given to the fact that a variety of natively or intrinsically disordered proteins [119–122] are substrates for TGase *in vivo*, including synuclein [123], tau protein [124] and huntingtin [125]. These proteins are involved in severe diseases as those of Alzheimer, Parkinson and Huntington and consequently a role of TGase in these diseases has been proposed [126–131].

Local unfolding of the protein substrate dictates the selective attack by TGase

Numerous studies have been conducted on the substrate specificity of TGases using synthetic peptides, as well as proteins, with the aim of unraveling the pattern of amino acid sequences around the reactive Gln residues [132–138]. However, the results of these studies revealed that there is no obvious consensus sequence around the Gln residues modified by TGase and that both the amino acid sequence and local conformation around the Gln residues contribute to the specificity of the enzyme [132, 136]. Consequently, it was concluded that other factors should govern the TGase reactivity of protein-bound Gln residues. It was proposed that surface accessibility of a reactive Gln could dictate the TGase-catalyzed modification [132]. However, it was also recognized that surface exposure alone does not justify the often made observation that several exposed Gln residues in native globular proteins of known 3D-structure do not react with TGase [34, 132].

Here, we have illustrated the results of recent experimental studies that demonstrate that TGase can mediate site-specific PEGylation of globular protein substrates, a feature that has important utility and useful applications for the development of safer and long lasting protein drugs. We have shown that a protein-bound Gln residue can be attacked by TGase only if it is encompassed in a flexible or disordered region of the protein chain. Indeed, there is a correlation between sites of specific TGase's attack with sites or regions displaying enhanced chain flexibility or even disorder, as evidenced by the analysis of the B-factor profile along the polypeptide chain of the protein substrate (see Figs 3 and 4, as well as [34]). We conclude that the main features dictating the site-specific modification of a protein-bound Gln residue by TGase in a globular protein is the local unfolding of the chain region encompassing that Gln residue. We emphasize here again that surface exposure of the site of enzymatic attack is not sufficient to explain site-specific TGase attack, since in a protein there are often quite numerous surface exposed Gln residues, but only the ones that are encompassed in a flexible or unfolded region can be attacked by TGase [34].

In several cases it has been found that the same region of the polypeptide chain of a globular protein that reacts with TGase is also the region prone to limited proteolysis phenomena (see [34] for additional data). We anticipate that the biorecognition phenomenon between TGase or a protease and a polypeptide substrate shares analogous mechanism of action, in keeping with the idea that TGase can be considered a reverse-protease and makes use of the same triad of active site residues of a Cys-protease (see Introduction). On this basis, we may propose similar substrate's binding phenomena occurring with

TGase and a protease. A widely used nomenclature to describe the interaction of a polypeptide substrate at the protease's active site is that introduced by Schechter and Berger [139]. For TGase, as shown in Figure 5, it is proposed that the amino acid residues of the polypeptide substrate bind at subsites of the TGase's active site. These subsites on the enzyme are called S (for subsites) and the substrate's amino acid residues are called P (for peptide). The amino acid residues of the N-terminal side of the site of TGase reaction are numbered P3, P2, P1 and those residues of the C-terminal side are numbered P1', P2', P3'...The P1 or P1' residues are those residues located near the site of TGase's attack. The subsites on TGase that complement the substrate binding residues are numbered S3, S2, S1, S1', S2', S3'... The enzyme-substrate binding scheme outlined in Figure 5 is in line with the view that a polypeptide substrate should suffer extensive unfolding and conformational adaptation for an efficient and productive binding at the enzyme's active site. Indeed, modeling studies have indicated that the protease-peptide/protein substrate interaction involves a stretch of up to 12 amino acid residues [140, 141].

Chain flexibility or local unfolding is proposed here for explaining the TGase's selectivity, as well as before for the site-specific limited proteolysis of globular proteins [142–145]. A mechanism of local unfolding of the site of enzymatic attack appears to be the critical parameter dictating a site-specific enzymatic reaction. The key role of flexibility in the digestion of a polypeptide substrate is substantiated by the recent systematic analysis of the recognition mechanism of proteases for polypeptide substrates and inhibitors [146, 147]. This analysis, made possible by the recent availability of many protease-inhibitor crystallographic structures, revealed that an almost universal recog-

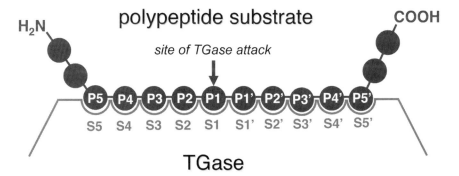

Figure 5. Schematic representation of the binding of a polypeptide substrate at the active site of TGase. A 10–12 residue segment of a polypeptide chain interacts with its side-chain residues (P) at a series of subsites (S) of TGase. The interaction of the substrate at the TGase's active requires a specific stereochemical adaptation of the substrate and thus likely a significant degree of chain mobility. It is suggested that the polypeptide substrate binds at the TGase's active site in an extended conformation (see text). The P1 side-chain residue interacting with the S1 binding site of TGase is the carboxamido side-chain group of a Gln residue. The figure and nomenclature are adapted from the representation of a similar binding of a polypeptide substrate at the protease's active site introduced by Schechter and Berger [139].

nition mechanism by all proteolytic enzymes implies that the binding of a polypeptide substrate at the protease's active site occurs in an extended β-strand conformation [147]. Consequently, we may understand why a folded and quite rigid globular protein is usually rather resistant to an enzymatic attack and we may anticipate that only a flexible or disordered polypeptide can have the structural plasticity to adopt the extended conformation that enables a stereospecific binding at the enzyme's active site [142–145].

It is tempting to propose that the molecular features herewith emphasized for a productive interaction between TGase or a protease with a polypeptide substrate can be extended also to other enzymatic reactions occurring with proteins, i.e., to other post-translational protein modifications, such as phosphorylation, glycosylation, acetylation, methylation, hydroxylation, amidation, sulfation and other reactions. In fact, the importance of protein disorder in a variety of biomolecular recognition processes of proteins has been recently emphasized [120–122]. For example, it is intriguing to observe that the TGase-mediated PEGylation of G-CSF occurs at Gln134, while *in vivo* glycosylation occurs at the nearby Thr133 (see above Fig. 4). We may add here that phosphorylation sites in proteins mostly occur in protein regions characterized by an intrinsic disorder. The reader is referred to Iakoucheva et al. [148] for a discussion on the importance of disorder for protein phosphorylation. Indeed, in a number of X-ray structures of kinases bound to peptide substrates or inhibitors [149], it has been found that the ligands in their protein-bound state have extended, irregular conformations, in analogy to the X-ray structures of proteases with their peptide inhibitors [147] (see also in PDB the kinases-inhibitor structures 1ATP, 1IR3, 1O6I, 1PHK, 1O6K, 1GY3, 1CDK and 1JBP). Therefore, the polypeptide substrate should be disordered and not intramolecularly hydrogen bonded in a regular secondary structure (helices, β-strands) in order to be able to bind at the enzyme's active site [142, 147]. It seems appropriate to consider here that the implications of these crystallographic data in terms of disordered chain segments as a requirement for the enzymatic reactions involving polypeptide substrates clearly have been overlooked. Summing up, we hypothesize that this mechanism of local unfolding of the site of enzymatic reaction applies to TGases and proteases, as well as likely to other enzymes that act on protein or polypeptide substrates, thus leading to a variety of enzymatic post-translational modifications of proteins.

TGase acting on flexible peptide substrates

Here, we have emphasized the importance of local unfolding in the TGase's reaction when globular proteins are used as substrates. In this case, the flexibility of the polypeptide chain is clearly the most critical parameter in dictating site-specific reactions [34]. However, when short flexible peptides are used in a TGase's reaction, chain flexibility is no more the critical parameter controlling the rate of reaction. In this case, the amino acid sequence near a Gln-residue is expected to play an important role in dictating the proper and favorable interaction between the enzyme and the peptide substrate, as it can be deduced from the scheme shown in Figure 5. Thus, the kinetics of the enzymatic reaction will be influenced by the amino acid residues neighboring the Gln-residue. Indeed, in a recent study the preferred peptide substrate sequences of microbial TGase have been identified by using a phage-display library and indeed some peptide sequences react faster with TGase than others [138]. Considering the features of the microenvironment of the active site of microbial TGase (see Introduction), it can be anticipated that charge and hydrophobicity of a polypeptide substrate will play a role in the enzyme's kinetics. In particular, since the active site is negatively charged by the presence of several carboxylate groups near the reactive Cys64 [46], likely negatively or positively charged residues near a Gln-residue will hinder or facilitate the enzymatic reaction on a polypeptide substrate, respectively. Indeed, the flexible peptidyl linkers containing a positively charged arginine (Arg) residue near the Gln-residue exhibited the highest reactivity in a TGase-mediated protein crosslinking reaction [102]. Similarly, in our laboratory we have found that a negatively charged glutamic acid (Glu) residue near a Gln-residue strongly hinders the TGase reaction in a Gln-peptide substrate. Finally, in agreement with the fact that the active site of microbial TGase is surrounded by hydrophobic residues [46], it was found that introducing into a Gln-peptide substrate a hydrophobic amino acid residue substantially accelerated the TGase's reaction [134]. These observations regarding the role of amino acid sequences of polypeptide substrates in the TGase's reaction can be related to analogous results obtained in studying the substrate specificities of proteolytic enzymes. For example, it is known that trypsin hydrolyses selectively Lys-X and Arg-X peptide bonds, but that tryptic cleavages at these basic residues are strongly inhibited by nearby negatively charged amino acid residues, so that for example a Lys–Glu peptide bond is not cleaved (see the website http://www.expasy.ch/tools/peptidecutter/).

Finally, we may recognize that herewith we have not discussed differences in the enzymatic activity between microbial or mammalian TGases. Even if microbial and mammalian TGases share the same catalytic triad (see Introduction) and often various TGases behave similarly with some peptide or protein substrates [114], likely differences in their enzymatic activities and kinetics are expected [137, 138].

Outlook

Considering the increasing relevance of protein pharmaceuticals and the high regulatory demands for their approval [21, 150], it can be anticipated that innovative methods for the site-specific PEGylation of proteins will be further investigated for the years to come [14, 21]. The broad applicability and comparably low cost of PEG will maintain this polymer in a leading position and the PEGylation reaction will find more and more useful applications for the successful development of pharmaceutical drugs. The site-specific TGase-mediated PEGylation likely will be further explored [50] and applied for preparing more homogeneous protein conjugates than those that can be obtained by using other reactive derivatives of PEG. Here, we have presented data that indicate that the main features dictating the site-specific modification of a protein-bound Gln residue by TGase in a globular protein is the flexibility or local unfolding of the chain region encompassing the reactive Gln residue [34]. In particular, TGase appears to act on a polypeptide substrate in analogy to a protease [142–145] and, in fact, often the same region of the polypeptide chain of a globular protein that suffers limited proteolysis is also the site of specific TGase attack [34]. Therefore, it is possible to predict the site(s) of TGase-mediated modification of a protein on the basis of its 3D-structure and dynamics and, consequently, the likely effects on its physicochemical and functional properties.

Acknowledgments
This research was supported in part by the Italian Ministry of University and Research (PRIN-2006 No. 03035 and FIRB-2003 No. RBNEOPX83). The authors are grateful to BioKer Srl (Pula, Italy) for a generous supply of recombinant protein samples of G-CSF and hGH. The expert typing of the manuscript by Ms. Barbara Sicoli is also gratefully acknowledged.

References

1. Leader B, Baca QJ, Golan DE (2008) Protein therapeutics: A summary and pharmacological classification. *Nature Rev Drug Discov* 7: 21–39
2. Frokjaer S, Otzen DE (2005) Protein drug stability: A formulation challenge. *Nature Rev Drug Discov* 4: 298–306
3. Pavlou A, Reichert J (2004) Recombinant protein therapeutics: Success rates, market trends and values to 2010. *Nature Biotechnol* 22: 1513–1519
4. Harris JM, Chess R (2003) Effect of PEGylation on pharmaceuticals. *Nature Rev Drug Discov* 2: 214–221
5. Davis F (2002) The origin of PEGnology. *Adv Drug Deliv Rev* 54: 457–458
6. Abuchowski A, McCoy JR, Palczuk NC, van Es T, Davis FF (1977) Effect of covalent attachment of poly(ethylene glycol) on immunogenicity and circulating life of bovine liver catalase. *J Biol Chem* 252: 3582–3586
7. Abuchowski A, van Es T, Palczuk NC, Davis FF (1977) Alteration of immunological properties of bovine serum albumin by covalent attachment of polyethylene glycol. *J Biol Chem* 252: 3578–3581
8. Harris JM (ed.) (1991) *Poly(ethylene glycol) chemistry: Biotechnological and biomedical applications.* Plenum Press, New York
9. Pasut G, Guiotto A, Veronese FM (2004) Protein, peptide and non-peptide drug PEGylation for therapeutic applications. *Expert Opin Ther Pat* 14: 1–36
10. Malik DK, Baboota S, Ahuja A, Hasan S, Ali J (2007) Recent advances in protein and peptide drug delivery systems. *Curr Drug Deliv* 4: 141–151
11. Veronese FM, Pasut G (2005) PEGylation, successful approach to drug delivery. *Drug Discovery Today* 10: 1451–1458
12. Harris JM, Veronese FM (eds): (2002) Peptide and protein PEGylation. *Adv Drug Deliv Rev* 54: 453–610
13. Harris JM, Veronese FM (eds): (2003) Peptide and protein PEGylation II: Clinical evaluation. *Adv Drug Deliv Rev* 55: 1259–1350

14. Harris JM, Veronese FM (eds): (2008) Peptide and protein PEGylation III: Advances in chemistry and clinical applications. *Adv Drug Deliv Rev* 60: 1–88
15. Duncan R (2003) The dawning era of polymer therapeutics. *Nature Rev Drug Discov* 2: 347–360
16. Thordarson P, Le Droumaguet B, Velonia K (2006) Well-defined protein-polymer conjugates: Synthesis and potential applications. *Appl Microbiol Biotechnol* 73: 243–254
17. Zalipsky S (1995) Chemistry of poly(ethylene glycol) conjugates with biologically active molecules. *Adv Drug Deliv Rev* 16: 157–182
18. Roberts MJ, Bentley MD, Harris JM (2002) Chemistry for peptide and protein PEGylation. *Adv Drug Deliv Rev* 54: 459–476
19. Pasut G, Veronese FM (2006) PEGylation of proteins as tailored chemistry for optimized bioconjugates. *Adv Polym Sci* 192: 95–134
20. Veronese FM (2001) Peptide and protein PEGylation: A review of problems and solutions. *Biomaterials* 22: 405–417
21. Reichert JM (2003) Trends in development and approval times for new therapeutics in the United States. *Nature Rev Drug Discov* 2: 695–702
22. Gentle I, DeSouza I, Baca M (2004) Direct production of proteins with N-terminal cysteine for site-specific conjugation. *Bioconjug Chem* 15: 658–663
23. Goodson RJ, Katre NV (1990) Site-directed PEGylation of recombinant interleukin-2 at its glycosylation site. *Biotechnology* 8: 343–346
24. Doherty DH, Rosendahl MS, Smith DJ, Hughes JM, Chilpala EA, Cox GN (2005) Site-specific PEGylation of engineered cysteine analogs of recombinant human granulocyte-macrophage colony-stimulating factor. *Bioconjug Chem* 16: 1291–1298
25. Wetzel R, Halualani R, Stults JT, Quan C (1990) A general method for highly selective crosslinking of unprotected polypeptides *via* pH-controlled modification of N-terminal α-amino groups. *Bioconjug Chem* 1: 114–122
26. Wang YS, Youngster S, Grace M, Bausch J, Bordens R, Wyss DF (2002) Structural and biological characterization of PEGylated recombinant interferon 2b and its therapeutic implications. *Adv Drug Deliv Rev* 54: 547–570
27. Lee H, Jang H, Ryu S, Park T (2003) N-Terminal site-specific mono-PEGylation of epidermal growth factor. *Pharm Res* 20: 818–825
28. Kinstler O, Molineux G, Treuheit M, Ladd D, Gegg C (2002) Mono-N-terminal poly(ethylene glycol)-protein conjugates. *Adv Drug Deliv Rev* 54: 477–485
29. Gaertner HF, Offord RE (1996) Site-specific attachment of functionalized poly(ethylene glycol) to the amino terminus of proteins. *Bioconjug Chem* 7: 38–44
30. Sato H, Ikeda M, Suzuki K, Hirayama K (1996) Site-specific modification of interleukin-2 by the combined use of genetic engineering techniques and transglutaminase. *Biochemistry* 35: 13072–13080
31. Sato H, Yamamoto Y, Hayashi E, Takahara Y (2000) Transglutaminase-mediated dual and site-specific incorporation of poly(ethylene glycol) derivatives into a chimeric interleukin-2. *Bioconjug Chem* 11: 502–509
32. Sato H, Hayashi E, Yamada N, Yatagai M, Takahara Y (2001) Further studies on the site-specific protein modification by microbial transglutaminase. *Bioconjug Chem* 12: 701–710
33. Sato H (2002) Enzymatic procedure for site-specific PEGylation of proteins. *Adv Drug Deliv Rev* 54: 487–504
34. Fontana A, Spolaore B, Mero A, Veronese FM (2008) Site-specific modification and PEGylation of pharmaceutical proteins mediated by transglutaminase. *Adv Drug Deliv Rev* 60: 13–28
35. Folk JE (1980) Transglutaminases. *Annu Rev Biochem* 49: 517–531
36. Lorand L, Conrad SM (1984) Transglutaminases. *Mol Cell Biochem* 58: 9–35
37. Folk JE (1983) Mechanism and basis for specificity of transglutaminase-catalyzed ε-(γ-glutamyl)lysine bond formation. *Adv Enzymol Relat Areas Mol Biol* 54: 1–56
38. Gorman JJ, Folk JE (1980) Structural features of glutamine substrates for human plasma factor XIIIa (activated blood coagulation factor XIII). *J Biol Chem* 255: 419–427
39. Gorman JJ, Folk JE (1984) Structural features of glutamine substrates for transglutaminases: Role of extended interactions in the specificity of human plasma factor XIIIa and of the guinea pig liver enzyme. *J Biol Chem* 259: 9007–9010
40. Griffin R, Casadio R, Bergamini CM (2002) Transglutaminases: Nature's biological glues. *Biochem J* 368: 377–396
41. Folk JE, Finlayson JS (1977) The ε-(γ-glutamyl)lysine crosslink and the catalytic role of trans-

glutaminases. *Adv Protein Chem* 31: 1–133

42. Mero A, Spolaore B, Veronese FM, Fontana A (2009) Transglutaminase-mediated PEGylation of proteins: Direct identification of the sites of protein modification by mass spectrometry using a novel monodisperse PEG. *Bioconjug Chem* 20: 384–389

43. Ando H, Adachi M, Umeda K, Matsuura A, Nonaka M, Uchio R, Tanaka H, Motoki M (1989) Purification and characterization of a novel transglutaminase derived from microorganisms. *Agric Biol Chem* 53: 2613–2617

44. Washizu K, Ando K, Koiked S, Hiros S, Matsuura A, Akagi H, Motoki M, Takeuchi K (1994) Molecular cloning of the gene for microbial transglutaminase from *Streptoverticillium* and its expression in *Streptomyces lividans*. *Biosci Biotechnol Biochem* 58: 82–87

45. Kanaji T, Ozaki H, Takao T, Kawajiri H, Ide H, Motoki M, Shimonishi Y (1993) Primary structure of microbial transglutaminase from *Streptoverticillium* sp. strain s-8112. *J Biol Chem* 268: 11565–11572

46. Kashiwagi T, Yokoyama K, Ishikawa K, Ono K, Ejima D, Matui H, Suzuki E (2002) Crystal structure of microbial transglutaminase from *Streptoverticillium mobaraense*. *J Biol Chem* 277: 44252–44260

47. Berman HM, Westbrook J, Feng Z, Gilliland G, Bhat TN, Weissig H, Shindyalov IN, Bourne PE (2000) The protein data bank. *Nucleic Acids Res* 28: 235–242

48. Yee VC, Pedersen LC, Le Trong I, Bishop PD, Steukamp RE, Teller DC (1994) Three-dimensional structure of a transglutaminase: Human blood coagulation factor XIII. *Proc Natl Acad Sci USA* 91: 7296–7300

49. Menéndez O, Rawel H, Schwarzenbolz U, Henle T (2006) Structural changes of microbial transglutaminase during thermal and high-pressure treatment. *J Agric Food Chem* 54: 1716–1721

50. Zhu Y, Tramper J (2008) Novel applications for microbial transglutaminase beyond food processing. *Trends Biotechnol* 26: 559–565

51. Zhu Y, Rinzema A, Tramper J, Bol J (1995) Microbial transglutaminases: A review of its production and application in food processing. *Appl Microbiol Biotechnol* 44: 277–282

52. Yokohama K, Nio N, Kikuchi Y (2004) Properties and applications of microbial transglutaminases. *Appl Microbiol Biotechnol* 64: 447–454

53. Mariniello L, Porta R (2005) Transglutaminases as biotechnological tools. In: K Mehta, R Eckert (eds): *Transglutaminase*. *Prog Exp Tum Res*, Basel, Karger, 38: 174–191

54. Waldmann TA (2006) The biology of interleukin-2 and interleukin-15: Implications for cancer therapy and vaccine design. *Nature Rev Immunol* 6: 595–601

55. Malek TR (2008) The biology of interleukin-2. *Annu Rev Immunol* 26: 453–479

56. Brandhuber BJ, Boone T, Kenney WC, McKay DB (1987) Three-dimensional structure of interleukin-2. *Science* 238: 1707–1709

57. Cohen FE, Kosen PA, Kuntz ID, Epstein LB, Ciardelli TL, Smith KA (1986) Structure-activity studies of interleukin-2. *Science* 234: 349–352

58. Mott HR, Baines BS, Hall RM, Cooke RM, Driscoll PC, Weir MP, Campbell ID (1995) The solution structure of the F42A mutant of human interleukin-2. *J Mol Biol* 247: 979–994

59. Arkin MA, Randal M, DeLano WL, Hyde J, Luong TN, Oslob JD, Raphael DR, Taylor L, Wang J, McDowell RS et al. (2003) Binding of small molecules to an adaptive protein-protein interface. *Proc Natl Acad Sci USA* 100: 1603–1608

60. Frauenfelder H, Petsko GA, Tsernoglou D (1979) Temperature-dependent X-ray diffraction as a probe of protein structural dynamics. *Nature* 280: 558–563

61. Sternberg MJE, Grace DEP, Phillips DC (1979) Dynamic information from protein crystallography: An analysis of temperature factors from refinement of the hen egg-white lysozyme. *J Mol Biol* 130: 231–253

62. Ringe D, Petsko GA (1985) Mapping protein dynamics by X-ray diffraction. *Prog Biophys Mol Biol* 45: 197–235

63. Ringe D, Petsko GA (1986) Study of protein dynamics by X-ray diffraction. *Methods Enzymol* 131: 389–433

64. Kundu S, Melton JS, Sorensen DC, Phillips Jr GN (2002) Dynamics of proteins in crystals: Comparison of experiment with simple models. *Biophys J* 83: 723–732

65. Smith DK, Radivojac P, Obradovic Z, Dunker AK, Zhu G (2003) Improved amino acid flexibility parameters. *Protein Sci* 12: 1060–1072

66. Radivojac P, Obradovic Z, Smith DK, Zhu G, Vucetic S, Brown CJ, Lawson JD, Dunker AK (2004) Protein flexibility and intrinsic disorder. *Protein Sci* 13: 71–80

67. Akbarzadeh S, Layton JE (2001) Granulocyte colony-stimulating factor receptor: Structure and function. *Vitam Horm* 63: 159–194
68. Zink T, Ross A, Luers K, Cieslar C, Rudolph R, Holak TA (1994) Structure and dynamics of the human granulocyte colony-stimulating factor determined by NMR spectroscopy: Loop mobility in a four-helix-bundle protein. *Biochemistry* 33: 8453–8463
69. Hill CP, Osslund TD, Eisenberg D (1993) The structure of granulocyte-colony-stimulating factor and its relationship to other growth factors. *Proc Natl Acad Sci USA* 90: 5167–5171
70. Morstyn G, Dexter TM (1994) *Neopogen (r-metHuG-CSF) in clinical practice*. M. Dekker, New York
71. Welte K, Gabrilove J, Bronchud MH, Platzer E, Morstyn G (1996) Filgrastim (r-metHuG-CSF): The first 10 years. *Blood* 88: 1907–1929
72. Lubenau H, Bias P, Maly AK, Siegler KE, Mehltretter K (2009) Pharmacokinetic and pharmacodynamic profile of new biosimilar filgrastim XM02 equivalent to marketed filgrastim Neupogen: Single-blind, randomized, crossover trial. *BioDrugs* 23: 43–51
73. Herman AC, Boone TC, Lu HS (1996) Characterization, formulation, and stability of Neupogen (Filgrastim), a recombinant human granulocyte-colony stimulating factor. *Pharm Biotechnol* 9: 303–328
74. Molineux G (2004) The design and development of pegfilgrastim (PEG-rmetHuG-CSF, Neulasta). *Curr Pharm Des* 10: 1235–1244
75. Piedonte DM, Treuheit MJ (2008) Formulation of Neulasta (pegfilgrastim). *Adv Drug Deliv Rev* 60: 50–58
76. Veronese FM, Mero A, Caboi F, Sergi M, Marongiu C, Pasut G (2007) Site-specific PEGylation of G-CSF by reversible denaturation. *Bioconjug Chem* 18: 1824–1830
77. Tonon G, Orsini G (2008) *G-CSF site-specific mono-conjugates*. Patent WO/2008/7017603, Int. Application No. PCT/EP2007/057824
78. Li CH (1982) Human growth hormone: 1974–1981. *Mol Cell Biochem* 46: 31–41
79. Clark R, Olson K, Fuh G, Marian M, Mortensen D, Teshima G, Chang S, Chu H, Mukku V, Canova-Davis E et al. (1996) Long-acting growth hormones produced by conjugation with poly(ethylene glycol). *J Biol Chem* 271: 21969–21977
80. Cox GN, Rosendahl MS, Chlipala EA, Smith DJ, Carlson SJ, Doherty DH (2007) A long-acting mono-PEGylated human growth hormone analog is a potent stimulator of weight gain and bone growth in hypophysectomized rats. *Endocrinology* 148: 1590–1597
81. Dorwald F, Johansen N, Iversen L (2006) *Transglutaminase-mediated conjugation of growth hormone*. Patent WO/2006/134148, Int. Application No. PCT/EP2006/063246
82. de Vos AM, Ultsch MH, Kossiakoff AA (1992) Human growth hormone and extracellular domain of its receptor: Crystal structure of the complex. *Science* 255: 306–312
83. Ultsch MH, Somers W, Kossiakoff AA, de Vos AM (1994) The crystal structure of affinity-matured human growth hormone at 2 Å resolution. *J Mol Biol* 236: 286–299
84. Spolaore B, Polverino de Laureto P, Zambonin M, Fontana A (2004) Limited proteolysis of human growth hormone at low pH: Isolation, characterization and complementation of the two biologically relevant fragments 1–44 and 45–191. *Biochemistry* 43: 6576–6586
85. Polverino de Laureto P, Toma S, Tonon G, Fontana A (1995) Probing the structure of human growth hormone by limited proteolysis. *Int J Pept Prot Res* 45: 200–208
86. Jelkmann W (2007) Erythropoietin after a century of research: Younger than ever. *Eur J Haematol* 78: 183–205
87. Egrie JC, Dwyer E, Browne JK, Hitz A, Lykos MA (2003) Darbepoetin alfa has a longer circulating half-life and greater *in vivo* potency than recombinant human erythropoietin. *Exp Hematol* 31: 290–299
88. Pool CT (2004) *Formation of novel erythropoietin conjugates using transglutaminase*. Patent WO/2004/148667, Int. Application No. PCT/US2004/016670
89. Syed RS, Reid SW, Li C, Cheetham JC, Aoki KH, Liu B, Zhan H, Osslund TD, Chirino AJ, Zhang J et al. (1998) Efficiency of signalling through cytokine receptors depends critically on receptor orientation. *Nature* 395: 511–516
90. Cheetham JC, Smith DM, Aoki KH, Stevenson JL, Hoeffel TJ, Syed RS, Egrie J, Harvey TS (1998) NMR structure of human erythropoietin and a comparison with its receptor bound conformation. *Nature Struct Biol* 5: 861–866
91. Evans SV, Brayer GD (1990) High-resolution study of the three-dimensional structure of horse heart metmyoglobin. *J Mol Biol* 213: 885–897

92. Eliezer D, Wright PE (1996) Is apomyoglobin a molten globule? Structural characterization by NMR. *J Mol Biol* 263: 531–538

93. Eliezer D, Yao J, Dyson HJ, Wright PE (1998) Structural and dynamic characterization of partially folded states of apomyoglobin and implications for protein folding. *Nature Struct Biol* 5: 148–155

94. Fontana A, Zambonin M, Polverino de Laureto P, De Filippis V, Clementi A, Scaramella E (1997) Probing the conformational state of apomyoglobin by limited proteolysis. *J Mol Biol* 266: 223–230

95. Picotti P, Marabotti A, Negro A, Musi V, Spolaore B, Zambonin M, Fontana A (2004) Modulation of the structural integrity of helix F in apomyoglobin by single amino acid replacements. *Protein Sci* 13: 1572–1585

96. Musi V, Spolaore B, Picotti P, Zambonin M, De Filippis V, Fontana A (2004) Nicked apomyoglobin: A noncovalent complex of two polypeptide fragments comprising the entire protein chain. *Biochemistry* 43: 6230–6240

97. Brooks CL (1992) Characterization of "native" apomyoglobin by molecular dynamics simulation. *J Mol Biol* 227: 375–380

98. Tirado-Rives J, Jorgensen WL (1993) Molecular dynamics simulations of the unfolding of apomyoglobin in water. *Biochemistry* 32: 4175–4184

99. Hirst JD, Brooks CL (1995) Molecular dynamics simulations of isolated helices of myoglobin. *Biochemistry* 34: 7614–7621

100. Onufriev A, Case DA, Bashford D (2003) Structural details, pathways and energetics of unfolding apomyoglobin. *J Mol Biol* 325: 555–567

101. Taki M, Shiota M, Taira K (2004) Transglutaminase-mediated N- and C-terminal fluorescein labelling of a protein can support the activity of the modified protein. *Protein Eng Des Select* 17: 119–126

102. Tanaka T, Kamiya N, Nagamune T (2004) Peptidyl linkers for protein heterodimerization catalyzed by microbial transglutaminase. *Bioconjug Chem* 15: 491–497

103. Meusel M (2004) Synthesis of hapten-protein conjugates using microbial transglutaminase. *Methods Mol Biol* 283: 109–123

104. Kamiya N, Tanaka T, Suzyuki T, Takazawa T, Takeda S, Watanabe K, Nagamune T (2003) S-Peptide as a potent peptidyl linker for protein crosslinking by microbial transglutaminase from *Streptomyces mobaraensis*. *Bioconjug Chem* 14: 351–357

105. Kim E, Motoki M, Seguro K, Muhlrad A, Reisler E (1995) Conformational changes in subdomain 2 of G-actin: Fluorescence probing by dansyl-ethylenediamine attached to Gln-41. *Biophys J* 69: 2024–2032

106. Mornet D, Ue K (1984) Proteolysis and structure of skeletal muscle actin (limited proteolysis/organization of G-actin). *Proc Natl Acad Sci USA* 81: 3680–3684

107. Moraczewska J, Wawro B, Seguro K, Strzelecka-Golaszewska H (1999) Divalent cation-, nucleotide- and polymerization-dependent changes in the conformation of subdomain 2 of actin. *Biophys J* 77: 373–385

108. Khaitlina SY, Moraczewska J, Strzelecka-Golaszewska H (1993) The actin/actin interactions involving the N-terminus of the DNase-I-binding loop are crucial for stabilization of the actin filament. *Eur J Biochem* 218: 911–920

109. Borovikov YS, Moraczewska J, Khoroshev MI, Strzelecka-Golaszewska H (2000) Proteolytic cleavage of actin within the DNase-I-binding loop changes the conformation of F-actin and its sensitivity to myosin binding. *Biochim Biophys Acta* 1478: 138–151

110. Klenchin VA, Allingham JS, King R, Tanaka J, Marriott G, Rayment I (2003) Trisoxazole macrolide toxins mimic the binding of actin-capping proteins to actin. *Nature Struct Biol* 10: 1058–1063

111. Matsumura Y, Yuporn C, Kumazawa Y, Ohtsuka T, Mori T (1996) Enhanced susceptibility to transglutaminase reaction of α-lactalbumin in molten globule state. *Biochim Biophys Acta* 1292: 69–76

112. Gu YS, Matsumura Y, Yamaguchi S, Mori T (2001) Action of protein-glutaminase on α-lactalbumin in the native and molten globule states. *J Agric Food Chem* 49: 5999–6005

113. Nieuwenhuisen WF, Dekker HL, De Koning LJ, Groneveld T, De Koster CG, De Jong GA (2003) Modification of glutamine and lysine residues in holo and apo α-lactalbumin with microbial transglutaminase. *J Agric Food Chem* 51: 7132–7139

114. Lee DS, Matsumoto S, Matsumura Y, Mori T (2002) Identification of the ε-(γ-glutamyl)lysine

crosslinking sites in α-lactalbumin polymerized by mammalian and microbial transglutaminases. *J Agric Food Chem* 50: 7412–7419

115. Kuwajima K (1996) The molten globule state of α-lactalbumin. *FASEB J* 10: 102–109
116. Schulman BA, Kim PS, Dobson CM, Redfield C (1997) A residue-specific NMR view of the non-cooperative unfolding of a molten globule. *Nature Struct Biol* 4: 630–634
117. Polverino de Laureto P, De Filippis V, Di Bello M, Zambonin M, Fontana A (1995) Probing the molten globule state of α-lactalbumin by limited proteolysis. *Biochemistry* 34: 12596–12604
118. Polverino de Laureto P, Frare E, Gottardo R, Fontana A (2002) Molten globule of bovine α-lactalbumin at neutral pH induced by heat, trifluoroethanol and oleic acid: A comparative analysis by circular dichroism spectroscopy and limited proteolysis. *Proteins: Struct Funct Genet* 49: 385–397
119. Dyson HJ, Wright PE (2002) Coupling of folding and binding for unstructured proteins. *Curr Opin Struct Biol* 12: 54–60
120. Dunker AK, Brown CJ, Lawson LD, Iakoucheva LM, Obradovic Z (2002) Intrinsic disorder and protein function. *Biochemistry* 41: 6573–6582
121. Uversky VN (2002) Natively unfolded proteins: A point where biology waits for physics. *Protein Sci* 11: 739–756
122. Tompa P (2002) Intrinsically unstructured proteins. *Trends Biochem Sci* 27: 527–533
123. Junn E, Ronchetti RD, Quezabo MM, Kim SY, Mouradian MM (2003) Tissue transglutaminase-induced aggregation of α-synuclein: Implications for Lewy body formation in Parkinson's disease and dementia with Lewy bodies. *Proc Natl Acad Sci USA* 100: 2047–2052
124. Prasana Murthy SN, Wilson JH, Lukas TJ, Kuret J, Lorand L (1998) Crosslinking sites of the human tau protein probed by reactions with human transglutaminase. *J Neurochem* 71: 2607–2614
125. Karpuj MV, Garren H, Slunt H, Price DL, Gusella J, Becker MW, Steinman L (1999) Transglutaminase aggregates huntingtin into non-amyloidogenic polymers and its enzymatic activity increases in Huntington's disease brain nuclei. *Proc Natl Acad Sci USA* 96: 7388–7393
126. Karpuj M, Steinman L (2004) The multifaceted role of transglutaminase in neurodegeneration. *Amino Acids* 26: 373–379
127. Karpuj MV, Becker MW, Steinman L (2002) Evidence for a role for transglutaminase in Huntington's disease and the potential therapeutic implications. *Neurochem Int* 40: 31–36
128. Lesort M, Chun W, Johson GVW, Ferrante RJ (1999) Tissue transglutaminase is increased in Huntington's disease brain. *J Neurochem* 73: 2018–2027
129. Selkoe DJ, Abraham C, Ihara Y (1982) Brain transglutaminase: *In vitro* crosslinking of human neurofilament proteins into insoluble polymers. *Proc Natl Acad Sci USA* 79: 6070–6074
130. Johnson GV, Cox TM, Lockar JP, Zimmerman MD, Miller ML, Powers RE (1997) Transglutaminase activity is increased in Alzheimer's disease in brain. *Brain Res* 75: 323–329
131. Konno T, Morii T, Hirata A, Sato S, Oiki S, Ikura K (2005) Covalent blocking of fibril formation and aggregation of intracellular amyloidogenic proteins by transglutaminase-catalyzed intramolecular crosslinking. *Biochemistry* 44: 2072–2079
132. Coussons PJ, Price NC, Kelly SM, Smith B, Sawyer L (1992) Factors that govern the specificity of transglutaminase-catalysed modification of proteins and peptides. *Biochem J* 282: 929–930
133. Case A, Smith RL (2003) Kinetic analysis of the action of tissue transglutaminase on peptide and protein substrates. *Biochemistry* 42: 9466–9481
134. Ohtsuka T, Ota M, Nio N, Motoki M (2000) Comparison of substrate specificities of transglutaminases using synthetic peptides as acyl donors. *Biosci Biotechnol Biochem* 64: 2608–2613
135. Sugimura Y, Hosono M, Wada F, Yoshimura T, Maki M, Hitomi K (2006) Screening for the preferred substrate sequence of transglutaminase using a phage-displayed peptide library: Identification of peptide substrates for TGase 2 and Factor XIIIA. *J Biol Chem* 281: 17699–17706
136. Facchiano F, Facchiano A (2005) Transglutaminases and their substrates. *Prog Exp Tumor Res* 38: 37–57
137. Facchiano A, Facchiano F (2009) Transglutaminases and their substrates in biology and human diseases: 50 years of growing. *Amino Acids* 36: 599–614
138. Sugimura Y, Yokoyama K, Nio N, Maki M, Hitomi K (2008) Identification of preferred substrate sequences of microbial transglutaminase from *Streptomyces mobaraensis* using a phage-displayed peptide library. *Arch Biochem Biophys* 477: 379–383
139. Schechter I, Berger A (1967) On the size of the active site in proteases. I. Papain. *Biochem*

Biophys Res Commun 27: 157–162
140. Hubbard SJ, Eisenmenger F, Thornton JM (1994) Modelling studies of the change in conforma-
     tion required for cleavage of limited proteolytic sites. *Protein Sci* 3: 757–768
141. Hubbard SJ (1998) The structural aspects of limited proteolysis of native proteins. *Biochim
     Biophys Acta* 1382: 191–206
142. Fontana A, Fassina G, Vita C, Dalzoppo D, Zamai M, Zambonin M (1986) Correlation between
     sites of limited proteolysis and segmental mobility in thermolysin. *Biochemistry* 25: 1847–1851
143. Fontana A, Polverino de Laureto P, De Filippis V, Scaramella E, Zambonin M (1999) Limited
     proteolysis in the study of protein conformation. In: EE Sterchi, W Stöcker (eds): *Proteolytic
     Enzymes: Tools and Targets*. Springer Verlag, Heidelberg, 257–284
144. Fontana A, Polverino de Laureto P, De Filippis V, Scaramella E, Zambonin M (1997) Probing the
     partly folded states of proteins by limited proteolysis. *Folding Des* 2: R17–R26
145. Fontana A, Polverino de Laureto P, Spolaore B, Frare E, Picotti P, Zambonin M (2004) Probing
     protein structure by limited proteolysis. *Acta Biochim Pol* 51: 299–321
146. Tyndall JDA, Fairlie DP (1999) Conformational homogeneity in molecular recognition by pro-
     teolytic enzymes. *J Mol Recognit* 12: 363–370
147. Tyndall JDA, Nall T, Fairlie DP (2005) Proteases universally recognize beta strands in their
     active site. *Chem Rev* 105: 973–999
148. Iakoucheva LM, Radivojac P, Brown CJ, O'Connor TR, Sikes JG, Obradovic Z, Dunker AK
     (2004) The importance of intrinsic disorder for protein phosphorylation. *Nucleic Acid Res* 32:
     1037–1049
149. Zheng J, Trafny EA, Knighton DR, Xuong NH, Taylor SS, Ten Eyck LF, Sowadski JM (1993) A
     refined crystal structure of the catalytic subunit of cAMP-dependent protein kinase complexed
     with MnATP and a peptide inhibitor. *Acta Crystallogr* 49: 362–365
150. Reichert JM (2006) Trends in US approvals: New biopharmaceuticals and vaccines. *Trends
     Biotechnol* 24: 293–298

Protein conjugates purification and characterization

Conan J. Fee

Department of Chemical & Process Engineering, University of Canterbury, Private Bag 4800, Christchurch 8040, New Zealand

Abstract

Methods for separation and characterization of PEGylated proteins are reviewed in this chapter. It is explained that these methods are challenging because PEG itself is a relatively inert, neutral, hydrophilic polymer and the starting point for PEGylation is a pure protein. Other than changes to molecular weight and size, differences between the properties of the PEGylated forms of a pure protein are relatively small, since they arise only from the addition to the protein of relatively inert, neutral polymer chains, which tend to shield interactions.
Physicochemical properties that are routinely used to characterize and purify proteins are discussed with regard to their applications for PEGylated proteins, including molecular mass, size and shape (mass spectrometry, size exclusion chromatography, membranes, capillary electrophoresis, gel electrophoresis), electrostatic charge (cation and anion exchange chromatography, isoelectric point gel electrophoresis, capillary electrophoresis) and relative hydrophobicity (hydrophobic interaction, reversed phase).

Introduction

As described in other chapters in this book, PEG-protein conjugates, or PEGylated proteins, are an important class of modern therapeutic drugs. It follows from the stringent regulatory requirements surrounding the need for proven clinical efficacy and safety that they must be characterized and purified before use. Other than changes to molecular weight and size, differences between the properties of the PEGylated forms of a pure protein are relatively small, since they arise only from the addition of relatively inert, neutral polymer chains to the protein. Although some changes to the surface properties of the protein (i.e., charge and hydrophobicity) are inevitable upon conjugation, these changes are generally small and the large, heavily hydrated PEG chains tend to shield and weaken the strength of the protein's surface-related interactions. This imparts exploitable differences between the native protein and its PEGylated forms but successive PEG chain attachments result in rapidly diminishing changes to physicochemical properties and there are significant challenges in separating the PEGylated forms from one another.

Methods for characterization and purification are linked, in that the unique properties of the PEGylated target molecule must be exploited in each case to achieve a distinction between the target and other molecules in analytical or preparative techniques. Often the same properties can be used both to characterize the molecule and to purify it from contaminants. Of particular interest to PEGylated proteins for therapeutic use, given that the starting point for their PEGylation will normally be the highly purified native protein (where "native" pertains to its un-PEGylated form, as opposed to its wild *versus* recombinant or engineered forms) are the removal of PEGylation reaction byproducts and unconverted reactants, PEGylation extent (the number of PEG adducts on each molecule, N), and positional isomerism (the positions of the PEG adducts on each molecule). Potential changes to bioactivity and protein folding are also of interest. Advantages in terms of reduced plasma clearance rates and increased clinical efficacies have been reported for branched-PEG adducts over linear PEG adducts [1] so methods by which these forms can be characterized may be of interest during research and development activities.

Physicochemical properties that are routinely used to characterize and purify proteins can also be applied with varying effectiveness to PEGylated proteins, including molecular mass, size and shape (mass spectrometry, size exclusion (gel permeation) chromatography, membranes, capillary electrophoresis, gel electrophoresis, light scattering), protein folding (circular dichroism), electrostatic charge (cation and anion exchange chromatography, isoelectric focussing, isoelectric point gel electrophoresis, capillary electrophoresis), relative hydrophobicity (hydrophobic interaction and reversed phase chromatography), diffusivity, immunoassays (ELISA, Western blotting, surface plasmon resonance (SPR)) and relative solubility (aqueous two-phase systems). Additional characterization information can be obtained from bioassays and pharmacokinetic studies, such as plasma clearance rates, toxicity, immunogenicity and clinical efficacy but these will not be discussed in this chapter.

Molecular mass, size and shape

Commercial PEGylation reagents are identified by nominal molecular weight but this usually underestimates the true molecular weight. For example, PEG reagents with nominal molecular weights of 5 kDa (NOF Corporation, Japan), 10 kDa (Nektar Therapeutics, Alabama, USA) and 20 kDa (NOF Corporation, Japan) were found by mass spectroscopy to have actual molecular masses of 5589 ± 56 Da, $11,555 \pm 116$ Da and $21,910 \pm 219$ Da, respectively [1]. In the case of branched PEGs, which are manufactured by combining two identical PEG molecules, the discrepancy between nominal and actual molecular masses can be larger than that for linear PEG chains.

Molecular mass of both native PEGs and PEGylated proteins can be detected accurately by matrix-assisted laser desorption ionisation time of flight (MALDI-TOF) mass spectrometry, as first reported by Watson et al. [2], a technique subsequently used by many authors (see, for example, [3–7]). Molecular mass indicates directly the PEGylation extent, since the total molecular mass equals approximately the sum of the native protein and total conjugated PEG.

Specific operating conditions depend on the model of MALDI-TOF used but an acceleration voltage of 20 kV with linear detection has been found to be suitable, with samples prepared on various matrices, including sinapic acid [8], dihydroxybenzoic acid [1] and α-cyano-4-hydroxycinnamic acid [3].

Molecular size changes significantly and by large quanta with each PEG chain added to a protein. The viscosity radii of random coil PEG molecules in solution are correlated to their molecular masses by $R_{h,PEG} = 0.1912 \, M_r^{0.559}$ [8], while the viscosity radii of globular proteins is $R_{h,prot} = (0.82 \pm 0.02) \, M_r^{0.333}$ [9]. Using these correlations, Table 1 shows the radii and equivalent protein molecular masses of some commonly used PEG nominal molecular weights.

Table 1. Equivalent protein molecular weights for common PEG molecules

PEG Nominal M_r (kDa)	Viscosity radius (nm)	Equivalent globular protein M_r (kDa)
2	1.34	4.4
5	2.24	20.4
10	3.29	65.4
20	48.5	209.5
40	71.5	670.7

Polyacrylamide gel electrophoresis (PAGE), routinely used for protein molecular weight and purity determinations can also be used also for analysis of PEGylated proteins. However, unlike the case for native proteins where standard protein ladders can be used to make quantitative determinations of molecular weight, no correlation currently exists for determining PEGylated protein molecular weight by this method. The migration rates of PEGylated proteins through porous gels are slowed by the large, heavily hydrated and uncharged PEG chains attached to the proteins, so their apparent molecular masses cannot be determined from their positions relative to standard protein ladders. Often, PEGylation extent can be inferred qualitatively from the band positions in PAGE gels, since the latter differ significantly with each PEG chain added to the protein, but this can provide only circumstantial evidence of PEGylation extent. For example, if site-specific PEGylation is obtained through maleimide-PEG reacting with a single, free available cysteine residue on a protein, then a significant shift in the PAGE band position after PEGylation would support the conclusion that PEGylation was successful. On the other hand, if one was using a non-specific chemistry such as lysine con-jugation, the PEGylation extents of the migrating species can only be guessed by assuming that bands occur in the order of native protein, mono-PEGylated, di-PEGylated and tri-PEGylated protein, etc., with no missing values of N. In the absence of a native protein band showing among the reaction products it would be dangerous to make assumptions regarding the values of N obtained by this method.

PEG tends to interact with SDS under reducing conditions, creating a clump of negatively-charged chains that move in the gel, despite PEG having no charge. Native SDS-PAGE appears to eliminate PEG interaction with the gel, providing sharper bands and giving consistent band positions depending on molecular weight of conjugated PEG, e.g., a di-PEGylated 10-kDa PEG-protein having a band at the same position as the mono-PEGylated 20 kDa PEG-protein [10].

PEGylation extent and protein concentration can be determined by size exclusion chromatography (SEC), provided that the native protein fits well on the standard protein calibration curve. It is important to remember that elution volumes in SEC are related to molecular *size* and not molecular weight, despite the normal practice in protein chromatography of calibrating SEC columns in terms of the latter. If molecular weights are used to create calibration curves, protein and PEG standards lie on distinctly different curves. However, if molecular size, in terms of viscosity radius, is used to calibrate the column, protein and PEG standards fall on the same curve. The viscosity radius of a PEGylated protein depends only on the total molecular weight of conjugated PEGs, i.e., the viscosity radius is independent of N, the number of PEG chains attached [8]. Figure 1 shows the elution volumes for cytochrome C PEGylated to various extents with PEGs of differing nominal molecular weight. Clearly, the PEGylated protein peaks overlap at positions where the

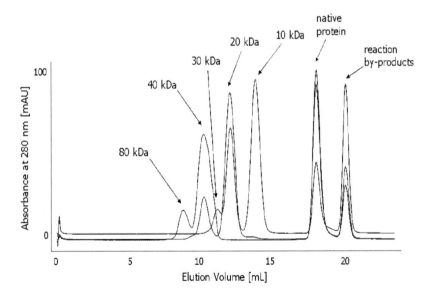

Figure 1. Size exclusion chromatogram for cytochrome C, PEGylated to varying extents with various nominal molecular weight PEGs. The peak labeled "10 kDa" is mono-PEGylated; the peaks labeled "20 kDa" are di-PEGylated with 10 kDa PEG and mono-PEGylated with 20 kDa PEG; the peak labeled "30 kDa" is tri-PEGylated with 10 kDa PEG; the peaks labeled 40 kDa are di-PEGylated with 20 kDa PEG and mono-PEGylated with 40 kDa PEG; and the peak labeled "80 kDa" is di-PEGylated with 40 kDa PEG.

total molecular weight of PEG adducts is equal, regardless of the number of PEG chains used to reach the total.

Using a model in which PEG chains surround the protein molecule as a shell and the observation that the PEG layer apparently expands in volume to attain a surface area to volume ratio equal to that of a native PEG molecule with the same total molecular weight in solution, the viscosity radius of a PEGylated protein can be predicted using equations (1) and (2) (where $R_{h,PEG}$ is calculated using the total PEG M_r attached to the protein) [8].

$$R_{h,PEGprot} = \frac{A}{6} + \frac{2}{3A} R_{h,PEG}^2 + \frac{1}{3} R_{h,PEG} \tag{1}$$

where

$$A = \left[108 R_{h,prot}^3 + 8 R_{h,PEG}^3 + 12 \left(81 R_{h,prot}^6 + 12 R_{h,prot}^3 R_{h,PEG}^3 \right)^{1/2} \right]^{1/3} \tag{2}.$$

Thus, by proper calibration using viscosity radii, the PEGylation extent can be determined quantitatively by SEC. Furthermore, since PEG itself is invisible to UV at 280 nm and typical amine reactive groups such as succinimide do not contain UV active chromophores, UV adsorption at 280 nm may be used to quantify the concentration of PEGylated species. However, some chromophores, such as maleimide or fmoc, do have UV active components. For conjugates with these types of linkers, it may be necessary to determine the extinction coefficient after a non-UV method, such as amino acid analysis, has been used to determine the protein concentration.

Because of the large hydrodynamic radii of PEGylated proteins, SEC can be used in desalting mode for removal of the small molecule by-products of PEGylation reactions, e.g., succinic acid, the main by-product of lysine conjugation chemistry. It can also be used efficiently for the separation of native and PEGylated forms of a protein generally up to a maximum of tri-PEGylated species, while removing small molecules at the same time. The general rule of thumb for efficient SEC separation of proteins at preparative scale is that the proteins should differ in molecular weight by no less than two times to ensure good peak separation. Because of their sizes, even for large PEG adducts ($M_r \geq 20$ kDa), it is not efficient to separate species with a PEGylation extent of N from those with an extent N-1 when $N \geq 3$ [11]. For 5 kDa PEG adducts, the upper limit is N = 2, while for 2 kDa adducts, SEC can be used efficiently for separating mono-PEGylated species from native protein but not for separating species with higher values of N.

A number of authors have used ultrafiltration for purification of PEGylated proteins [12–15]. More recently, it has been found, unlike the case in SEC, that it is not only the total PEG molecular weight that affects transport through ultrafiltration membranes but also N [16]. Molek and Zydney [16] measured a two-fold decrease in the sieving coefficient for ovalbumin (43 kDa) PEGylated with two 5-kDa PEGs compared to PEGylation with one 10-kDa PEG. α-lactalbumin (14.2 kDa) tetra-PEGylated with 5 kDa-PEG had a sieving coefficient more than an order of magnitude smaller than that di-PEGylated with 10-kDa PEG. This dependence on N may be due to deformation and/or elongation of PEG chains caused by convective flow into the membrane pores. One might picture a single conjugated PEG chain as a flexible entity, streaming out in advance or trailing behind a globular molecule as it moves through a membrane pore. A higher number of conjugated chains would add more to the effective molecular diameter because of the constraints to movement at the conjugation sites. Such a view of chain flexibility aiding pore transport is consistent with the observation that PEG molecules have greater sieving coefficients than proteins with an equivalent hydrodynamic radius [16] and similar observations regarding differences between PEG and Ficoll (a highly branched, cross-linked, almost spherical polysaccharide) [1].

Molek and Zydney used a two-stage combination of conventional and charged ultrafiltration membranes to achieve separation of PEGylated α-lactalbumin from reaction byproducts and unreacted precursors [17]. Conventional (neutral charge) ultrafiltration membranes were used first to remove unreacted native protein and small reaction byproducts from the reaction mixture by diafiltration, taking advantage of the greatly increased hydrodynamic radius of the PEGylated species. In the second stage, a higher molecular weight cut-off, negatively charged membrane was used to repel the PEGylated protein, while allowing the (charge neutral) unreacted PEG to pass through, again in diafiltration mode. In diafiltration processes, there is of course a trade-off between purity and yield, which increase and decrease, respectively, with diafiltration volume. In practice, one cannot completely remove unwanted components by diafiltration but their concentrations can be considerably reduced before final purification using more selective processes.

Capillary electrophoresis can separate PEGylated proteins on the basis of size, shape and surface activity (charge, hydrophobicity) and can thus, unlike SEC and membranes, separate positional isomers in addition to determining PEGylation extent. Li et al. [18] separated positional isomers of PEGylated ribonuclease A and lysozyme in a semi-aqueous capillary electrophoresis technique in which a high molecular weight polyethylene oxide layer was first adsorbed to the internal surface of the capillary, followed by electrophoresis in an acetonitrile-water solvent at pH 2.5. Na et al. [19] were able to differentiate between PEGylated α-interferon with very small differences in conjugated PEG molecular weight (40 kDa, 43.5 kDa and 47 kDa) using a sodium dodecyl sulfate-capillary gel electrophoresis system. Characterisation of the PEGylation site in positional isomers can also be carried out by proteolytic digestion of a PEGylated protein and comparing fragments with those obtained with native protein digestion. For example, using the endo-proteolytic enzyme Lys-C, differences between native and PEGylated parathyroid hormone 1-34 were examined by capillary electorphoresis and PEGylation sites assigned using the assumption that steric hindrance prevented cleavage near the conjugation site [20].

Electrostatic charge

Ion exchange chromatography is the most commonly used technique for purification of PEGylated proteins, with many examples available in the literature, e.g., [21–30]. PEG may affect the charge properties of proteins in three ways. First, PEG may shield the surface charges of a protein and weaken electrostatic interactions. Second, the isoelectric point (pI) may be altered by conjugation of PEG to charged residues (effectively neutralization of a single charge with each PEG group conjugated). Third, PEGs may hydrogen bond with some groups to raise their pKa, thus affecting their charge *versus* pH behavior.

The most common observation regarding ion exchange separation of PEGylated proteins is that more heavily PEGylated species (whether from higher N or higher molecular weight PEG chains [31]) elute at lower ionic strengths than their less heavily PEGylated counterparts. In some cases, highly PEGylated forms are contained in the flowthrough fractions of ion exchange, while native and less heavily PEGylated forms bind [21, 11]. Piquet et al. [27] obtained a 97% purity of mono-PEGylated growth hormone release factor using a step elution ion exchange process at gram scale. Similar to the earlier success of Kinstler et al. with cation exchange [24], Lee et al. were able to separate mono-, di- and tri-PEGylated forms of a recombinant granulocyte colony stimulating factor using anion exchange [32], obtaining a good purity of N-terminal mono-PEGylated product. Yamamoto et al. showed that not only the PEG:protein molecular weight ratio but also the site of conjugation is important in determining ion exchange behavior, as they were able to separate three mono-PEGylated forms of lysozyme by cation exchange chromatography [33]. However, the differences in the strengths of electrostatic interactions between positional isomers are small and cannot be exploited effectively at the preparative scale.

Aside from charge considerations, ion exchange capacity will be affected by the size of the molecules, lowering diffusivity and access to internal pores in chromatography media [31]. From equation (1), a 20 kDa protein, PEGylated with a 20 kDa PEG, has viscosity radii of native, mono- and di-PEGylated forms equivalent to 20 kDa, 265 kDa and 730 kDa protein molecular weights, respectively. These size differences are significant and one would expect associated differences in mobility to affect access to ion exchange binding sites. A macro-porous cation exchange resin has been shown to achieve higher binding capacities than a conventional resin for PEGylated species [34]. It should also be noted here that PEG is commonly used to repel proteins from surfaces [35, 36], so there is a possibility that PEGylated molecules, once bound on the exterior of a chromatography resin, could repel further molecules from approaching, particularly larger, slower-moving, more heavily PEGylated species.

PEGylated proteins, of course, assume net negative and positive charges above and below their isoelectric points, respectively. PEGylation itself alters the isoelectric point (pI) of a protein because each PEG chain binds to and neutralizes a charged residue such as lysine on the protein surface. The change in pI with PEGylation will depend on the number of charged residues on the protein and the shape of the net charge *versus* pH curve near the isoelectric point itself. The theoretical charge *versus* pH curves, calculated using PropKa [37] and sequence data from the Swiss Protein Database, are shown in Figure 2 for BSA (PDB ID: 1HK1, [38]) as a function of PEGylation extent. The large number of charged residues in BSA results in only a small difference to the net charge with each PEG chain added. Also, the pI for native BSA occurs in a region where there is a relatively high slope in the charge *versus* pH curve. Thus little relative shifts in total charge and pI occur with each successive PEG chain addition.

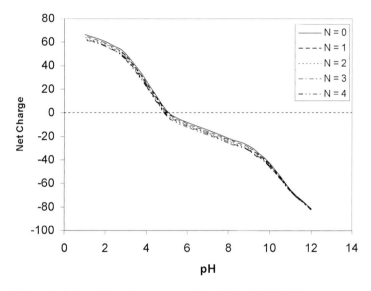

Figure 2. Theoretical change in net charge *versus* pH for BSA with PEGylation extent.

For smaller proteins, each charged residue that is neutralized by PEGylation represents a significant fraction of the total charge. Combined with the pI of the protein being located in a region where the charge *versus* pH curve has a low slope, PEGylation can, at least in theory, significantly shift the pI. For example, Figure 3 shows the theoretical charge-pH curves for native and PEGylated α-interferon 2A (Protein Data Bank (PDB) ID: 2HIE, [39]. The pI lies in a region of low slope on the charge *versus* pH curve and there are a low number of charged residues, thus PEGylation shifts the calculated pI by two points from that of the native protein (N = 0) with a pI of 6.6 to that of tetra-PEGylated protein (N = 4) with a pI of 4.6, with the largest single shift (0.8 pH points) occurring between the native and mono-PEGylated forms.

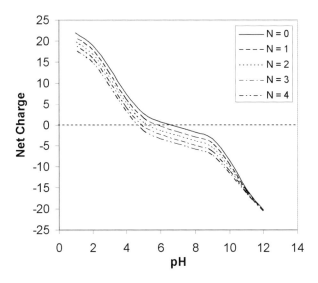

Figure 3. Theoretical change in net charge *versus* pH for α-interferon 2A with PEGylation extent.

Hydrophobicity

Little work has been done on hydrophobic interactions of PEGylated proteins, other than its use as the first step in a purification process for growth hormone by Clark et al. [40], to separate PEGylated forms of growth hormone antagonist by Nijs et al. [41], and the observation of Vincintelli et al. [42] that a PEGylated form of β-lactoglobulin bound more strongly to a hydrophobic interaction resin than either the free PEG or the native protein. The explanation offered for the latter behavior was that surface deformability of the protein was enhanced by conformational changes to the protein upon PEGylation. Thiol PEGylation has been used with hydrophobic interaction chromatography to aid the separation of plant enzymes that were otherwise difficult to resolve but the objective was the removal of contaminants by selective modification with PEG rather than purification of the PEGylated protein *per se* [43, 44]. There appears to be little, if any, resolution in hydrophobic interaction chromatography between various PEGylated forms of a protein, either on the basis of PEGylation extent or PEG chain length.

Reversed phase chromatography can be used for analytical as well as preparative (if the protein or peptide is stable) separation of PEGylated proteins both for purity assays and for separation of positional isomers [32, 45–47]. For example, Lee et al. [48] separated mono-PEGylated salmon calcitonins first by size exclusion and then used reversed phase chromatography to separate them into positional isomers. Johnson et al. [49] developed a quantitative reversed phase assay for use during manufacturing and for stability studies of a PEGylated staphylokinase mutant for use as a thrombolytic agent.

Their assay was able to detect and quantitate de-PEGylation events as well as the presence of host cell protein contaminants.

Summary

Separation and analysis of PEGylated proteins is challenging because PEG itself is a relatively inert, neutral, hydrophilic polymer and the starting point for PEGylation is a pure protein. Thus, other than molecular weight and size, differences in the physicochemical properties between PEGylated forms of a protein tend to be rather small. The usual properties of electrostatic charge and molecular weight (size) form the basis of the most commonly used separation techniques, particularly ion exchange chromatography, size exclusion chromatography and ultrafiltration. The main effect of PEGylation on ion exchange separations is to shield the electrostatic charges on the protein surface and to reduce the strength of interactions with higher PEG chain molecular weight or higher PEGylation extent. Thus, ion exchange can be used very effectively to separate on the basis of PEGylation extent. Separation of positional isomers is possible by ion exchange at analytical scale but it is problematic at preparative scale due to the small size of the differences in electrostatic interactions between isomers.

PEGylation imparts quantum changes in molecular weight with each chain added to a protein, so MALDI-TOF mass spectrometry is a logical choice for determining PEGylation extent. There are corresponding quantum increases in molecular size, so size exclusion chromatography and ultrafiltration (and dialysis) are very effective methods for separating native and PEGylated proteins. However, the relative size difference between variants with PEGylation extent N and $(N + 1)$ reduces with N so that efficient size exclusion chromatography separation between PEGylated species differing by one PEG chain is not achievable at the preparative scale for $N > 3$ even for large PEG chains ($M_r \geq 20$ kDa). For small PEG chains ($M_r = 2$ kDa), only native and PEGylated species can be separated effectively. At the analytical scale, with proper calibration, size exclusion chromatography can provide valuable information on PEGylation extent. Membranes can be used to reduce the concentration of smaller molecular weight species by dialysis but cannot fully remove them and require an operational trade-off between purity and yield. Gel electrophoresis can confirm PEGylation reactions have proceeded and indicate the relative purity of products but it cannot confirm PEGylation extent. The main drawback of separations based solely upon molecular size is that they cannot differentiate between positional isomers. Capillary electrophoresis is an exception, quantitatively combining any or all of size, shape, conformational freedom and small differences in protein surface properties to separate by both PEGylation extent and positional isomerism.

Relative hydrophobicity is a useful property for analytical separations using reversed phase chromatography but hydrophobic interaction chromatography, which is used routinely for production-scale purification of proteins, does not appear to be particularly useful for separation of PEGylated species.

References

1. Fee, C.J., *Size comparison between proteins PEGylated with branched and linear poly(ethylene glycol) molecules.* Biotechnology and Bioengineering, 2007. 98(4): p. 725–31.
2. Watson, E., Shah, B., DePrince, R., Hendren, R.W., and Nelson, R., *Matrix-assisted laser desorption mass spectrometric analysis of a pegylated recombinant protein.* Biotechniques, 1994. 16(2): p. 278–80.
3. Basu, A., Yang, K., Wang, M., Liu, S., Chintala, R., Palm, T., Zhao, H., Peng, P., Wu, D., Zhang, Z. et al., *Structure-function engineering of interferon-α-1b for improving stability, solubility, potency, immunogenicity, and pharmacokinetic properties by site-selective mono-PEGylation.* Bioconjugate Chem., 2006. 17(3): p. 618–30.
4. Foser, S., Schacher, A., Weyer, K.A., Brugger, D., Dietel, E., Marti, S., and Schreitmüller, T., *Isolation, structural characterization, and antiviral activity of positional isomers of monopegylated interferon [alpha]-2a (PEGASYS).* Protein Expression and Purification, 2003. 30(1): p. 78–87.
5. Lee, K.C., Moon, S.C., Park, M.O., Lee, J.T., Na, D.H., Yoo, S.D., Lee, H.S., and DeLuca, P.P., *Isolation, characterization, and stability of positional isomers of mono-PEGylated salmon calcitonins.* Pharmaceutical Research, 1999. 16(6): p. 813–18.
6. Li, X.-Q., Lei, J.-D., Su, Z.-G., and Ma, G.-H., *Comparison of bioactivities of monopegylated rhG-CSF with branched and linear mPEG.* Process Biochemistry, 2007. 42(12): p. 1625–31.
7. Youn, Y.S., Na, D.H., Yoo, S.D., Song, S.-C., and Lee, K.C., *Chromatographic separation and mass spectrometric identification of positional isomers of polyethylene glycol-modified growth hormone-releasing factor (1-29).* Journal of Chromatography A, 2004. 1061(1): p. 45–49.
8. Fee, C.J. and Van Alstine, J.M., *Prediction of viscosity radius and size exclusion chromatography behavior of PEGylated proteins.* Bioconjugate Chemistry, 2004. 15(6): p. 1304–13.
9. Hagel, L., *Gel Filtration*, in *Protein Purification*, J.-C. Janson and Rydén, L., Editors. 1998, John Wiley & Sons: New York.
10. Zheng, C.Y., Ma, G.H., and Su, Z.G., *Native PAGE eliminates the problem of PEG-SDS interaction in SDS-PAGE and provides an alternative to HPLC in characterization of protein PEGylation.* Electrophoresis, 2007. 28(16): p. 2801–7.
11. Fee, C.J. and Van Alstine, J.M., *PEG-proteins: Reaction engineering and separation issues.* Chemical Engineering Science, 2006. 61(3): p. 924–39.
12. Bailon, P. and Berthold, W., *Polyethylene glycol-conjugated pharmaceutical proteins.* Pharmaceutical Science & Technology Today, 1998. 1(8): p. 352–56.
13. Edwards, C.K., Martin, S.W., Seely, J., Kinstler, O.B., Buckel, S., Bendele, A.M., Cosenza, M.E., Feige, U., and Kohno, T., *Design of PEGylated soluble tumour necrosis factor receptor type I (PEG sTNF-RI) for chronic inflammatory diseases.* Advanced Drug Delivery Reviews, 2003. 55: p. 1315–36.
14. Maeda, N., Kimura, M., Sasaki, I., Hirose, Y., and Konno, T., *Toxicity of bilirubin and detoxification by PEG-bilirubin oxidase conjugate*, in *Poly(ethylene glycol) chemistry: Biotechical and biomedical applications*, J.M. Harris, Editor. 1992, Plenum Press: New York. p. 153–69.
15. Tan, Y., Sun, X., Xu, M., An, Z., Tan, X., Han, Q., Miljkovic, D.A., Yang, M., and Hoffman, R.M., *Polyethylene glycol conjugation of recombinant methioninase for cancer therapy.* Protein Expression and Purification, 1998. 12(1): p. 45–52.
16. Molek, J.R. and Zydney, A.L., *Ultrafiltration characteristics of pegylated proteins.* Biotechnology and Bioengineering, 2006. 95(3): p. 474–82.
17. Molek, J.R. and Zydney, A.L., *Separation of PEGylated alpha-lactalbumin from unreacted precursors and byproducts using ultrafiltration.* Biotechnology Progress, 2007. 23(6): p. 1417–24.
18. Li, W., Zhong, Y., Lin, B., and Su, Z., *Characterization of polyethylene glycol-modified proteins by semi-aqueous capillary electrophoresis.* Journal of Chromatography A, 2001. 905(1–2): p. 299–307.
19. Na, D.H., Park, E.J., Jo, Y.W., and Lee, K.C., *Capillary electrophoretic separation of high-molecular-weight poly(ethylene glycol)-modified proteins.* Analytical Biochemistry, 2008. 373(2): p. 207–12.

20. Na, D.H. and Lee, K.C., *Capillary electrophoretic characterization of PEGylated human parathyroid hormone with matrix-assisted laser desorption/ionization time-of-flight mass spectrometry.* Analytical Biochemistry, 2004. 331(2): p. 322–28.

21. Brumeanu, T.-D., Zaghouani, H., and Bona, C., *Purification of antigenized immunoglobulins derivatized with monomethoxypolyethylene glycol.* Journal of Chromatography A, 1995. 696: p. 219–25.

22. Esposito, P., Barbero, L., Caccia, P., Caliceti, P., D'Antonio, M., Piquet, G., and Veronese, F., *Pegylation of growth hormone-releasing hormone GRF analogues.* Advanced Drug Delivery Reviews, 2003. 55(10): p. 1279–91.

23. He, X.H., Shaw, P.C., and Tam, S.C., *Reducing the immunogenicity and improving the in vivo activity of trichosanthin by site-directed pegylation.* Life Sciences, 1999. 65(4): p. 355–68.

24. Kinstler, O.B., Brems, D.N., Lauren, S.L., Paige, A.G., Hamburger, J.B., and Treuheit, M.J., *Characterization and stability of N-terminally PEGylated rhG-CSF.* Pharmaceutical Research, 1996. 13: p. 996–1002.

25. Koumenis, I.L., Shahrokh, Z., Leong, S., Hsei, V., Deforge, L., and Zapata, G., *Modulating pharmacokinetics of an anti-interleukin-8 F(ab')(2) by amine-specific PEGylation with preserved bioactivity.* 2000. 198(1): p. 83–95.

26. Manjula, B.N., Tsai, A., Upadhya, R., Perumalsamy, K., Smith, P.K., Malavalli, A., Vandegriff, K., Winslow, R.M., Intaglietta, M., Prabhakaran, M et al., *Site-specific PEGylation of hemoglobin at cys-93(b): correlation between the colligative properties of the PEGylated protein and the length of the conjugated PEG chain.* Bioconjugate Chem., 2003. 14(2): p. 464–72.

27. Piquet, G., Gatti, M., Barbero, L., Traversa, S., Caccia, P., and Esposito, P., *Set-up of large laboratory-scale chromatographic separations of poly(ethylene glycol) derivatives of the growth hormone-releasing factor 1-29 analogue.* Journal of Chromatography A, 2002. 944(1–2): p. 141–48.

28. Reddy, K.R., Modi, M., and Pedder, S., *Use of PEGinterferon alfa-2a (40 kD) (Pegasys) for the treatment of hepatitis C.* Advanced Drug Delivery Reviews, 2002. 54: p. 571–86.

29. Sato, H., *Enzymatic procedure for site-specific pegylation of proteins.* Advanced Drug Delivery Reviews, 2002. 54(4): p. 487–504.

30. Wang, Y.-S., Youngster, S., Grace, M., Bausch, J., Bordens, R., and Wyss, D.F., *Structural and biological characterization of PEGylated interferon alpha-2b and its therapeutic implications.* Advanced Drug Delivery Reviews, 2002. 54(4): p. 547–70.

31. Pabst, T.M., Buckley, J.J., Ramasubramanyan, N., and Hunter, A.K., *Comparison of strong anion-exchangers for the purification of a PEGylated protein.* Journal of Chromatography A, 2007. 1147(2): p. 172–82.

32. Lee, D.L., Sharif, I., Kodihalli, S., Stewart, D.I.H., and Tsvetnitsky, V., *Preparation and characterization of monopegylated human granulocyte-macrophage colony-stimulating factor.* Journal of Interferon and Cytokine Research, 2008. 28(2): p. 101–12.

33. Yamamoto, S., Fujii, S., Yoshimoto, N., and Akbarzadehlaleh, P., *Effects of protein conformational changes on separation performance in electrostatic interaction chromatography: Unfolded proteins and PEGylated proteins.* Journal of Biotechnology, 2007. 132(2): p. 196–201.

34. Fee, C.J., Bergstrom, J., Stadler, J., Magnusson, R., and Van Alstine, J.M., *Challenges related to the processing of PEG-modified proteins,* in *International Conference on Biopartitioning and Purification (BPP 2005)* 2005: Delft, Netherlands.

35. Harris, J.M., ed. *Poly(ethylene glycol) Chemistry: Biotechnical and Biomedical Applications.* Topics in Applied Chemistry, ed. A.R. Katritzky and Sabongi, G.J. 1992, Plenum Press: New York.

36. Harris, J.M. and Zalipsky, S., eds. *Poly(ethylene glycol): Chemistry and Biological Applications.* ACS Symposium Series. Vol. 680. 1997, American Chemical Society: Washington D.C.

37. Li, H., Robertson, A.D., and Jensen, J.H., *Very fast empirical prediction and rationalization of protein pKa values.* Proteins, 2005. 61: p. 704–21.

38. Petitpas, I., Petersen, C.E., Ha, C.E., Bhattacharya, A.A., Zunszain, P.A., Ghuman, J., Bhagavan, N.V., and Curry, S., *Structural basis of albumin-thyroxine interactions and familial dysalbuminemic hyperthyroxinemia.* Proceedings of National Academic Science, USA, 2003. 100: p. 6440–45.

39. Murgolo, N.J., Windsor, W.T., Hruza, A., TReichert, P., Tsarbopoulos, A., Baldwin, S., Huang, E., Pramanik, S., Ealick, P., and Trotta, P., *A homology model of human interferon alpha-2.* Proteins, 1993. 17: p. 62–74.

40. Clark, R., Olson, K., Fuh, G., Marian, M., Mortensen, D., Teshima, G., Chang, S., Chu, H., Mukku, V., Canova-Davis, E et al., *Long-acting growth hormones produced by conjugation with poly(ethylene glycol).* Journal of Biological Chemistry, 1996. 271(21): p. 969–77.

41. Nijs, M., Azarkan, M., Smolders, N., Brygier, J., Vincentelli, J., Vries, G.M.P., Duchateau, J., and Looze, Y., *Preliminary characterization of poly(ethylene glycol)ylated human growth hormone antagonist*, in *Poly(ethylene glycol): Chemistry and Biological Applications*, J.M. Harris and Zalipsky, S., Editors. 1997, American Chemical Society: Washington, D.C. p. 170–81.

42. Vincentelli, J., Paul, C., Azarkan, M., Guermant, C., El Moussaoui, A., and Looze, Y., *Evaluation of the polyethylene glycol-KF-water system in the context of purifying PEG-protein adducts.* International Journal of Pharmaceutics, 1999. 176(2): p. 241–49.

43. Azarkan, M., El Moussaoui, A., van Wuytswinkel, D., Dehon, G., and Looze, Y., *Fractionation and purification of the enzymes stored in the latex of carica papaya.* Journal of Chromtography B, 2003. 790(1–2): p. 229–38.

44. Azarkan, M., Maes, D., Bouckaert, J., Thi, M.-H.D., Wyns, L., and Looze, Y., *Thiol PEGylation facilitates purification of chymopapain leading to diffraction studies at 1.4 A resolution.* Journal of Chromatography A, 1996. 749(1–2): p. 69–72.

45. Lee, H.S. and Park, T.G., *Preparation and characterization of mono-PEGylated epidermal growth factor: evaluation of in vitro biologic activity.* Pharmaceutical Research, 2002. 19(6): p. 845–51.

46. Lee, L.S., Conover, C., Shi, C., Whitlow, M., and Filpula, D., *Prolonged circulating lives of single-chain Fv proteins conjugated with polyethylene glycol: a comparison of conjugation chemistries and compounds.* Bioconjugate Chemistry, 1999. 10(6): p. 973–81.

47. Veronese, F., Sacca, B., Laureto, P.P.d., Sergi, M., Caliceti, P., Schiavon, O., and Orsolini, P., *New PEGs for peptide and protein modification, suitable for identification of the PEGylation site.* Bioconjugate Chemistry, 2001. 12(1): p. 62–70.

48. Lee, K.C., Tak, K.K., Park, M.O., Lee, J.T., Woo, B.H., Yoo, S.D., Lee, H.S., and DeLuca, P.P., *Preparation and characterization of polyethylene-glycol-modified salmon calcitonins.* 1999. 4(2): p. 269–75.

49. Johnson, C., Royal, M., Moreadith, R., Bedu-Addo, F., Advant, S., Wan, M., and Conn, G., *Monitoring manufacturing process yields, purity and stability of structural variants of PEGylated staphylokinase mutant SY161 by quantitative reverse-phase chromatography.* Biomedical Chromatography, 2003. 17(5): p. 335–44.

PEG and PEG conjugates toxicity: towards an understanding of the toxicity of PEG and its relevance to PEGylated biologicals

Rob Webster[1], Victoria Elliott[2], B. Kevin Park[2], Donald Walker[1], Mark Hankin[3] and Philip Taupin[3]

[1] *Pharmacokinetics, Dynamics and Metabolism, Pfizer Global Research and Development, Kent, CT13 9NJ, UK*
[2] *University of Liverpool, MRC Centre for Drug Safety Science, Department of Pharmacology and Therapeutics, Liverpool, L69 3BX, UK*
[3] *DSRD, Pfizer Global Research and Development, Kent, CT13 9NJ, UK*

Abstract

PEG is used to improve pharmacokinetic properties of biologicals. Concern has been expressed about the toxicological effect and/or fate of the PEG. This paper reviews the available toxicity, metabolism and clearance data of PEG and PEGylated products in order to place such concerns in to appropriate context. The available data demonstrates that PEG itself only shows toxicity at high, parenteral doses and the usual target organ is the kidney as this is the route of excretion for unchanged PEG. A large therapeutic window (approximately 600-fold) exists between the maximum PEG burden from a current biological agent and the doses of PEG associated with human toxicity. Pathological changes which results in no functional deficit, PEG containing vacuoles in cells, have been observed with PEGylated biologicals. There is evidence that these PEG vesicle can resolve with time. In conclusion the doses used clinically for current and many future PEGylated biologicals are low and will result in exposures to PEG significantly lower than that required to elicit PEG toxicity. In all cases the routine regulatory toxicology studies would identify relevant pathology should it occur.

Introduction

PEGylation, the addition of PEG molecules to a protein, peptide or other molecule has become a widely used methodology to improve the pharmacokinetics of the moiety (by reducing clearance and increasing half-life) in humans, increasing dosing interval and increasing patient convenience compared to the non PEGylated product. PEGylation is generally considered to improve the pharmacokinetics of a conjugated molecule by either decreasing glomerular filtration, decasing enzymatic degradation or by decreasing affinity for the target receptor and therefore reducing target mediated clearance [1]. It has also been proposed that PEGylation can reduce the immunogenicity of conjugated proteins [2].

Polyethylene glycols are polymers of ethylene oxide with a formula of $HO-(CH_2-CH_2-O)_n-H$ where n is the number (average) of ethylene glycol units present in the molecule. The number of ethylene glycol units present in a molecule of PEG can range from 4 to greater than 400. PEGs, including those used to synthesise PEGylated biologicals are not single chemical entities but are mixtures of various polymer chain lengths and for some of the higher molecular weight materials they can also be branched [3]. PEGs now generally have low polydispersion due to a chromatographic purification that minimise the dispersion of molecular weights in the product.

Also the PEG used to PEGylate biological products is routinely capped with a methyl group at the hydroxyl group not used for conjugation to the biological. These PEGs are generally described as methoxy PEGs [3].

The PEG is chemically activated prior to the addition to the protein, the technologies used to carry out this linkage are discussed in more detail elsewhere [4, 5]. The process of PEGylation generally involves the addition of one or more high molecular weight PEGs to the molecule of interest. Commercially, PEGylated proteins were initially linked with multiple 5 kD PEGs in a relatively non-specific manner, targeting reactive amino acids in the molecule, these can include amines (lysines) and thiols. As the technology has evolved the PEGylation has frequently become more targeted and it has become more common to add a single large, potentially branched, PEG with a molecular weight in the 20–40 kD range. This has led to a far more targeted approach to PEGylation [6–8].

The choice of PEG as a molecule with which to increase the dosing interval of biologicals seems to have been largely driven by its lack of toxicity, except at very high doses and its immune-modulatory properties. PEG is generally considered to have low toxicity by all routes of administration. PEG is not a strong ocular or mucosal irritant but can be slightly irritant to skin. While this is a very brief overview of a complex area, generally PEG has been shown to be very safe in a large battery of tests and using a large range of routes and molecular weights [9]. PEGs are widely used and are present in many products used in daily life, these include toothpaste which can be up to 10% PEG [9]. PEG is also widely used as an excipient in many medicines administered by the parenteral, topical, ophthalmic, oral and rectal routes. Examples of intravenous medicines that contain PEG include Vespesid® and Ativan®. The World Health Organisation (WHO) has set an estimated acceptable daily dose of PEG of up to 10 mg/kg [9]. However, this WHO limit is likely to refer to the oral route and also does not specify molecular weight, but a molecular weight range up to 10 kD.

This paper discusses the toxicity observed with PEG and attempts to place this toxicity in a context that will be useful to those developing PEGylated biologicals in the future and faced with question around the toxicity of PEG portion of the molecule.

Toxicity of PEG in animals

The toxicity of PEG has been widely evaluated and has been thoroughly reviewed in several papers including most recently that of Fruijtier-Polloth et al. [9]. However, the majority of the toxicology studies with PEG have been performed by the oral route. While these oral data are of interest to the understanding of the target organ toxicity of PEG they are not discussed in this paper. Pharmacokinetic data clearly demonstrate that as the molecular weight of PEG increases oral absorption decreases [1]. Based on the available data it is therefore impossible to separate changes in toxicity from changes in absorption. This paper has therefore focused on PEGs administered by either the intravenous, subcutaneous or intra peritoneal routes, allowing a direct comparison of the toxicology exposure relationship. The majority of toxicology studies with PEG have been carried out with 'normal' PEG (i.e., both hydroxyl groups are present). Generally there is little or no toxicity associated with these molecules, when toxicity is observed it is most frequently associated with the kidneys and can result in ultra structural vacuolation of the proximal renal tubules [10]. The available toxicological information is summarised in Table 1. The data presented in Table 1 have been generated in the rabbit which is not a commonly used toxicology species. However, it is believed that the rat and rabbit are likely to yield similar results, especially as in both cases renal excretion of PEG is likely to be the predominant route of clearance (see below, '*Metabolism and excretion of PEG, does this provide an understanding of the toxicological mechanism?*'). The mechanism underlying the renal toxicity of PEG is not clear.

As stated above PEG appears to be largely devoid of toxicity except at very high parental doses and when observed is most often associated with renal related findings. The data presented above relates to PEG. However, as stated, the majority of PEGylated agents are routinely conjugated to a methoxy PEG. Also the majority of the PEGs toxicity data presented in the literature is of low molecular weights, up to 7,500. PEGs with molecular weights in the region of 20–40 kD are routinely used to PEGylate biologicals. There is a small amount of data available on the toxicity of methoxy PEG of molecular weight similar to those used for PEGylating biologicals, these data are summarised in Table 2. The doses used in the toxicological testing of biologicals are generally lower

Table 1. Summary of the toxicity of acute and chronically administered PEGs by the subcutaneous and intravenous routes (adapted from [9])

Type of PEG	Species	Result
Acute toxicity		
PEG-750 (IV)	Rabbit	LD50: 10,000 mg/kg: the two animals survived to termination (day 14), one animal had a renal tubular swelling
Chronic toxicity		
PEG-200, 300, 400, 1000, 1540, 3600, 7500 (IV)	Rabbit	350 mg/kg/day for 5 weeks: cloudy swellings of renal tubules and hepatic parenchyma amongst all groups
PEG-200 (SC)	Rabbit	1,000 mg/kg/day, 6 days/week for 60 days: In some animals increased urea. Two out of six rabbits died on day 6 and 21. 3/3 rabbits died that were fed on rolled barley diet

Table 2. Summary of the toxicity of acute and chronically administered methoxy PEG by the subcutaneous and intravenous routes [11]

Type of PEG	Approximate molecular weight of the PEG	Species	Result
Acute toxicity			
Mpeg	Assumed 12 kD	Mice and rats	IV and SC doses at 6.48 mg/m^2, no toxicity
Chronic toxicity			
mPEG	Assumed 12 kD	Rats	SC administered twice weekly at 2.276 mg/m^2/week for 13 weeks. No mPEG related macro or microscopic findings
mPEG	Assumed 12 kD	Cynomolgus monkeys	SC administration twice weekly at 2.276 mg/m^2/week for 13 weeks (followed by 4 week recovery period). No mPEG related macro or microscopic findings

than those used to study the toxicity of PEG alone [11]. However, these doses are relevant to the biological burdens of PEG administered during the administration of PEGylated biologicals.

These data demonstrate that the main target organ for PEG toxicity noted in toxicological studies is the kidney. By these routes PEG is a very safe molecule requiring very high doses to achieve a toxicological effect. The data presented above on methoxy PEG also indicate that again this form of PEG has little or no toxicity at relevant biological doses. The observations reported above are in keeping with the toxicity observed by other routes (e.g., orally) where again the predominant target organ is the kidney [9].

Does molecular weight impact the safety of PEG in animals?

There are very little data that allow the examination of the impact of molecular weight on the toxicity of PEG. The majority of the data available is for PEG administered by the oral route, as discussed above, this is not suitable for direct comparison as absorption also changes as molecular weight increases. However, Table 3 summarises the available data where a consistent route has been used in a single species to study the single dose toxicity of PEG [12]. Following single dose administration there is no strong evidence of a molecular weight dependency on the toxicity of PEG. However, it is possible that the higher molecular weight PEGs are more acutely toxic. Again the doses described in these acute studies are much higher than the PEG burden seen for currently available PEGylated biologicals (g/kg *versus* µg/kg). However, as the use of PEG for biologicals expands it is possible that this window will narrow,

Table 3. Acute toxicity (LD_{50}) of PEG in the mouse and rat following intra peritoneal administration [12]

	Mouse LD_{50} (g/kg)	Rat LD_{50} (g/kg)
PEG200	11.8	-
PEG300	10.4	17.0
PEG400	12.9	-
PEG600	10.2	-
PEG1000	3.1	15.6
PEG4000	10.7	13.0
PEG6000	5.9	6.8
PEG9000	5.9	-

but it is likely that a window will still exist as biological doses in the g/kg range are unlikely.

A similar comparison cannot be carried out for the chronic toxicity of PEG due to a lack of consistent data. However, in the literature, data that are available for PEGs with molecular weight similar to those used for PEGylating biologicals demonstrate its safety up to very high doses. Therefore, while understanding this trend is of scientific interest, it does not provide an improved understanding of the safety of PEG.

Reports of PEG toxicity in humans

The majority of the toxicology reports related to PEG in humans relate to findings by the oral route, while not ideal in terms of establishing an exposure/molecular weight/response relationship in human these data are crucial to the understanding of the toxicity in human as they demonstrate the target organ/s of toxicity.

PEG3750 is widely used for colon cleansing: no major side effects have been reported with the exception of a single case of acute renal insufficiency (reversible) in a patient that received 18L of PEG solution [13]. Also ingestion of PEG 200 (13% solution) contained in lava lamp liquid induced renal toxicity in a 65 year old man [14].

Similar renal toxicity has also been observed for nitrofurantoin cream (63% PEG300/5% PEG1000/32% PEG 4000) in burns patients. In these patients estimates were also made of the systemic concentrations associated with these adverse events. These plasma concentrations were in the region of 30–70 mM [15, 16] and demonstrate the high concentrations of PEG required to result in renal toxicity. In these patients there was also evidence of acidosis believed to be due to the acid metabolites of PEG, possibly including glycolic acid.

Finally, on a small number of occasions toxicity has been observed following the administration of PEG as an excipient by the intravenous route.

Intravenous nitrofurantoin [17] has been shown to cause acute renal necrosis, oliguria and azotemia in patients dosed for 3–5 days with a total of 121–220 g of PEG. One patient who received multiple administrations of Lorazepam (Ativan®) presented with acute renal tubular necrosis after a cumulative dose of approximately 240 g of PEG400 [18]. The doses of PEG associated with these toxicities, the relationships between these doses and those seen for biologicals are discussed below in '*PEG exposures for biological products, excipient and associated with toxicity*'. However, it is clear that at high PEG doses toxicity is seen in human, predominately limited to target organ effects on the kidney and is associated with high concentrations potentially in the 10's of mM range.

Comparisons of the toxicity of PEG in animals and humans

In both animals and humans the major target organ for PEG toxicity is the kidney. Generally this toxicity is seen as some form of either acute renal failure or some evidence of renal kidney cell/tubular changes (vacuolation). Animal studies have been carried out with a methoxy PEG and demonstrate that methoxylation of the PEG does not dramatically change the toxicity of PEG with the molecule being well tolerated. These observations are in keeping with the wide range of PEG like products discussed in the review of PEG toxicity carried out by Fruijtier-Polloth et al. [9] where despite the various PEG like structures toxicity was similar and generally only seen at very high doses.

As the kidney is a key clearance organ for PEG [1] in animals and human there is almost certainly a link between this observation and the observed toxicity of PEG. Clearance by the kidney will mean that this organ is likely to be exposed to the highest burden of PEG and therefore it is this organ that should be considered when reviewing toxicology data for evidence of PEG specific findings. However, this does not preclude toxicology occurring in other organs (see below, '*Evidence of PEG specific toxicology for PEGylated biologicals*').

Metabolism and excretion of PEG, does this provide an understanding of the toxicological mechanism?

Predominately PEGs of all molecular weights are excreted unchanged in the urine, increasing molecular weight generally gives rise to an increase in the residence time in the body rather than a fundamental shift in the major excretory route [1]. The urinary excretion processes acting on unchanged PEG would appear to be largely glomerular filtration, a purely passive process. The metabolism of PEG has been discussed in some detail by Webster et al. [1]. Data clearly demonstrate that metabolism of PEG does occur and that this metabolism involves the metabolism of alcohol group/s on the PEG to the corresponding acid or di acid metabolite. These metabolites have been reported in the plasma and urine of burns patients and rabbits and in the bile of cats. The phase 1 metabolism of PEG is a molecular weight dependent process with up to 25% of the dose of PEG400 metabolised in humans. This is rapidly reduced as the molecular weight increases such that by molecular weights of 6000 less than 4% of the dose is potentially metabolised [1]. The phase 1 metabolism that has been observed for PEG would appear to be mediated via alcohol dehydrogenase [19]. Other enzymes have also been implicated in the metabolism of PEG, these include P450 and sulphur transferases. When the picture of metabolism for PEG is considered there is no clear evidence of any major species differences in metabolites formed and the phase 1 metabolites seen in human are formed in animals.

Metabolism studies with PEG have demonstrated that there is no evidence of ethylene glycol formation in animals and humans [20]. However, while the presence of ethylene glycol has not been demonstrated *in vivo* there is evidence that minor amounts of oxalic acid are liberated after the metabolism of PEG [9]. Oxalic acid is a key metabolite in the degradation of ethylene glycol and the calcium salt of oxalic acid is believed to be responsible for the renal toxicity of ethylene glycol [21, 22]. It is therefore possible that oxalic acid or similar metabolites give rise to the renal toxicology of PEG. The metabolism of ethylene glycol is detailed in Figure 1.

In a similar vein the acid metabolites of ethylene glycol are thought to be responsible for the acidosis seen for this molecule following high dose admin-

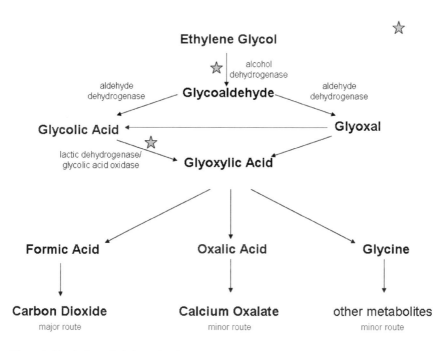

Figure 1. Metabolism of ethylene glycol *in vivo*

istration. Also as stated in the section entitled '*Reports of PEG toxicity in human*' acidosis is also seen in patients during PEG overdose. This toxicity is believed to be due to the acid metabolites of PEG, possibly including glycolic acid. Again the metabolism and toxicities seen for PEG are similar to those seen for ethylene glycol and could be due to similar metabolites.

If the toxicity seen for PEG is driven through similar pathways as that seen for ethylene glycol, two factors will act to limit the toxicity of PEG associated with biologicals: Firstly, as molecular weight increases the fraction of PEG metabolised decreases. It is likely that as the molecular weight of PEG increases the risk of chronic toxicity should decrease; there are however no data to support this suggestion as insufficient comparable data exist (see section '*Does molecular weight impact the safety of PEG in animals?*'). Secondly, it is likely that the methoxy capping of PEGs used in PEGylation will act to reduce the toxicity of PEG as the hydroxyl groups required to initiate this metabolism/toxicity are not present or would at least require O-demethylation to yield a free OH group prior to metabolism by alcohol dehydrogenase.

Problems associated with bioanalytical measurement of PEG disposition and metabolism

The limiting factor in determining the disposition of PEG is the measurement of PEG and its metabolites *in vivo*. PEG is transparent, has no UV chromophore, is non-fluorescent and hence is relatively difficult to detect by itself. For small organic molecules, radio-labelling is the most commonly used methodology to define the metabolic fate of a compound, when administered to animals and humans. Depending on the modification of the PEG, a range of radiolabels can be incorporated onto the PEG molecule including ^{3}H, ^{125}I, ^{14}C, ^{18}F, and ^{111}In [1]. These methods are generally not applicable to PEGylated proteins as the radio-labelling is performed during PEG reagent synthesis and the radiolabel is added near the terminal hydroxyl groups of the PEG molecule and these groups are frequently either methylated or activated in order to provide the a linker to the protein. Methods involving radio-labelling can also result in the modification of the PEG structure, limiting their usefulness for metabolic and pharmacokinetic investigation.

Mass spectrometry based upon matrix-assisted laser desorption/ionisation (MALDI)-mass spectrometry has been utilised to characterise PEGylated proteins in terms of both the extent of PEGylation [23] and the location of PEGylation site [24]. PEG itself can be measured by MALDI, with a 1 ng loading being an approximate detection limit of PEG 4000 and 6000. However with increasing molecular weight PEG becomes difficult to ionise without the aid of metal cations [25]. Furthermore identifying metabolites of PEG by this methodology would be difficult because the PEGs used have polydispersed molecular weight [3] as shown in Figure 2 and hence subtle changes in mass due to metabolism will be undetectable by mass spectrometry.

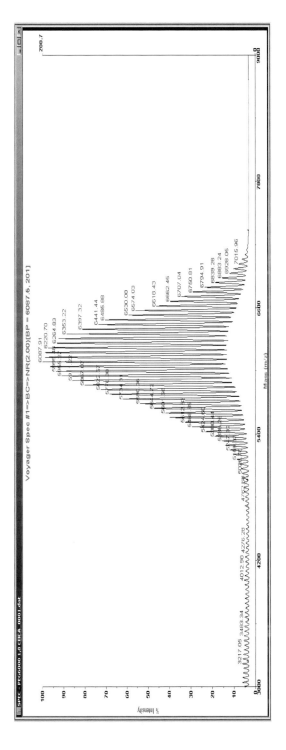

Figure 2. MALDI-TOF-MS spectrum of PEG 6000 showing the polydispersed nature of PEG

Additionally, work by Weaver et al. has shown that PEG can suppress ionisation in electrospray mass spectrometry which may increase the uncertainty
surrounding metabolite identification and quantification [26].

An alternative approach would be to measure small molecular weight
derivatives of PEG. In the previous section, '*Metabolism and excretion of
PEG, does this provide an understanding of the toxicological mechanism?*',
glycolic acid and oxalic acid were identified as potential biomarkers of PEG
metabolism. Both glycolic acid and oxalic acid are amenable to detection by
mass spectrometry as demonstrated in work by Hunt et al. [27] and Keevil et
al. [28]. Measurement of these metabolites in serum and urine following a

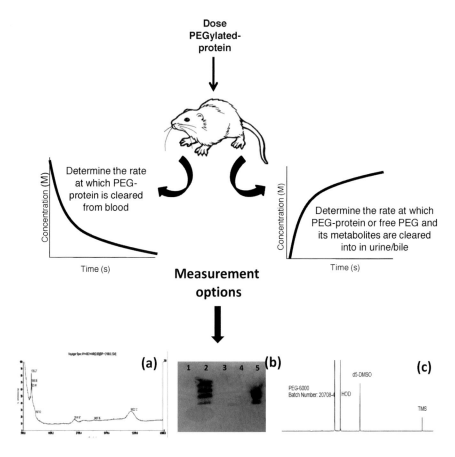

Figure 3. Bioanalytical approaches to determine the disposition and potential metabolism of PEG. (a)
MALDI–TOF-MS spectra of 1 mg/ml 40 K-PEGylated-Insulin (MW ~50,000). (b) Western Blot –
proteins and PEGylated proteins were electrophoresed in a 12% SDS-PAGE gel, transferred to nitro-
cellulose and probed with an anti-PEG antibody – (1) 10 µg superoxide dismutase, (2) 10 µg [5K]PEG-
superoxide dismutase, (3) rat plasma, (4) rat plasma spiked with 10 µg superoxide dismutase and (5)
rat plasma spiked with 10 µg [5K]PEG-superoxide dismutase. (c) ^{1}H NMR of 1 mg/ml PEG 6000, with
all protons from PEG producing a single peak

toxic dose of PEG may further confirm the hypothesis that the metabolism of PEG is closely linked to the metabolism pathway of ethylene glycol and furthermore whether metabolism of PEG decreases with an increase in molecular weight.

Antibody-based methods have recently been reported for the measurement of PEGylated biologicals based upon Western blotting and enzyme linked immunosorbent assay (ELISA). The studies utilised antibodies raised against PEG itself, with two raised against the backbone of PEG [29] and one against the terminal methoxy group [30]. However, in our experience these antibodies are not suitable for detecting hydroxy or methoxy terminated PEG (unpublished observations), although they are capable of detecting a variety of PEGylated proteins, see Figure 3. As PEG is thought to be non-immunogenic [31] and the three reported antibodies have been raised against PEG while it is coupled to a protein, it is likely that the epitope to which the antibody is raised involves the linker which conjugates the PEG moiety to the protein. Nevertheless, these anti-PEG antibodies may be useful to determine levels of PEGylated proteins, if the conjugate remains intact after disposition.

SDS electrophoresis has also been utilised for analysis of PEG and PEG-proteins, but due to steric hindrance by the PEG conjugates, their migration is hindered through the gel, giving a higher than anticipated MW for the conjugate, when compared to protein MW markers. Traditional Coomassie stain for proteins cannot be applied to visualise PEG. However alternative methodology based on the formation of a barium iodide complex with the polyethylene glycol molecule has been demonstrated by Bailon et al. [32] and could offer alternative measurement possibilities to determine disposition.

NMR polymer analysis can be used to determine molecular weight, polymer chain branching and molecular end-groupings. NMR has also been utilised to characterise PEGylated proteins [33] and could prove to be a useful tool in providing a 'simple' assay to measure the disposition of PEG and potential products of metabolism. Furthermore it will be possible to integrate all PEG moieties offering direct quantitation.

Evidence of PEG specific toxicology for PEGylated biologicals

There are a range of PEGylated biological products that are available as medicines. The major products are detailed in Table 4. All of these agents are used clinically via the subcutaneous or intravenous route and therefore provide an opportunity to look for examples of PEG related toxicities [11, 34–36].

Macugen, a PEGylated oligonucleotide, has not been included as this molecule is directly injected in the eye and data indicate that while absorption into the systemic circulation is essentially complete the molecule only slowly diffuses out of the vitreous humour. This product has no toxicity in the eye and due to limitations of dose size and injection volume no maximum tolerated dose could be reached in animal toxicology studies. Macugen® can be used to

Table 4. Toxicity associated with currently marketed PEGylated biological products (information extracted from the relevant EPARs)

	Toxicology findings in animals	Toxicology findings in humans
Pegasys® [34] (Pegylated interferon α2a)	Single dose toxicity. Mild transient increases in liver transaminases, anaemia, leucopenia and foci of sub acute inflammation in the liver parenchyma (SC) at approximately 2,000 times the clinical dose. By the IV route no treatment related toxicities were reported besides bruising at the injection site and elevated AST. Following repeat dose generally Pegasys® was well tolerated with no mortality and only rare serious side effects. The characteristic pattern of interferon alpha toxicity was observed. The lack of specific studies to assess the toxicity of the PEG moiety was acceptable taking into account the toxicity studies did not reveal any toxic signs that could be due to this moiety and there is clinical experience with other PEGylated products.	The safety profile of Pegasys® plus ribavirin® was generally similar to that of interferon α2a plus ribavirin®.
PegIntron® [11] (Pegylated interferon α2b)	Single dose toxicity studies in mice, rats and monkeys indicate a low order of toxicity in animals. Repeat dose toxicity studies were performed in the cynomolgus monkey (SC) every day for 1 month. The findings observed with PegIntron® were similar in nature to those produced by intron A®.	The side effect profile observed with Peg Intron® was quantitatively similar to that of Intron A®. The most common adverse events were those typically associated with interferon use.
Neulasta® [35] (Pegfilgrastim)	In single dose studies in rat (IV 100–1,000 µg/kg) the compound was well tolerated and caused the expected pharmacological effects. In repeat dose toxicology studies in rats and cynomolgus monkeys pegfilgrastim produced a range of changes that reflected an exagger ated pharmacological response, or reaction to the primary response.	The most frequently/occurring adverse events were associated with the administration of chemotherapy and the incidence was similar in the filgrastim and combined pegfilgrastim group. The nature of these adverse events was also similar in the pegfligrastim and filgrastim groups.
Somavert® [36] (Pegvisomant)	No non PEGylated comparator available. However, acute studies in mice (IV and SC – 10 mg/kg) did not induce any functional or toxic effects. Pre-clinical testing revealed no special hazard for human based on repeat dose toxicity studies in rat and monkey.	The overall incidence of treatment emergent signs and symptoms was similar in the treat ment and place-bo groups. There were 23 adverse events that occurred ≥5% more commonly in the treatment group. The most likely explanation for this difference is the longer dura-tion of observation in the pegviso-mant group (up to 34.6 *versus* 3 months).

(Continued on next page)

Table 4. (Continued)

	Toxicology findings in animals	Toxicology findings in humans
Conclusions	Toxicity of currently available PEGylated products is driven by the toxicity of the protein portion, super pharmacological effects and not by the PEG. These data demonstrate that to date PEG incorporated into a PEGylated biological has been well tolerated and has not been associated with any adverse events.	

demonstrate the excellent local, ocular, safety of PEG attached to PEGylated products [37].

The toxicity associated with PEGylated products administered by routes that will achieve systemic exposure (intravenous, subcutaneous and intra peritoneal) in animals and humans are summarised in Table 4. In summary generally all biological products marketed to date have no evidence of toxicity associated with the administered PEG. However, it is also apparent that the biological products PEGylated to date have marked toxicity themselves and as such this could limit the evidence of PEG toxicity, but, given the low toxicity of PEG discussed above the possibility of masked PEG toxicity is considered unlikely. These data also demonstrate that the PEGylated products used clinically to date have relatively low doses and this has limited the systemic exposure to PEG and so PEG toxicity is unlikely to occur. This area is discussed in more detail in section 'PEG exposures for biological products, excipient and associated with toxicity'.

While reports of PEG related toxicity are rare for marketed PEGylated biologicals there are a number of reports in the literature of organ specific vacuolation occurring in animals for several molecules. Tumour necrosis factor (TNF) binding protein PEGylated with 20K PEG has been shown to cause vacuolation in the kidneys on single intravenous doses of 20–40 mg/kg and at lower doses on chronic dosing (10 mg/kg) in the rat [38]. Further experiments demonstrated that these vacuoles are not seen when similar doses of TNF binding protein or PEG itself is administered, indicating that this phenomenon is specific to the PEG protein conjugate. The size of the vacuoles formed is also related to the dose. At low cumulative doses (i.e., on single dose) small vacuoles are formed that do not distort the epithelial cells or distort the nucleus. High doses lead to the formation of large single or coalescing vacuoles which distort the tubular profile and compress the nuclei. However, necrosis was not observed. Reversibility experiments demonstrated that the small vacuoles are reversible over a two month period. However, those that had progressed to multilocular stage were not reversible. Similar vacuoles have been seen with the administration of PEG alone [10]. However, higher doses of PEG are required to demonstrate this effect. It has also been noted that even in the presence of the most striking morphological alterations of kidney tubular cells following administration of PEGylated TNF binding protein there is no evidence of tubular functional abnormalities, lysozomal damage or changes in biomarkers of renal damage including BUN, creatinine and NAG.

Histopathology examination of these vacuoles demonstrated that at the end of the dosing period they were immuno-positive for TNF binding protein. However, at the end of the recovery period (2 months) the vacuoles were no longer immuno-positive for TNF binding protein and it is assumed that the vacuoles contained PEG only. Similar experiments conducted with TNF binding protein conjugated to a 50 kD PEG indicated minimal or absence of vacuolation in rat kidneys. This is believed to be due to a reduction in the glomerular filtration of large PEG conjugates and therefore the reduction in PEG taken up by the kidney tubular cells. The same author [38] also reports similar vacuoles are formed with IGF-bp1.

Vacuoles have also been observed frequently in monkeys administered PEGylated haemoglobin. Vacuoles have been reported in the liver, renal tubules and macrophages in the bone marrow, spleen and lymph nodes [39]. The primary clearance mechanism for haemoglobin is by sequestration in cells of the reticulo-endothelial system [40] and it is these cells that show evidence of vacuolation, indicating the link between clearance mechanism/organ and the distribution of the vacuoles [41, 42]. In studies with PEGylated haemoglobin [42] vacuolation was most severe in splenic macrophages after 7 days of dosing, but had resolved 30 days after dosing. Renal tubular vacuoles were less severe, but persisted 30 days after dosing with reduced severity. However, by 90 days post dose the vacuoles were also absent. For PEGylated haemoglobin at the doses tested the vacuolation findings were dose dependent, transient and without toxic effect. It has also been reported that PEGylated super oxide dismutase caused vacuolation of splenic macrophages in rats in a dose dependent manner (dose up to 1,000 Units/kg) [43].

These data demonstrate that the toxicity of marketed PEGylated biologicals is mediated by the pharmacology of the protein portion and not by that of the PEG. For some PEGylated biologicals there is evidence of vacuole formation that is specific to the PEGylated biologicals. The formation of these vacuoles is linked to the clearance mechanism of the PEGylated protein and can be seen in the major organs of clearance for the specific biological. Information indicates that while these vacuoles can be large they do not appear to be associated with marked adverse events or dysfunction. Small vacuoles have been shown to resolve within two months. However, large vacuoles have been shown to persist for at least two months. It is possible that these vacuoles would resolve over time by a similar mechanism to the small vesicles. However, there are no data available in the literature to support this statement. It is also clear that PEG can be concentrated in particular organs or cell types that are responsible for the clearance of the biological portion of the molecule. Even when this is reported to occur there is no evidence of functional or adverse events associated with this finding. This indicates the safety of PEG and the sequestered nature of the PEGylated material.

Metabolism of PEGylated biological; how and where will PEG be released?

While there are no data in the literature demonstrating the route by which PEG is released from PEGylated biologicals, there is clear evidence that PEGylated proteins are degraded in animals and human, probably by proteases [36] and this metabolism should lead to the release of PEG. Experiments with Pegasys® demonstrate that following the administration of Pegasys® there was no free PEG or interferon circulating in rats [34]. Extensive metabolism studies have also been Modi et al. [44] on Pegasys® in the rat. Experiments were carried out using ^{14}C lysyl labelled PEG (40 kDa) interferonα2a. Following administration of the radiolabel samples of injection sites, blood, liver, kidney and spleen were collected at 0.5, 24 and 336 h post dose and assayed with a range of technologies including immuno-staining, SDS PAGE, western blots and immunoprecipitation. These data demonstrated that the liver was the primary organ responsible for the metabolism of the Pegasys® in the rat. Pegasys® was metabolised in the rat to fragments of the interferon molecule attached to the PEG. These metabolites of Pegasys® were subsequently excreted in the urine. Based on these observations it is believed that the mechanism by which PEG is released from the protein is likely to be by degradation of the protein portion of the molecule by proteases and similar enzymes until the PEG itself is cleaved from the amino acid backbone of the molecule possibly by hydrolysis of the linker moiety.

Work carried out with TNF binding protein supports the hypothesis that PEGylated biologicals are cleared in this manner [38]. Experiments have demonstrated the presence of vacuoles formed from lyzosomes. Immunohistochemical analysis of these vesicles demonstrated that immediately after dosing TNF binding protein is present in these vacuoles. However after the 2 month reversibility period these vacuoles are no longer positive for TNF binding protein and are thought to contain just the PEG. These vacuoles are formed in the kidney and are believed to be a product of the target clearance of the PEGylated conjugate. Pegylated (20 kD) TNF binding protein is filtered at the glomerulus and then reabsorbed in the kidney tubules. Following readsorption the molecule is targeted to the lysozome for degradation. In this compartment the TNF binding protein is degraded by proteases leaving the PEG. As PEG of a molecular weight of 20 kD is highly resistant to metabolism the PEG persists in these vesicles. It is likely that over time the PEG in these vesicles will be released back to the systemic circulation either by diffusion, the redistribution of the vesicle to the cell surface or by apotosis of the cell. Evidence for the persistence of these vacuoles indicates that this redistribution process is very slow with small vacuoles resolving in 2 months. However large vacuoles persist for more than 2 months in the rat [38]. It is not clear whether this indicates that these vacuoles are permanent or whether 2 months is insufficient time for the PEG to redistribute out of the cell. Higher molecular weight PEGylated TNF binding protein conjugates (50 kD PEG) demonstrated less vacuole formation. This was believed to be due to the reduced role of the kidney in the clearance of the PEGylated biological or perhaps a reduced rate of presentation. This indicates the link between the clearance mechanism and the formation of these vesicles.

These vacuoles have also been reported in other tissues including liver and spleen and in other cell types such as macrophages [39]. Again there is evidence that the appearance of these vacuoles in these tissues is driven by the role of these tissues, cells or organs in the clearance of the PEGylated biological. It is therefore highly likely that the mechanism underlying the formation and disappearance of these vacuoles in these tissues is essentially the same as that detailed in kidney above.

The clearance mediated disposition of PEG from PEGylated biologicals may alter the site/s of PEG disposition in the body, especially if the clearance mechanism targets PEGylated biologicals to organs other than the kidney. While this may initially appear to be a concern, it is likely clearance organs of the majority of PEGylated biologicals will be conserved across species. Generally two mechanisms are responsible for the clearance of biologicals. These are non specific processes that act on the molecule and are highly species conserved. There are also clearance mechanisms that are related to binding to the target pharmacological receptor. While this can show marked species differences it is also a key component of the selection of toxicology species to ensure that interaction and distribution of the target receptor are the same in animals and humans. Given the above it is likely that the clearance mechanism of a PEGylated biological in toxicology species and in human will be highly conserved.

PEG exposures for biological products, excipient and associated with toxicity

In order to put in to context the exposure of PEG from PEGgylated biological Webster et al. [1] reviewed the relative exposures of a range of PEG containing products. In carrying out this analysis it has been assumed that the toxicity and targeting of PEG released from a biological and that seen for free PEG will be similar. While there is evidence of changes in PEG distribution related to clearance mechanism (PEG containing vacuoles) these changes appear to have little toxicological consequence and have not been considered in this analysis. It is also likely that in the long-term PEG targeted to different organs will either leave the body largely unchanged via the kidneys (predominantly) or liver in a similar manner to PEG itself. The products used in this analysis include:

1. PEGylated biologicals, Somavert®, Neulasta®, Pegasys® and PEG-Intron®.
2. A range of intravenous medicines which contain PEG as excipients; Busuflex®, Vepesid®, Dipyrimidamole® and Ativan®.
3. Products that contain PEG as an impurity Venoglobin-s® and Aralast®.

4. Finally exposures of PEG that are associated with toxicity from the two most relevant literature example for the medicines nitrofurantoin and lorazopam (Ativan®).

These data are presented graphically in Figure 4 and clearly demonstrate the large safety margins that exist between the exposures seen for current PEGylated biologicals and those associated with toxicity. These data also demonstrate that at no point do the potential exposures of PEG from PEGylated products ever become greater than that achieved for an array of products that have PEG impurities or as excipients, again demonstrating the safety of PEGylated biological products.

Overall, for PEGylated products the PEG exposures lie below the estimated safe doses of PEG as presented by the World Health Organisation [9] and are no higher than the exposure multiples for PEG use as an excipient or as an impurity. Finally, the exposure of PEG from PEGylated biologicals is approximately 600-fold lower than the PEG exposures associated with toxicity in human. These data indicate that given the low exposures of PEG that arise from the administration of PEGylated biologicals and the excellent safety seen for PEG *in vivo*, it is unlikely that adverse events will occur in human specifically as a result of PEG from these products.

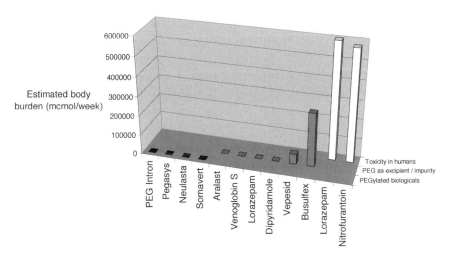

Figure 4. Comparisons of the exposures of PEG for various PEGylated and PEG containing products; comparison of those associated with toxicity

Conclusions

PEG is frequently used to improve pharmacokinetic properties of biological (proteins, peptides, aptimers, etc.). Concern has been expressed about the toxicological effect and/or fate of the PEG. This paper reviews the available toxicity, metabolism and clearance data of PEG and PEGylated products in order to place such concerns in to appropriate context.

Review of the available data demonstrates that PEG itself only shows toxicity at high, parenteral, doses and the usual target organ is the kidney as this is the route of excretion for unchanged PEG. Therefore, if PEG is removed from a biological it is likely to be predominately excreted unchanged and in this case renal toxicity is the most likely result, if toxicity occurs. Assuming the simple peptide degradation, PEG excretion relationship described above can be shown then data indicate that a therapeutic window of approximately 600-fold exists between the maximum PEG burden from a current biological agent and the doses of PEG associated with human toxicity. This window is large and will not readily be eroded. However, there are examples where pathological changes, PEG containing vacuoles, have been demonstrated to be due to the PEGylated biological (not PEG alone or the biological alone). To date these vesicles have not resulted in functional deficit of the affected organ. It is likely that the target organ for vesicle formation will depend on the pharmacological action of the compound, due to target mediated receptor uptake or due to specific uptake mechanisms. Initially the vacuoles contain the PEGylated biological but in time, following degradation of the biological component contains PEG alone. There is evidence that these PEG vesicles can resolve in time indicating that PEG may be removed from the cell over a prolonged period. This PEG is then likely to again be excreted largely in the kidneys in which case the earlier arguments about PEG toxicology and elimination apply. Large vesicles have not been shown to resolve, whether this is due to these vesicles being a permanent change to the cell or inadequate time allowed for resolution to occur is not clear. To be better able to define disposition of PEGylated biologicals, new bioanalytical techniques are required, this would enable extrapolation from animal to man for risk assessment purposes. To date, it has been found that immunochemical and mass spectrometric techniques cannot be used for such *in vivo* analyses.

In conclusion the doses used clinically for current PEGylated biologicals and for many future PEGylated biologicals will remain low and will result in exposures to PEG significantly lower than that required to elicit PEG toxicity. In all cases the routine regulatory toxicology studies would identify relevant pathology should it occur.

References

1. Webster R, Didier E, Harris P, Siegel N, Stadler J, Tilbury L, Smith D (2007) PEGylated proteins: evaluation of their safety in the absence of definitive metabolism studies. *Drug Metab Dispos* 35:9–16
2. Mehvar R (2000) Modulation of the pharmacokinetics and pharmacodynamics of proteins by polyethylene glycol conjugation. *J Pharm Sci* 3:125–136
3. Veronese FM, Pasut G (2005) PEGylation, successful approach to drug delivery. *Drug Discov*

Today 10:1451–1458
4. Roberts MJ, Bentley MD, Harris JM (2002) Chemistry for peptide and protein PEGylation. *Adv Drug Deliv Rev* 54:459–476
5. Morpurgo M, Veronese FM (2004) Conjugates of peptides and proteins to polyethylene glycols. *Methods Mol Biol* 283:45–70
6. Fontana A, Spolaore B, Mero A, Veronese FM (2008) Site-specific modification and PEGylation of pharmaceutical proteins mediated by transglutaminase. *Adv Drug Deliv Rev* 60:13–28
7. Veronese FM, Harris JM (2008) Peptide and protein PEGylation III: advances in chemistry and clinical applications. *Adv Drug Deliv Rev* 60:1–2
8. Veronese FM, Harris JM (2002) Introduction and overview of peptide and protein pegylation. *Adv Drug Deliv Rev* 54:453–456
9. Fruijtier-Polloth C (2005) Safety assessment on polyethylene glycols (PEGs) and their derivatives as used in cosmetic products. *Toxicology* 214:1–38
10. Alden CL, Frith CH (1991) Urinary system. In: Haschek WM, Rousseaux CG (eds): Handbook of Toxicology Pathology. Acedemic Press, New York, 339–341
11. *PegIntron EPAR*. Available at: http://www.emea.europa.eu/humandocs/PDFs/EPAR/Pegintron/024400en6.pdf. (Accessed 07 April 2008)
12. *Polyethylene glycols (Who Food Additives Series 14)*. Available at: http://www.inchem.org/documents/jecfa/jecmono/v14je19.htm. (Accessed 07 April 2008)
13. Descamps C, Cabrera G, Depierreux M, Deviere J (2000) Acute renal insufficiency after colon cleansing. *Endoscopy* 32:S11
14. Erickson TB, Aks SE, Zabaneh R, Reid R (1996) Acute renal toxicity after ingestion of Lava light liquid. *Ann Emerg Med* 27:781–784
15. Bruns DE, Herold DA, Rodeheaver GT, Edlich RF (1982) Polyethylene glycol intoxication in burn patients. *Burns Incl Therm Inj* 9:49–52
16. Herold DA, Rodeheaver GT, Bellamy WT, Fitton LA, Bruns DE, Edlich RF (1982) Toxicity of topical polyethylene glycol. *Toxicol Appl Pharmacol* 65:329–335
17. McCabe W, Jackson G, Hans G (1959) Treatment of chronic pyelonephritis. *Arch Intern Med* 104:710–719
18. Laine GA, Hossain SM, Solis RT, Adams SC (1995) Polyethylene glycol nephrotoxicity secondary to prolonged high-dose intravenous lorazepam. *Ann Pharmacother* 29:1110–1114
19. Herold DA, Keil K, Bruns DE (1989) Oxidation of polyethylene glycols by alcohol dehydrogenase. *Biochem Pharmacol* 38:73–76
20. Schaffer C, Critchfield F, Nair J (1950) The absorption and excretion of a liquid polyethylene glycol. *J Am Pharm Assoc Sci Ed* 39:340–344
21. Cruzan G, Corley RA, Hard GC, Mertens JJ, McMartin KE, Snellings WM, Gingell R, Deyo JA (2004) Subchronic toxicity of ethylene glycol in Wistar and F-344 rats related to metabolism and clearance of metabolites. *Toxicol Sci* 81:502–511
22. Guo C, McMartin KE (2005) The cytotoxicity of oxalate, metabolite of ethylene glycol, is due to calcium oxalate monohydrate formation. *Toxicology* 208:347–355
23. Na DH, Lee KC (2004) Capillary electrophoretic characterization of PEGylated human parathyroid hormone with matrix-assisted laser desorption/ionization time-of-flight mass spectrometry. *Anal Biochem* 331:322–328
24. Dou H, Zhang M, Zhang Y, Yin C (2007) Synthesis and purification of mono-PEGylated insulin. *Chem Biol Drug Des* 69:132–138
25. Tanaka K, Waki H, Ido Y, Akita S, Yoshida Y, Yohida T (1988) Protein and polymer analyses up to m/z 100,000 by laser ionization time of flight mass spectrometry. *Rapid Communications in Mass Spectroscopy* 2:151–153
26. Weaver R, Riley RJ (2006) Identification and reduction of ion suppression effects on pharmacokinetic parameters by polyethylene glycol 400. Rapid Commun *Mass Spectrom* 20:2559–2564
27. Hunt DF, Giordani AB, Rhodes G, Herold DA (1982) Mixture analysis by triple-quadrupole mass spectrometry: metabolic profiling of urinary carboxylic acids. *Clin Chem* 28:2387–2392
28. Keevil BG, Thornton S (2006) Quantification of urinary oxalate by liquid chromatography-tandem mass spectrometry with online weak anion exchange chromatography. *Clin Chem* 52:2296–2299
29. Tsai NM, Cheng TL, Roffler SR (2001) Sensitive measurement of polyethylene glycol-modified proteins. *Biotechniques* 30:396–402
30. Wunderlich DA, Macdougall M, Mierz DV, Toth JG, Buckholz TM, Lumb KJ, Vasavada H (2007)

Generation and characterization of a monoclonal IgG antibody to polyethylene glycol. *Hybridoma (Larchmt)* 26:168–172

31. Chapman AP (2002) PEGylated antibodies and antibody fragments for improved therapy: a review. *Adv Drug Deliv Rev* 54:531–545

32. Bailon P, Palleroni A, Schaffer CA, Spence CL, Fung WJ, Porter JE, Ehrlich GK, Pan W, Xu ZX, Modi MW, Farid A, Berthold W, Graves M (2001) Rational design of a potent, long-lasting form of interferon: a 40 kDa branched polyethylene glycol-conjugated interferon alpha-2a for the treatment of hepatitis C. *Bioconjug Chem* 12:195–202

33. Kellner K, Tessmar J, Milz S, Angele P, Nerlich M, Schulz MB, Blunk T, Gopferich A (2004) PEGylation does not impair insulin efficacy in three-dimensional cartilage culture: an investigation toward biomimetic polymers. *Tissue Eng* 10:429–440

34. *Pegasys EPAR*. Available at: http://www.emea.europa.eu/humandocs/PDFs/EPAR/pegasys/199602en6.pdf. (Accessed 07 April 2008)

35. *Neulasta EPAR*. Available at: http://www.emea.europa.eu/humandocs/PDFs/EPAR/neulasta/296102en6.pdf. (Accessed 07 April 2008)

36. *Somavert EPAR*. Available at: http://www.emea.europa.eu/humandocs/PDFs/EPAR/somavert/486302en6.pdf (Accessed 07 April 2008)

37. *Macugen EPAR*. Available from at: http://www.emea.europa.eu/humandocs/PDFs/EPAR/macugen/H-620-en6.pdf. (Accessed 07 April 2008)

38. Bendele A, Seely J, Richey C, Sennello G, Shopp G (1998) Short communication: renal tubular vacuolation in animals treated with polyethylene-glycol-conjugated proteins. *Toxicol Sci* 42:152–157

39. Young MA, Malavalli A, Winslow N, Vandegriff KD, Winslow RM (2007) Toxicity and hemodynamic effects after single dose administration of MalPEG-hemoglobin (MP4) in rhesus monkeys. *Transl Res* 149:333–342

40. Schaer DJ, Schaer CA, Buehler PW, Boykins RA, Schoedon G, Alayash AI, Schaffner A (2006) CD163 is the macrophage scavenger receptor for native and chemically modified hemoglobins in the absence of haptoglobin. *Blood* 107:373–380

41. Shum KL, Leon A, Viau AT, Pilon D, Nucci M, Shorr RG (1996) The physiological and histopathological response of dogs to exchange transfusion with polyethylene glycol-modified bovine hemoglobin (PEG-Hb). *Artif Cells Blood Substit Immobil Biotechnol* 24:655–683

42. Conover C, Lejeune L, Linberg R, Shum K, Shorr RG (1996) Transitional vacuole formation following a bolus infusion of PEG-hemoglobin in the rat. Artif Cells *Blood Substit Immobil Biotechnol* 24:599–611

43. Viau AT, Abuchowski A, Greenspan S, Davis FF (1986) Safety evaluation of free radical scavengers PEG-catalase and PEG-superoxide dismutase. J *Free Radic Biol Med* 2:283–288

44. Modi MW, Fulton JS, Buckmann DK (2000) Clearance of PEGylated (40KDa) interferone alfa-2a (PEGASYS) is primarily hepatic. *Hepatology* 32:371A

The occurrence, induction, specificity and potential effect of antibodies against poly(ethylene glycol)

Jonathan K. Armstrong

Department of Physiology and Biophysics, Keck School of Medicine, University of Southern California, Los Angeles, California 90033, USA

Abstract

Specific antibodies against poly(ethylene glycol) (anti-PEG) were induced in animals following expo-sure to PEG-conjugated proteins and particles, resulting in rapid clearance of PEG-conjugated agents. In humans, induction of anti-PEG was observed following exposure to a PEG-conjugated drug, and pre-existing anti-PEG was identified in over 25% the healthy population. In clinical studies, the pres-ence of anti-PEG was strongly associated with rapid clearance of PEG-asparaginase and PEG-uricase. PEGylation of therapeutic agents will continue to be of significant value in medicine to reduce immunogenicity, antigenicity and toxicity as well as markedly reducing renal clearance, while main-taining drug efficacy. It is important to recognize that PEG itself may possess antigenic and immuno-genic properties. Further comprehensive studies are warranted to fully elucidate the effect of anti-PEG on PEG-conjugated agents and if confirmed in a prospective trial, patients should be screened and monitored for anti-PEG, and strategies developed to overcome the potential negative effect of anti-PEG on drug clearance to improve the effectiveness of therapy.

Introduction

Many therapeutic agents are highly immunogenic, antigenic, toxic or rapidly cleared by the host posing significant hurdles for the effective and safe treat-ment of a disease. One very successful method to overcome these hurdles is the covalent attachment of a synthetic nonionic polymer, poly(ethylene gly-col), to the drug. Poly(ethylene glycol) (PEG) is one of the most widely used biocompatible polymers with a long history of use in the food, cosmetic and pharmaceutical industries [1–4].

Thirty years ago, Abuchowski and colleagues [5, 6] were the first to describe the application of covalent attachment of PEG to exogenous proteins, the process that is now commonly referred to as 'PEGylation'. They demon-strated a significant increase in the circulatory time and that immunogenicity and antigenicity was almost abolished for PEGylated proteins *versus* the unmodified protein [5, 6].

The covalent attachment of PEG to a wide range of therapeutic agents is now commonly employed including PEG-asparaginase (PEG-ASNase, [Pegaspargase, Oncaspar®]) for leukemia and lymphoma [7–9], PEG-interfer-on alpha 2a (PEGASYS®) and 2b (PEGINTRON™) for chronic hepatitis C virus [10–12], PEG-filgrastim (Neulasta®) for neutropenia [13–15] and PEG-adenosine deaminase (Adagen®) for severe combined immunodeficiency syndrome [16, 17]. In addition, numerous other PEGylated agents are in clinical use or development [18–34].

PEG has long been considered as an inert biocompatible polymer, however antibodies against PEG (anti-PEG), either pre-existing or induced in humans and induced in rabbits, were first described by Richter and Åkerblom 25 years ago, suggesting that PEG is both antigenic and immunogenic [35, 36]. Studies in animal models demonstrated that induced anti-PEG was associated with rapid clearance of PEG-conjugated particles and drugs [37–53], and, in humans, both pre-existing and induced anti-PEG was associated with the rapid clearance of PEG-conjugated drugs [54–56].

While PEGylation is clearly of great value to medicine, and has significantly improved treatments for many diseases by achieving sustained and effective activity of drugs that would otherwise be rapidly cleared from a patient or give unacceptable toxicity, the emergence of anti-PEG and its potential impact on the effectiveness of treatment with PEGylated therapeutics requires careful attention. With the growing body of studies regarding pre-existence and induction of anti-PEG, the purpose of this chapter is to summarize the published studies, to discuss the implication of anti-PEG on PEGylated therapeutic agents and the possible approaches to ensure effective therapy for patients identified as positive for anti-PEG.

PEG-conjugated drugs – current market

In 1990, PEG-adenosine deaminase (Adagen®) was the first FDA approved drug for treatment of severe combined immune deficiency (SCID) [17]. Subsequently many PEG-conjugated drugs are in widespread use and clinical development [7–34].

From available company annual reports [57–61], in 2007, sales of seven FDA-approved PEG-conjugated drugs were $5.8 billion. This class of drugs is dominated by PEG-filgrastim (Neulasta) for the treatment of febrile neutropenia ($3 billion [57]) and PEG-interferon (PEGINTRON™ and PEGASYS®) for the treatment of Hepatitis-C ($2.5 billion [58, 59]), followed by a PEG-liposome formulation of doxorubicin (CAELYX®) for the treatment of breast/ovarian cancer and AIDS-related Kaposi's sarcoma ($257 million [58]), PEG-asparaginase (Oncapsar®) for the treatment of acute lymphoblastic leukemia ($38.7 million [60]), PEG-adenosine deaminase (Adagen®) for the treatment of SCID ($24.5 million [60]) and PEG-anti-VEGF aptamer (pegaptanib, Macugen®) for the treatment of age-related macular degeneration ($18 million [61]). Over the past 3 years, sales of PEG-conjugated drugs have increased between 10–15% per annum, and when combined with the recent approval of PEG-erythropoietin (Mircera®, FDA approval in November 2007), the development of other potential blockbuster drugs, e.g., PEG-insulin [25], and the broader application of existing PEG-conjugated drugs (e.g., FDA approval in July 2006 for PEG-asparaginase as a front-line treatment for acute lymphoblastic leukemia), the market for PEG-conjugated drugs will likely markedly increase in the near future.

Animal models of anti-PEG induction

Induction of anti-PEG has been achieved with PEG-conjugated proteins [36, 37, 41, 42, 62–64], PEG-liposomes [44–53] and PEG-conjugated red blood cells (PEG-RBCs) [39, 40] in mice, rats and rabbits. A variety of techniques have been employed to detect anti-PEG and determine antibody specificity including passive hemagglutination, gel diffusion, ELISA, serology and flow cytometry.

   Richter and Åkerblom performed a comprehensive series of studies to characterize anti-PEG induced in rabbits. Ovalbumin, superoxide dismutase and ragweed pollen extracts, conjugated with mPEG of molecular mass 3,000 g/mol and 10,000 g/mol with 5–127 PEG chains per molecule were administered intramuscularly to rabbits with complete Freund's adjuvant [36]. Anti-PEG was induced in 18 of 34 rabbits dependent on the base protein and degree of modification. For example, modification of ovalbumin with 6 PEG chains of 10,000 g/mol produced both anti-PEG and anti-ovalbumin, whereas modification of ovalbumin with 20 PEG chains predominantly conferred non-immunogenic properties [36]. Two techniques were used to detect the presence of anti-PEG: (i) passive hemagglutination, where mPEG-coated red blood cells (mPEG-RBCs) were incubated with serial dilutions of sera. After 2 h, settling patterns were observed and recorded as the reciprocal of the highest serum dilution, giving complete hemagglutination, defined as anti-PEG positive for titers of 32 and above; and (ii) single, double and reversed single radial gel immunodiffusion. In the latter technique, PEG 40,000 g/mol was incorporated in the gel for antibody specificity studies [36]. To confirm specificity of anti-PEG, inhibition of passive hemagglutination of mPEG-RBCs was investigated by pre-incubation of rabbit sera with PEG of 4,000, 15,000 and 5.9 million g/mol over a range of concentrations, and readings of settling patterns were recorded at 2 and 24 h. Inhibition of passive hemagglutination was observed with all PEGs investigated irrespective of molecular weight [36]. Cross-reactivity was also observed as Pluronic® F68 (a PEG-containing block copolymer) gave precipitate lines with anti-PEG positive rabbit sera in the gel double diffusion assay which completely fused with those produced by PEG 6,000 g/mol [36]. Epitope studies were performed using double gel diffusion by observation of precipitate formation between rabbit sera and PEG of 300 g/mol to 6 million g/mol. No precipitate formation was observed with PEG 300 or 2,000 g/mol, but precipitate formation was observed with PEGs of ≥4,000 g/mol with the strongest bands observed for PEG 40,000 g/mol [36]. Additional epitope studies were performed using reversed single radial gel immunodiffusion, following the formation of precipitate lines with rabbit sera and PEG 40,000 g/mol, a solution of PEG of 300 or 2,000 g/mol was poured on the gel and left for 24 h, which resulted in complete dissolution of the bands. From these combined results, the authors concluded that following repeated challenges with methoxy-PEG conjugated proteins, anti-PEG induced in rabbits had an epitope equivalent to 6–7 repeat oxyethylene groups and bound to the PEG backbone [36]. The authors also sensitized guinea pigs with rabbit anti-PEG positive sera, followed by an intravenous challenge of 0.5 mg of PEG 40,000 g/mol after 24 h, resulting in typical signs of anaphylactic shock and death within 5 min of the challenge. Non-sensitized animals showed no adverse effects when injected with the same dose [36]. The effects of free, unconjugated PEG were also investigated and PEG of 10,000 and 100,000 g/mol were nonimmunogenic in rabbits, while PEG of 5.9 million g/mol was weakly immunogenic in mice [36].

A monoclonal anti-PEG IgM was induced in mice injected with PEG-conjugated β-glucuronidase-RH1 (where RH1 is a monoclonal antibody against AS-30D rat hepatoma cells) intravenously on days 1 and 54, and by intraperitoneal injection on days 7, 14 and 21 [41]. Following the final injection, spleen cells were excised, fused with myeloma cells, screened and then cloned. Monoclonal antibodies that were generated from the hybridoma cells, 1E8 (IgG1) and AGP3 (IgM), were further tested for specificity in ELISA assays, and showed that 1E8 was specific for β-glucuronidase, and AGP3 was specific for PEG [41]. Following a similar method as described before [41], a monoclonal anti-PEG IgG1 (E11) was induced in mice [42]. To determine the specificity of E11 and AGP3, immunoblots showed that AGP3 and E11 detected dansyl chloride-PEG-β-glucuronidase (PEG of molecular mass 3,400 g/mol), amine-PEG-amine of 1,500 g/mol and mPEG of 2,000 g/mol, but did not detect β-glucuronidase [42]. Specificity was also determined by ELISA assays with wells coated with methoxy-PEG-amine of 750 and 5,000 g/mol, amine-PEG-amine of 1,500 and 3,400 g/mol. Both E11 and AGP3 bound to all PEG coated wells except mPEG-amine 750 g/mol, suggesting that the binding epitope for both antibodies was 16 oxyethylene units [42]. Additional sandwich ELISA studies with E11 and AGP3 demonstrated the ability to quantitatively detect dansyl chloride-PEG-β-glucuronidase (PEG of molecular mass 3,400 g/mol), mPEG 20,000 g/mol, PEG 2,000 g/mol conjugated-quantum dots and PEG 2,000 g/mol-liposomes, confirming that both anti-PEG IgG and IgM bind to the PEG backbone [42].

Martinez and colleagues showed induction of anti-PEG in rabbits following repeated injection of a partially PEGylated recombinant uricase, (two chains of mPEG 10,000 g/mol per protein subunit, sufficient to yield solubility, but insufficient to suppress immunogenicity) [37, 64]. Employing an ELISA assay with bound mPEG conjugated to a structurally unrelated protein, the authors identified anti-PEG IgG in rabbit sera specific for the methoxy-group. The specificity of the induced anti-mPEG was confirmed in competition studies by pre-incubation of rabbit sera with methoxy-PEGs of various molecular weights. The data showed equivalent inhibition of binding with different mPEG on a molar basis and a minimal effect with dihydroxy-PEGs, confirming methoxy-PEG specificity [37].

Studies from our laboratories with PEG-RBCs in autologous transfusion studies in rabbits also demonstrated induction of anti-PEG [39, 40]. Blood was drawn from rabbits (n = 17), RBCs washed and then PEG coated using a succinimidyl propionate derivative of linear mPEG 20,000 g/mol over a range of concentrations from 2.5–20 mg/mL or cross-linked PEG coatings prepared by incubation of RBCs with an eight-multi-armed PEG-amine of 10,000 g/mol followed by addition of a succinimidyl propionate derivative of an eight-multi-armed PEG of 40,000 g/mol, with conditions carefully controlled to prevent gelation [40]. PEG-coated and uncoated (Control) RBCs were labeled with fluorescent lipophilic carbocyanine dyes (Vybrant DiI or DiD, (Invitrogen, Carlsbad, California, USA)), such that each rabbit received discretely labeled PEG-coated and uncoated RBCs to track RBC survival. Rabbits received up to seven consecutive infusions of PEG-coated RBCs, 60 days apart, alternating the DiI or DiD labeling for PEG-coated or control RBCs between each experiment. Plasma samples were taken at days 14–21 following infusion and screened for anti-PEG by serology with mPEG-coated rabbit RBCs and by flow cytometric analysis of 10 μm Tentagel®-OH beads, which are composed primarily of hydroxy-terminated PEG, incubated with sera, washed and stained for bound Igs by incubation with fluorescein isothiocyanate (FITC)-conjugated anti-rabbit IgG and R-phycoerthyrin (RPE)-conjugated anti-rabbit IgM. Samples with a mean fluorescence intensity of >100 for IgG and/or >50 for IgM were identified as positive for anti-PEG [40]. By the end of the study, 12 (71%) rabbits had developed anti-PEG. This was detected after the first infusion in six (35%) rabbits, after two infusions in four (24%) and after three or more infusions in two (12%). Two animals showed sustained (>1 year) anti-PEG (3+ and 2+ by direct agglutination); the remaining 10 showed a variable response that increased (up to 4+) following a challenge with PEG-RBCs. Five rabbits showed no evidence of anti-PEG induction, four of which received three infusions, and one seven infusions. The strongest anti-PEG responses were observed after multi-arm branched PEG-RBCs were infused. Rabbits with a 4+ response cleared PEG-RBCs of any type within two days; those with a weaker anti-PEG response (2+ or less) rapidly cleared all PEG-RBCs except the 2.5 mg/mL mPEG-coated cells [39, 40]. Figure 1 shows an example of the effect of anti-PEG on the survival of mPEG-coated autologous RBCs normalized by survival of uncoated (Control) autologous RBCs. Rabbit A, with a sustained anti-PEG (3+ by serology), shows rapid clearance of PEG-RBCs within three days. Rabbits B and C are naïve rabbits, rabbit B showed development of anti-PEG at day 7, resulting in rapid clearance of PEG-RBCs while rabbit C showed no evidence of anti-PEG or accelerated clearance of PEG-RBCs compared to uncoated RBCs.

After a single exposure to PEG-stabilized liposomes, rapid clearance of subsequent doses of PEG-liposomes was observed [44–53]. This was attributed to the induction of anti-PEG IgG and IgM [45, 46, 48–53]. However, it is

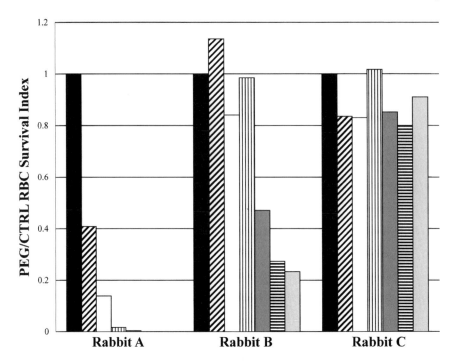

Figure 1. Effect of anti-PEG on the survival of mPEG-coated autologous RBCs normalized by uncoated (Control) RBCs at Days 0 (■), 1(▨), 2(☐), 3 (▥), 7 (■), 15 (▤) and 28 (▨). Rabbit A showed sustained anti-PEG (3+ by serology) with rapid clearance of PEG-RBCs within three days. Rabbits B and C are naïve rabbits, with rabbit B showing development of anti-PEG at day 7, resulting in rapid clearance of mPEG-RBCs and rabbit C with no evidence of anti-PEG shows no accelerated clearance of mPEG-RBCs compared to uncoated RBCs.

important to note that liposomes containing immunostimulatory agents are well known to act as potent adjuvants which can promote antibody responses against weakly immunogenic antigens [65–68]. Using mPEG 2,000 g/mol conjugated distearoylphosphatidylethanolamine (PEG2k-DSPE) [44, 46, 49–53], distearoylglycerol (PEG2k-DSG) [45], or dipalmitoylphospho-ethanolamine (PEG2k-DPPE) [48] as steric stabilizing materials for oligonu-cleotide-loaded liposomes, the authors noted an increasingly rapid clearance of PEG-liposomes upon repeated exposure in mice [44–46, 50–52], rats [49, 51–53] and rabbits [48] was attributed to generation of anti-PEG. An anaphylactic response was also observed in mice following a second exposure to PEG-liposomes with pronounced morbidity and mortality [44, 45] similar to that observed by Richter and Åkerblom following intravenous administration of free PEG to guinea pigs presensitized with anti-PEG rabbit sera [36]. Assays employed to detect induced antibodies included ELISA using immobilized lipid and PEG-lipid [45, 51], immobilized liposomes [44] as antigens, or by measuring adsorption of IgG by fresh PEG-liposomes incubated with sera [48]. Specificity of the induced antibodies towards PEG was confirmed by the absence antibody binding to unconjugated lipids [45, 51], inhibition of anti-PEG binding by pre-incubation of sera with free PEG [48] at extremely high concentrations (200 mg/mL of PEG 2,000 g/mol [48]) or lack of induced antibodies with non-PEG conjugated liposomes [44].

Interestingly, empty PEG-liposomes elicited no anti-PEG response in some studies [44, 45, 47] but induced anti-PEG in others [51]. As splenectomized rats did not show induction of anti-PEG [51, 69]; an anti-PEG IgM response was observed in the absence of T cells with immunodeficient athymic BALB/c mice [51, 53]; and as loaded liposomes are known to act as adjuvants, the induction of anti-PEG by PEG-conjugated liposomes was suggested as a T cell independent mechanism, and most likely that PEG-liposomes became trapped in the spleen by reactive B cells that triggered an IgM response [45, 51].

In summary, both anti-PEG and anti-mPEG IgG and IgM have been induced in naïve animals following exposure to PEG-conjugated agents, and animals with anti-PEG rapidly clear PEGylated agents. It is important to note that anti-PEG was readily and efficiently induced with PEG-conjugated particles (PEG-liposomes, PEG-RBCs) without the use of any co-administered adjuvant. Conversely, induction of anti-PEG with PEG-conjugated proteins required multiple exposures, usually combined with the use of Freund's adjuvant. This difficulty in anti-PEG induction was cited in one study [42]. Although the subsequent effect of induced anti-PEG is the same for PEG-particles and PEG-proteins, the mechanism of anti-PEG induction may likely be a T cell independent mechanism for PEG-particles following splenic sequestration, and a T cell dependent mechanism for PEG-proteins.

With the exception of one ELISA assay for the detection of anti-PEG [45], all other described ELISAs use Tween® 20 (0.05 or 0.1%) as a nonionic surfactant in the multiple washing steps following incubation with sera or incubation with secondary reagents [41, 42, 46, 48–53, 62, 63]. As Tween® 20 (polyoxyethylene-sorbitan monolaurate) contains three hydroxyl-terminated PEG chains and one PEG-laurate chain emanating from a sorbitan central group with an average of 5–6 oxyethylene groups per PEG chain, its use in the ELISA assays described may interfere with binding of anti-PEG to immobilized antigens.

Human studies – pre-existing anti-PEG

Only two studies to date investigated the occurrence of anti-PEG in the healthy blood donor population [35, 70]. In 1984, Richter and Åkerblom examined sera for the presence of anti-PEG from 453 healthy blood donors with approximately equal numbers from Japan, Germany and Italy and also from 92 allergic patients [35]. They employed a passive hemaggultination assay, where mPEG 6,000 g/mol-coated RBCs were incubated with serial dilutions of donor sera. After 2 h, settling patterns were observed and recorded as the reciprocal of the highest serum dilution giving complete hemagglutination and defined as anti-PEG positive for titers of 32 and above [35]. An occurrence of anti-PEG of 0.2% was determined in the healthy blood donor population and 3.3% in allergic patients. This was predominantly an anti-PEG IgM as near-complete abolition of agglutination was observed following treatment of sera with mercaptoethanol. No additional studies on the specificity or epitope identification were performed with anti-PEG positive sera from human subjects [35].

Some 20 years later, from studies in our own laboratories [38, 70] we determined an occurrence two orders of magnitude greater than previously shown, with anti-PEG detected in >25% of 350 healthy blood donors, predominantly of the IgG isotype [38, 39, 70], although anti-PEG IgM and both anti-PEG IgG and IgM were observed. Our findings of a 25% occurrence of anti-PEG is probably an underestimate as antibodies induced from environmental exposure to an antigen, such as anti-PEG, vary in response between donors and is likely to increase and decrease dependent on the last environmental challenge. As testing of a single serum sample reflects a snap-shot in time, it is possible that some anti-PEG positive samples were below the limits of detection for the assays we employed. In our studies, anti-PEG was determined using two techniques, serology with mPEG-RBCs and by flow cytometry with 10 μm diameter spherical PEG beads (Tentagel®-OH beads, RAPP Polymere GmbH, Tübingen, Germany). Donor plasma samples were incubated with mPEG-RBCs and uncoated (Control) RBCs. Agglutination of PEG-coated or uncoated (control) RBCs by each plasma sample was determined following centrifugation of the samples at $500 \times g$ for 1 min, and agglutination scored according to the 0–4+ scale [71]. By serology, 94 plasma samples (26.9%) agglutinated PEG-RBCs, 20 (5.7%) of which showed strong agglutination (3+ or 4+), with no agglutination observed with uncoated (Control) RBCs (Tab. 1). Plasmas were also incubated with 10 μm diameter spherical Tentagel®-OH beads, washed and stained for bound Igs by incubation with FITC-anti-human IgG and RPE-anti-human IgM, and examined by flow cytometry. Samples with a mean fluorescence intensity of >100 for IgG and/or >50 for IgM were identified as positive for anti-PEG. Similarly, flow cytometric analysis of Tentagel®-

Table 1. Serological testing for anti-PEG of normal donor sera with PEG-coated RBCs scored according to the 0–4+ agglutination scale [71]

Agglutination score	Number	Percentage
0	256	73.1
1+	31	8.9
2+w	26	7.4
2+	17	4.9
3+	12	3.4
4+	8	2.3
Negative	256	73.1
Positive	94	26.9
Total	350	100

Table 2. Flow cytometric analysis of TentaGel®-OH beads incubated in normal donor sera and stained for bound IgG and IgM. Mean fluorescence data were employed to identify the presence or absence of anti-PEG in test sera

Bound Igs	Number	Percentage
IgG only	67	19.1
IgM only	18	5.1
IgG + IgM	12	3.4
Negative	253	72.3
Positive	97	27.7
Total	350	100

OH particles showed 97 samples (27.7%) positive for IgG and/or IgM, of which 67 samples (19.1%) showed IgG binding only, 18 (5.1%) showed IgM only, and 12 (3.4%) showed both IgG and IgM uptake (Tab. 2). Further analysis of 11 anti-PEG IgG positive sera for IgG subtypes by flow cytometric analysis of TentaGel®-OH particles incubated in sera and stained for bound IgG-1, IgG-2, IgG-3 and IgG-4 with FITC-labeled antibodies, showed all anti-PEG sera to be positive for IgG-2, with one serum sample positive for IgG-1, IgG-2 and IgG-3, and no sera positive for IgG-4 [38, 70]. These findings are similar to anti-PEG induction following treatment with PEG-uricase where anti-PEG IgG-2 and both IgG-2 and IgG-3 were identified [55]. To confirm specificity, we further examined the ability of various polymers [PEGs, dextran, polyvinylalcohol (PvOH), poly(propylene glycol) (PPG)] and small ethers [di- to penta-(ethylene glycol); di- to tetra-(ethylene glycol) dimethyl ether] to inhibit agglutination of PEG-coated RBCs when pre-incubated with anti-PEG positive plasma [72]. Complete inhibition of agglutination of PEG-RBCs (from 4+ to 0) was observed in the presence of all PEGs (MW 300–20,000 g/mol), PPG (2,000 g/mol), tri- and tetra-(ethylene glycol)dimethyl ether and penta(ethylene glycol). Di(ethylene glycol)diethyl ether and tetra(ethylene glycol) reduced PEG-RBC agglutination (4+ to 2+). Dextran, PvOH, di- and tri-(ethylene glycol) had no effect [38, 72]. Comparison of the smallest inhibitors indicated that the minimum epitope required for binding of the PEG-antibody is a backbone of 4–5 repeat oxyethylene units (Tab. 3). This is in close agreement with Richter and Åkerblom for the anti-PEG induced in rabbits following injection of mPEG-conjugated proteins, showing an epitope of 6–7 repeat oxyethylene groups [36].

The remarkable increase in anti-PEG positive sera found in our studies [38, 39, 70] compared to the previous study in 1984 [35] may be explained by greater sensitivity of the techniques we employed. The passive hemagglutination method described by Richter and Åkerblom may only detect high-titer and/or high-affinity antibodies, and may be unable to detect weakly agglutinating IgGs as no amplification technique could be employed with this method (e.g., indirect antiglobulin test). Alternatively, the difference may reflect

Table 3. Agglutination of PEG-RBCs by three healthy blood donor plasma samples positive for anti-PEG (4+ reactive by serology with PEG-RBCs) in presence of 1% w/v PEG, PEG oligomers and analogs, poly(propylene glycol) (PPG), and irrelevant polymers (dextran 40 and polyvinyl alcohol (PvOH))

Polymer/compound added (all at 1% w/v)	Agglutination score			Ether/oligomer/polymer Backbone structure		
	Anti-PEG positive donor					
	A	B	C	◄——— Binding epitope ———►		
Ethlyene glycol	3+	4+	4+	HO–C ⋮ C–OH		
Diethylene glycol	3+	4+	4+	HO–C ⋮ C–O–CC–OH		
Triethylene glycol	2+	4+	3+	HO–C ⋮ C–O–CC–O–CC–OH		
Diethylene glycol dimethyl ether	*0+*	*2+*	*2+*	H₃ ⋮ C–O–CC–O–CC–O–CH₃		
Tetraethylene glycol	*1+*	*2+*	*1+*	HO–C ⋮ C–O–CC–O–CC–O–CC–OH		
Diethylene glycol diethyl ether	*0+*	*1+*	*1+*	H₃C ⋮ C–O–CC–O–CC–O–CCH₃		
Pentaethylene glycol	*1+*	*1+*	*0+*	HO–C ⋮ C–O–CC–O–CC–O–CC–O–CC–OH		
Triethylene glycol dimethyl ether	**0**	**0**	**0**	H₃ ⋮ C–O–CC–O–CC–O–CC–O–CH₃		
Tetraethylene glycol dimethyl ether	**0**	**0**	**0**	H₃ ⋮ C–O–CC–O–CC–O–CC–O–CC–O–C ⋮ H₃		
PEG 300 g/mol	**0**	**0**	**0**	HO–C ⋮ C–O–CC–O–CC–O–CC–O–CC–O–C ⋮ C–OH		
mPEG 5,000 g/mol	0	0	0	H₃C–O–C ⋮ C–O–(CC–O)₁₁₀–CC–O–CC–O–CC– ⋮ OH		
PEG 20,000 g/mol	0	0	0	HO–C ⋮ C–(O–CC)₄₅₀–O–CC–O–CC–O–CC– ⋮ OH		
PPG 2,000 g/mol	0	0	0	HO–C ⋮ C–(O–CC)₃₁–O–CC–O–CC–O–CC– ⋮ OH		
Dextran 40,000 g/mol	3+	4+	4+			
PvOH 25,000 g/mol	3+	4+	4+			
PBS (control)	4+	4+	4+			

increasing exposure to PEG and PEG-containing compounds due to their widespread use in agrochemicals (pesticides, herbicides and fertilizers), processed food products, cosmetics and pharmaceuticals.

Human studies – anti-PEG and PEG-conjugated drugs

Richter and Åkerblom investigated anti-PEG induced in patients following treatment with PEG-modified allergens (PEG-ragweed extract and PEG-bee venom) for hyposensitization therapy [35]. The presence of anti-PEG was detected using the passive agglutination assay described earlier. At baseline, 3.3% of allergic patients (total n = 92) were positive for anti-PEG. After one year of treatment with PEG-modified allergens (n = 58), the percentage of patients testing positive for anti-PEG jumped to 50%, and after two years of treatment (n = 28) the percentage of patients testing positive for anti-PEG dropped to 28.5%. The authors determined that the anti-PEG induced was an IgM isotype and as the response was moderate and not enhanced with subsequent exposure to PEG-modified allergens, they concluded that "*such an antibody response will most probably not interfere with the clinical usefulness of PEG-modified allergens in hyposenitization therapy*" [35]. This conclusion has been widely cited by other authors either directly or indirectly to apply to all PEG-conjugated drugs.

Recently, Ganson et al. [54] reported the induction of anti-PEG, detectable between 3–7 days after injection, in five of 13 (38.5%) patients treated for chronic refractory gout with mPEG-uricase, with no anti-uricase Igs detected in any of the 13 patients. Rapid clearance of mPEG-uricase was observed in the anti-PEG positive patients, identified as a low titer anti-PEG IgM detectable between 3–7 days, and anti-PEG IgG detectable between 7–14 days after the initial exposure to mPEG-uricase, characteristic of a T cell mediated immune response. An ELISA test was used to detect anti-PEG, with positives identified as those with Ig binding to immobilized mPEG-uricase and the absence of Ig binding to immobilized uricase. Confirmation of anti-PEG specificity was achieved by pre-incubation of sera with various agents showing complete inhibition of the ELISA with mPEG 10,000 g/mol, dihydroxy-PEG of 10,000 g/mol and PEG 10,000 g/mol conjugated with glycine; strong inhibition of the ELISA with a PEGylated purine nucleoside; and no effect on the ELISA with unmodified recombinant uricase [54]. Additionally, the effect of pre-incubation sera with mPEGs was investigated over a range of molecular weights and concentrations. Methoxy-PEGs inhibited the ELISA in the order 10,000 = 5,000 > 2,000 > 350 g/mol, which combined with the specificity experiments, indicate that the anti-PEG induced following treatment with mPEG-conjugated uricase binds to the backbone of the PEG irrespective of the end group moiety [54]. In a subsequent Phase I trial of a single intravenous dose of PEG-uricase [55], anti-PEG was induced in nine of 24 (37.5%) patients detected at 14 and 35 days after administration of the drug, showing an almost identical induction rate as described before [54]. No patients showed antibodies to either PEG or uricase prior to drug exposure [55]. The induced anti-PEG was predominantly IgG-2, and both IgG-2 and IgG-3 was observed in some patients, detected by employing the same ELISA assay as previously described [54] by binding of Igs to immobilized antigens, PEG-uricase and mPEG 10,000 g/mol. The authors did not observe any allergic responses to PEG-uricase in this study [55] as previously reported [54], however, as only a single infusion of PEG-uricase was studied no allergic reactions to PEG were anticipated [55].

In our own laboratory, we analyzed stored sera samples from 28 pediatric patients with acute lymphoblastic leukemia, treated with PEG-asparaginase (PEG-ASNase, Oncaspar) [56]. The samples were selected to include 15 subjects with undetectable ASNase activity after receiving PEG-ASNase. Sera were tested for the presence of anti-PEG by serology with mPEG-RBCs and by flow cytometry by detection of Igs bound to 10 µm diameter PEG-beads (as described earlier). All testing for anti-PEG was performed in a blinded manner, and the data was combined with ASNase activity results once all testing for anti-PEG was completed. Of the 15 sera from PEG-ASNase treated patients with undetectable ASNase activity, anti-PEG was detected in nine (60%) by serology and in 12 (80%) by flow cytometry [56]. Anti-PEG was also detected in one PEG-ASNase treated patient with lower ASNase activity (123 U/L). Twelve patients in the PEG-ASNase group had no documented prior exposure to PEG-ASNase, seven (58%) had undetectable ASNase activity of which four (33%) were positive for anti-PEG. All sera that were positive by serology were also positive for both IgG and IgM anti-PEG, and the observations were indicative of a pre-existing anti-PEG not induced by treatment. The presence of anti-PEG was shown to be a good predictor of low ASNase activity, and equally important, the absence of anti-PEG as a good predictor of high ASNase activity (positive/negative predictive value of anti-PEG for ASNase activity of 92%/80% and 100%/68% for flow cytometric and serological determinations respectively) with an overall accuracy of 86% for flow cytometric detection of anti-PEG and 79% and serological detection of anti-PEG (at the 95% confidence interval) [56].

Toxicity, safety and disposition of PEG

Although PEGs are widely considered as non-immunogenic and non-antigenic, the finding of a pre-existing anti-PEG in the healthy blood donor pool of 0.2% in 1984 [35] and over 100-fold higher some 20 years later from our own studies showing an occurrence of anti-PEG IgG and IgM of >25% would suggest the contrary [38, 39, 70]. In order to investigate the reasons for existence of antibodies against PEG in the healthy population who have not been exposed to PEG-conjugated drugs, it is of value to consider the environmental exposure to PEGs and PEG-containing compounds in everyday use, and the disposition of PEG-compounds in both animal models and human studies.

For over 60 years, the industrial use of PEG homopolymers, PEG- and polyoxyethylene-adducts, block and random copolymers as well as ethoxylated compounds is widespread in agrochemicals (pesticides, fungicides and fertilizers), in processed foodstuffs, household chemicals (e.g., laundry detergents), cosmetics and pharmaceuticals. The widespread application of oxyethylene-based polymers stems from their low toxicity, low irritability, inert properties, stability, versatility and solubility in wide range of solvents and also lack of color and flavor. Low molecular mass PEGs act as non-aqueous solvents and high molecular mass PEGs act as viscosity enhancing agents. PEG containing copolymers are used as viscosity stabilizers, emulsifying/demulsifying agents, foaming/defoaming agents, wetting agents, solubilizing agents, humectants, skin conditioning agents, cleaning and gelling agents, while ethoxylation of oils and hydrophobic compounds yields emulsifiability and solubility in an aqueous environment. Common examples of PEG-containing nonionic surfactants are PEG-conjugated alkyl chains and fatty acids, commonly known by the industrial trade names, Tween®, Triton®, Brij®, Myrj® and Span®. Common block copolymers containing PEG and poly(propylene glycol) are poloxamers (Pluronic® block copolymers), meroxapols (Pluronic-R® block copolymers) and poloxamines (Tetronic® block copolymers). The low acute and chronic toxicity, skin and eye irritability of PEGs [1–4, 73, 74] and PEG-containing compounds [4, 73, 75, 76] confirm their safe use in cosmetic products for external use, as additives for food products and excipients in pharmaceutical applications.

From disposition studies, the primary route of excretion is by renal filtration [77–80] for low molecular mass PEGs up to 6,000 g/mol, with a short circulatory half-life of 18 min [79]. With increasing molecular mass of PEG, the circulatory half life increases (e.g., 23 h for PEG of 170,000 g/mol), with increased phagocytosis and accumulation in the Kupffer cells above a PEG of molecular mass 50,000 g/mol [79]. It is also interesting to note the disposition of PEG-containing block copolymers (poloxamers), particularly poloxamer 188 (Pluronic® F68) that has been widely used in pharmaceutical preparations and in the clinical setting [81–83]. For Pluronic® F38 and Pluronic® F68, following intravenous injection, the primary route of excretion was via renal filtration (40–94% after 24 h) with a small amount in the bile and feces (2–6%) [83–85]. In human studies, Pluronic® F68 was well tolerated following intravenous administration at infusion levels of 100–150 g/day and was excreted, unchanged, primarily via the kidney [84, 86].

It is perhaps not surprising that antibodies against PEG are detected in healthy humans with no prior exposure to PEG-conjugated drugs [35, 38, 39, 70] considering the likely environmental exposure to PEG and PEG-containing compounds. PEG has long been considered non-antigenic and non-immunogenic as generation of anti-PEG in animal models has proved difficult, and in the studies cited in this chapter, induction of anti-PEG in animal models by PEG-conjugated proteins involved both multiple exposures to the PEGylated protein and, in most studies, the use of Freund's adjuvant (see earlier). The lack of generation of anti-PEG with unconjugated PEGs in animal models is likely caused by the rapid clearance of low molecular mass PEGs primarily via renal excretion [79]. However, conjugation of PEG to proteins and particles likely augments the generation of anti-PEG when compared to free, unconjugated PEGs, since PEGylated material is present in the circulation for longer periods of time and the route of excretion may be altered when compared to the unconjugated PEG [34, 87, 88]. There are few studies in the literature showing the fate and metabolism of PEG-conjugated drugs [89], which warrants further study with regards to induction of anti-PEG. In the case for simple mixtures of a PEG homopolymer or copolymer and a drug, where the polymer component may be defined as an excipient (e.g., the use of Pluronic® gels as drug delivery vehicles [90]), it is reasonable to assume that the disposition and metabolic fate of the mixture may be comparable to the individual components. However, in the case of PEG-conjugated agents, a similar rationale may not be appropriate [89], as one of the main reasons for PEG-conjugation is to reduce renal clearance and hence increase the circulatory time of the drug, (e.g., PEG-filgrastim [87]) suggesting that renal clearance is not the primary route of excretion as observed with unconjugated PEGs. Indeed, as early as 1924, Landsteiner showed that chemical modification of proteins induced anaphylaxis in dogs, but not the component unmodified compounds [91]. With our current knowledge, the observations by Landsteiner [91] reflect both altered bioavailability and presentation of a compound to the host's immune system following chemical modification, demonstrating that toxicological and metabolic studies on a component PEG and a candidate drug may not apply to a PEGylated drug as the presentation to the host and clearance of the PEGylated drug will likely differ from free PEG and unmodified drug [89].

Commercial sources of anti-PEG

Presently, there are four commercial sources of monoclonal anti-PEG with no cross-reactivity with other polymers or proteins, but with different reported specificity and sensitivity:

Epitomics Inc. (Burlingame, California, USA) and Novus Biologicals Inc. (Littleton, Colorado, USA) supply a rabbit monoclonal anti-PEG IgG (antibody clone PEG-B-47) with a specificity of the terminal methoxy group of mPEG and 16 oxyethylene repeat units (personal communication, Lee S, Epitomics Inc.), and report a capability to detect PEG concentrations as low as 8 ng/mL.

Silver Lake Research Corp. (Monrovia, California, USA) supplies two mouse monoclonal anti-PEG IgGs (both IgG1κ from antibody clones CH-2074 and CH-2076). The reagent specification summary states that the anti-PEG should not be used in the presence of Tween® 20. Using an ELISA assay, in competition experiments studying binding of mPEG-conjugated BSA to anti-PEG, co-incubation with Tween® 20 and linear dihydroxy PEGs of 1,500 to 20,000 g/mol interfered with the detection and quantification of mPEG-BSA (Personal communication, Geisberg M, Silver Research Corp.). These findings suggest that both mouse monoclonal anti-PEG IgGs have an epitope of <6 oxyethylene repeat units and specificity for the PEG backbone irrespective of the end group.

Institute of Biomedical Sciences, Academica Sinica (Taipei, Taiwan) supplies a mouse monoclonal anti-PEG IgG (E11 [42]) and a mouse monoclonal anti-PEG IgM (AGP3 [41, 42, 92]). Both antibodies have a specificity for 16 oxyethylene repeat units and bind to the PEG backbone irrespective of the end group or linker moiety [42]. In a sandwich ELISA using AGP3 for capture and AGP3-biotin for detection, PEG-conjugates were detected at concentrations as low as 15 ng/mL [92], and enhanced sensitivity was reported with capture/detection pairing of E11/AGP3 respectively with a reported capability to detect PEG conjugates as low as 1.2 ng/mL [42].

Uses of anti-PEG

The first *in vitro* application of anti-PEG, described by Richter and Åkerblom [36], for quantification of PEG and PEG-conjugates used a gel diffusion technique [36] with the ability to detect PEG-conjugates down to 1 µg/mL, and showed a linear correlation for PEG of molecular weight 40,000 g/mol over the range 2.5–40 µg/mL. With current methods of signal amplification (e.g., HRP, alkaline phosphatase, streptavidin for ELISA assays), detection of PEG and PEG conjugates using anti-PEG are possible down to single nanogram concentrations per mL (or femtomolar concentrations) [92]. For the commercially available anti-PEGs described above, the recommended *in vitro* uses are for ELISA, Western blots and immunohistochemistry for the detection and quantification of PEG and PEG-conjugated drugs.

In addition to the *in vitro* use of anti-PEG to quantifiably assay PEG-conjugates, potential clinical applications for administering anti-PEG have also been described [43, 63]. For antibody-directed enzyme prodrug therapy in the treatment of cancer, using a nude mouse model bearing xenografts of human colon adenocarcinoma, rapid serum clearance of a PEG-conjugated glucuronide prodrug occurred following administration of anti-PEG IgM [43]. A dramatic increase in the tumor/blood ratio from 3.9 to 29.6 occurred without decreasing the concentration of the prodrug in the target tumor, and delayed tumor growth when compared to the conjugate alone or conventional chemotherapy (β-glucuronidase or p-hydroxyaniline mustard) [43].

Intravenous administration of anti-PEG has been proposed as a method to enhance the clearance of a PEG-conjugated drug, and also as a method to potentially extend the circulatory time of a PEG-conjugated drug [63]. Additionally, the authors described the use of a labeled anti-PEG to detect PEG-conjugated drugs *in vivo*, by whole body imaging with scintigraphy or computed tomography [63].

Clinical significance and strategies to overcome anti-PEG

In the absence of a comprehensive prospective clinical trial to determine the effects of pre-existing or induced anti-PEG, currently it is not possible to confirm that the presence of anti-PEG in a patient would lead to an ineffective therapy with a PEG-conjugated drug. PEG-conjugated drugs have been successfully administered to many thousands of patients with no reports of any serious adverse reactions, and PEG-conjugated drugs have greatly improved the overall treatment success rates for several diseases. However, few treatments yield complete recovery in all patients, and, as most PEG-conjugated drugs have only recently been introduced, there are few long-term follow-up studies in the literature. In the treatment of SCID with PEG-adenosine deaminase, a decline in treatment effectiveness over time was observed in some patients [93], for the treatment of acute lymphoblastic leukemia with PEG-asparaginase the five year disease-free survival rates are 75–80% for children [94, 95] and 30–40% for adults [94], and eradication of Hepatitis C virus with PEG-interferons was achieved in approximately 50% of patients [96]. There are many reasons for unsuccessful treatments, ranging from poor patient compliance to severity of disease at the time of treatment to responsiveness of genotype variants to treatment (e.g., Hepatitis C). As most treatments require combination therapies with non-PEGylated drugs, the cause of treatment failure is exceedingly complex. In the absence of any testing for anti-PEG, it is not possible to confirm or refute whether the induction or pre-existence of anti-PEG plays a role in the 20–60% of patients for whom treatment with a PEG-conjugated drug was unsuccessful.

Accordingly, a cautious approach must be taken when considering alternative treatments to a PEG-conjugated drug, introduction of PEG-conjugated agents with decreased immunogenicity, or the application of methods to remove or reduce anti-PEG prior to or during treatment with a PEG-conjugated drug. However, considering the body of evidence from animal studies showing that anti-PEG rapidly clears PEG-conjugated drugs and particles [40–53], and the clinical studies correlating the presence of anti-PEG with clearance of a PEG-conjugated drug [54–56], it is of value to consider strategies that may be of value to remove or reduce anti-PEG to ensure an effective treatment.

In the simplest approach, if a pre-existing or induced anti-PEG is detected in a patient receiving a PEG-conjugated drug, then the clearance rate of the PEG-conjugated drug should be monitored. As required, adjustment of the dose may compensate for the increased clearance (by augmentation or more frequent dosing), or replacement with an alternative non-PEGylated drug may be considered to ensure an effective therapy.

Investigation of polymers with reduced immunogenicity and antigenicity as candidates for drug conjugation compared to mPEG is an alternative approach [37, 62, 64]. Using a mouse model, Caliceti and colleagues [62] investigated the immunogenicity and antigenicity of methoxy-PEG 5,000 g/mol, branched methoxy-PEG 10,000 g/mol, polyvinylpyrrolidone 6,000 g/mol and poly(N-acyloylmorpholine) 6,000 g/mol conjugated to uricase. For each polymer derivative investigated, approximately 40–47 polymer chains were covalently attached to each uricase (10–12 chains per protein subunit). Antibodies were induced against all polymers, but a weak response was observed (anti-PEG IgG and IgM) for both mPEG 5,000 and branched PEG 10,000 g/mol, with the branched PEG being the least immunogenic and antigenic [62]. A branched, hydroxyl-terminated PEG was investigated as a polymer with reduced immunogenicity and antigenicity compared to methoxy-terminated PEGs [63, 64]. Anti-mPEG was induced in rabbits following repeated injections of partially PEGylated–recombinant uricase (two strands of mPEG 10,000 g/mol per protein subunit, sufficient to solubilize the protein, but insufficient to suppress immunogenicity [64]). The induction of anti-PEG was confirmed by ELISA using 96-well plates coated with mPEG 10,000 g/mol conjugated to a structurally different protein, and end-group specificity of the induced anti-PEG was shown from competitive binding studies by pre-incubation of sera with mPEGs of various molecular mass, with equivalent anti-PEG binding inhibition shown on a molar basis [63]. In the same assay, hydroxyl-terminated PEGs and a t-butoxy terminated PEG showed a 100-fold lower binding affinity for the anti-mPEG, again confirming the methoxy-specificity of the induced anti-PEG [63, 64]. Additionally, rabbits injected with a hydroxyl-terminated branched PEG-uricase in a comparable manner to the mPEG-uricase showed only a 5% production of anti-PEG when compared to mPEG-uricase [63]. Further studies are required to confirm whether a hydroxyl-terminated PEG for covalent attachment to a therapeutic agent results in reduced clearance in the presence of a pre-existing anti-PEG or an induced anti-PEG with specificity for the PEG backbone [38, 54, 56, 72].

Infusion of a PEG-containing compound to block or suppress anti-PEG, prior to administration of a PEG-conjugated drug may be beneficial to ensure effective treatment,. Ideally, such a conjugate should be comprised of a core molecule with short PEG oligomers attached comparable in size to the anti-PEG minimum epitiope (i.e., 4–5 oxyethylene units) to avoid immune complex formation [56].

Finally, for acute treatment with a PEG-conjugated drug, techniques employed for antibody removal to suppress organ rejection [97] may be adapted to suppress or remove anti-PEG prior to treatment with a PEG-conjugated drug. Plasmapheresis or plasma exchange techniques may not be desirable methods to consider, as removal of physiological proteins and nutrients or the potential risk of infection may pose a high risk to the patient and outweigh the benefit of removal of anti-PEG. Preoperative removal of specific antibodies by immunoadsorption has shown value for organ transplantation. Plasma is separated from whole blood and then passed through a column containing an immobilized antigen before reinfusion [97]. Antibody titers must be reduced to a desirable level prior to organ transplantation, and may require several immunoadsorption session for patients with high antibody titers. Adaptation of this technique for removal of anti-PEG may be achieved by employing a column comprised of immobilized PEG (packed with the commercially available TentaGel®-OH particles for example) to specifically remove anti-PEG from plasma, although any additional procedure adds risk and also the cost of such a pre-treatment may be a consideration.

Conclusions

Antibodies against PEG have been induced in animal models following exposure to PEG-conjugated proteins, liposomes and red blood cells, resulting in rapid clearance of PEG-conjugated agents. In humans, induction of anti-PEG was observed following exposure to a PEG-conjugated drug, and pre-existing anti-PEG was identified in over 25% the healthy blood donor population, both anti-PEG IgG and/or IgM. In clinical studies, the presence of anti-PEG was strongly associated with rapid clearance of PEG-asparaginase and PEG-uricase, potentially rendering their treatment ineffective. The IgG subtype of induced and pre-existing anti-PEG was identified as predominantly IgG-2, although IgG-1 and IgG-3 was also observed. The epitope of anti-PEG varies from binding to the backbone of 4–5 repeat oxyethylene units irrespective of the end-group moiety to 16 oxyethylene units with methoxy-group specificity. Anti-PEG was shown to be useful for *in vitro* detection and quantitation of PEG and PEG-conjugated materials sensitive down to single nanogram quantities, and also may be useful, *in vivo*, for antibody-directed enzyme prodrug therapy in the treatment of cancer.

PEGylation of therapeutic agents is, and will continue to be, of significant value in medicine to reduce immunogenicity, antigenicity and toxicity as well as markedly reducing renal clearance of drugs while maintaining drug efficacy. It is important to recognize that PEG itself may possess antigenic and immunogenic properties. Based on the observations closely relating anti-PEG with the onset of rapid clearance of PEG-conjugated agents in a sub-set of patients, further comprehensive studies are warranted to fully elucidate the effect of induced and pre-existing anti-PEG on PEG-conjugated agents. If confirmed in a prospective trial, patients should be screened for pre-existing anti-PEG, routinely monitored for the development of anti-PEG throughout the course of treatment with any PEG-containing agent, and strategies developed to ensure effective treatment. To overcome the potential effect of anti-PEG on drug clearance, in the simplest approach, it may be necessary to augment or increase the frequency of dosing or by substitution with an effective non-PEGylated therapeutic agent. More complex strategies to overcome anti-PEG may encompass the development of less immunogenic PEGylating reagents, pre-infusion of a PEG-containing agent to block or suppress anti-PEG or removal of anti-PEG by immunoadsorption prior to administration of a PEG-conjugated agent.

Acknowledgements
The work from our laboratories described in this chapter was supported by National Institutes of Health research grants HL 15722, HL 70595 and HL 65637. I would like to thank Drs T.C. Fisher (Division of Plastic and Reconstructive Surgery, Department of Surgery, Keck School of Medicine at USC, Los Angeles, California, USA), G. Garratty (America Red Cross Blood Services, Southern California Region, Pomona, California, USA), and S.R. Roffler (Institute of Biomedical Sciences, Academica Sinica, Tapei, Taiwan) for their assistance in preparing this chapter.

References

1. Bailey FE, Koleske JV (eds) (1991) *Alkylene oxides and their polymers.* Marcel Dekker, New York
2. Bailey FE, Koleske JV (eds) (1976) *Poly(ethylene oxide).* Academic Press, New York
3. Pang SNJ (1993) Final report on the safety assessment of polyethylene glycols (PEGs) -6, -8, -32, -75, -150, -14M, -20M. *J Am Coll Toxicol* 12: 429–457
4. Fruijtier-Pölloth C (2005) Safety assessment on polyethylene glycols (PEGs) and their derivatives as used in cosmetic products. *Toxicol* 214: 1–38
5. Abuchowski A, van Es T, Palczuk NC, Davis FF (1977) Alteration and immunological properties

of bovine serum albumin by covalent attachment of polyethylene glycol. *J Biol Chem* 252: 3578–3581

6. Abuchowski A, McCoy JR, Palczuk NC, van Es T, Davis FF (1977) Effect of covalent attachment of polyethylene glycol on immunogenicity and circulating life of bovine liver catalase. *J Biol Chem* 252: 3582–3586

7. Pasut G, Sergi M, Veronese FM (2008) Anti-cancer PEG-enzymes: 30 years old, but still a current approach. *Adv Drug Deliv Rev* 60: 69–78

8. Graham ML (2003) Pegaspargase: a review of clinical studies. *Adv Drug Deliv Rev* 55: 1293–1302

9. Hawkins DS, Park JR, Thomson BG, Felgenhauer JL, Holcenberg JS, Panosyan EH, Avramis VI (2004) Asparaginase pharmacokinetics after intensive polyethylene glycol-conjugated L-asparaginase therapy for children with relapsed acute lymphoblastic leukemia. *Clin Cancer Res* 10: 5335–5341

10. Matthews SJ, McCoy C (2004) Peginterferon alfa-2a: A review of approved and investigational uses. *Clin Therapeut* 26: 991–1025

11. Koslowski A, Charles SA, Harris JM (2001) Development of pegylated interferons for the treatment of chronic hepatitis C. *BioDrugs* 15: 419–429

12. Luxon BA, Grace M, Brassard D, Bordens R (2002) Pegylated interferons for the treatment of chronic hepatitis C infection. *Clin Ther* 24: 1363–1383

13. Stathopoulos GP, Dimou E, Stathopoulos J, Xynotroulas J (2005) Therapeutic administration of pegfilgrastim instead of prophylatic use. *Anticancer Res* 25: 2445–2448

14. Molineux G (2004) The design and development of pegfilgrastim (PEG-rmetHuG-CSF, Neulasta®). *Curr Pharm Des* 10: 1235–1244

15. Piedmonte DM, Treuheit MJ (2008) Formulation of Neulasta® (pegfilgrastim). *Adv Drug Deliv Rev* 60: 50–58

16. Hershfield MS, Buckley RH, Greenberg ML, Melton AL, Schiff R, Hatem C, Kurtzberg J, Market ML, Kobayashi RH, Abuchowski A (1987) Treatment of adenosine deaminase deficiency with polyethylene glycol modified adenosine deaminase. *New Engl J Med* 316: 589–596

17. Hershfield MS (1997) Biochemistry and immunology of poly(ethylene glycol)-modified adenosine deaminase (PEG-ADA). In: JM Harris, S Zalipsky (eds): *Poly(ethylene glycol): chemistry and biological applications.* American Chemical Society, Washington DC, 145–154

18. Greenwald RB, Choe YH, McGuire J, Conover CD (2003) Effective drug delivery by PEGylated drug conjugates. *Adv Drug Deliv Rev* 55: 217–250

19. Francis GE, Fisher D, Delgado C, Malik F, Gardiner A, Neale D (1998) PEGylation of cytokines and other therapeutic proteins and peptides: the importance of biological optimisation of coupling techniques. *Int J Hematol* 68: 1–18

20. Delgado C, Francis GE, Fisher D (1992) The uses and properties of PEG-linked proteins. *Crit Rev Ther Drug Carrier Syst* 9: 249–304

21. Molineux G (2002) Pegylation: engineering improved pharmaceuticals for enhanced therapy. *Cancer Treat Rev* 28(suppl A): 13–16

22. Veronese FM, Morpurgo M (1999) Bioconjugation in pharmaceutical chemistry. *Farmaco* 54: 497–516

23. Monfardini C, Veronese FM (1998) Stabilization of substances in circulation. *Bioconj Chem* 9: 418–450

24. Veronese FM, Caliceti P, Schiavon O, Sergi M (2002) Polyethylene glycol-superoxide dismutase, a conjugate in search of exploitation. *Adv Drug Deliv Rev* 54: 587–606

25. Hinds KD, Kim SW (2002) Effects of PEG conjugation on insulin properties. *Adv Drug Deliv Rev* 54: 505–530

26. Roberts MJ, Harris MJ (1998) Attachment of degradable poly(ethylene glycol) to proteins has the potential to increase therapeutic efficacy. *J Pharm Sci* 87: 1440–1445

27. Gaberc-Porekar V, Zore I, Podobnik B, Menart V (2008) Obstacles and pitfalls in the PEGylation of therapeutic proteins. *Curr Opinin Drug Discov Develop* 11: 242–250

28. Chapman AP (2002) PEGylated antibodies and antibody fragments for improved therapy: A review. *Adv Drug Deliv Rev* 54: 531–545

29. Bjorkholm M, Fagrell B, Przybelski R, Winslow N, Young M, Winslow RM (2005) A phase I single blind clinical trial of a new oxygen transport agent (MP4), human hemoglobin modified with maleimide-activated polyethylene glycol. *Haematologica* 90: 505–515

30. Hinds KD, Kim SW (2002) Affects of PEG conjugation on insulin properties. *Adv Drug Deliv Rev*

54: 505–530
31. Parveen S, Sahoo SK (2006) Nanomedicine: Clinical application of polyethylene glycol conjugated proteins and drugs. *Clin Pharmacokinet* 45: 965–988
32. van Vlerken LE, Vyas TK, Amiji MM (2007) Poly(ethylene glycol)-modified nanocarriers for tumor-targeted and intracellular delivery. *Pharm Res* 24: 1405–1414
33. Dreborg S, Åkerblom E (1990) Immunotherapy with monomethoxypolyethylene glycol modified allergens. *Crit Rev Ther Drug Carrier Syst* 6: 315–365
34. Fishburn CS (2008) The pharmacology of PEGylation: Balancing PD with PK to generate novel therapeutics. *J Pharm Sci* 97: 4167–4183
35. Richter AW, Åkerblom E (1984) Polyethylene glycol reactive antibodies in man: titer distribution in allergic patients treated with monomethoxy polyethylene glycol modified allergens or placebo, and in healthy blood donors. *Int Arch Allergy Appl Immunol* 74: 36–39
36. Richter AW, Åkerblom E (1983) Antibodies against polyethylene glycol produced in animals by immunization with monomethoxy polyethylene glycol modified proteins. *Int Arch Allergy Appl Immunol* 74: 124–131
37. Martinez AL, Sherman MR, Saifer MGP, Williams LD (2004) Polymer conjugates with decreased antigenicity, methods of preparation and uses thereof. *Patent Cooperation Treaty* WO/2004/030617
38. Armstrong JK, Fisher TC (2008) Poly(ethylene glycol) anti-body detection assays and kits for performing thereof. *Patent Cooperation Treaty* WO/2008/063663
39. Garratty G (2008) Modulating the red cell membrane to produce universal/stealth donor red cells suitable for transfusion. *Vox Sang* 94: 87–95
40. Armstrong JK, Meiselman HJ, Wenby RB, Fisher TC (2003) *In vivo* survival of poly(ethylene glycol)-coated red blood cells in the rabbit. *Blood* 102: 94A
41. Cheng TL, Wu PY, Wu MF, Chern JW, Roffler SR (1999) Accelerated clearance of polyethylene glycol-modified proteins by anti-polyethylene glycol IgM. *Bioconj Chem* 10: 520–528
42. Cheng TL, Cheng CM, Chen BM, Tsao DA, Chuang KH, Hsiao SW, Lin YH, Roffler SR (2005) Monoclonal antibody-based quantitation of poly(ethylene glycol)-derivatized proteins, liposomes and nanoparticles. *Bioconj Chem* 16: 1225–1231
43. Cheng TL, Chen BM, Chern JW, Wu MF, Roffler SR (2000) Efficient clearance of poly(ethylene glycol)-modified immunoenzyme with anti-PEG monoclonal antibody for prodrug cancer therapy. *Bioconj Chem* 11: 258–266
44. Dams ETM, Laverman P, Oyen WJG, Storm G, Scherphof GL, van der Meer JWM, Corstens FHM, Boerman OC (2000) Accelerated blood clearance and altered biodistribution of repeated injections of sterically stabilized liposomes. *J Pharmacol Exp Ther* 292: 1071–1079
45. Judge A, McClintock K, Phelps JR, MacLachlan I (2006) Hypersensitivity and loss of disease site targeting caused by antibody responses to PEGylated liposomes. *Mol Ther* 13: 328–337
46. Semple SC, Harasym TO, Clow KA, Ansell SM, Klimuk SK, Hope MJ (2005) Immunogenicity and rapid blood clearance of liposomes containing polyethylene glycol-lipid conjugates and nucleic acid. *J Pharmacol Exp Ther* 312: 1020–1026
47. Laverman P, Carstens MG, Boerman OC, Dams ETM, Oyen WJG, van Rooijen N, Corstens FHM, Storm G (2001) Factors affecting accelerated blood clearance of polyethylene glycol-liposomes upon repeat injection. *J Pharmacol Exp Ther* 298: 607–612
48. Šroda K, Rydlewski J, Langner M, Kozubek A, Grzybek M, Sikorski AF (2005) Repeated injections of PEG-PE liposomes generate anti-PEG antibodies. *Cell Mol Biol Lett* 10: 37–47
49. Ishida T, Maeda R, Ichihara M, Irimura K, Kiwada H (2003) Accelerated clearance of PEGylated liposomes in rats after repeated injections. *J Control Release* 88: 35–42
50. Ishida T, Masuda K, Ichikawa T, Ichihara M, Irimura K, Kiwada H (2003) Accelerated clearance of a second injection of PEGylated liposomes in mice. *Int J Pharm* 255: 167–174
51. Ishida T, Kiwada H (2008) Accelerated blood clearance (ABC) phenomenon upon repeated injection of PEGylated liposomes. *Int J Pharm* 354: 56–62
52. Ishida T, Wang XY, Shimizu T, Nawat K, Kiwada H (2007) PEGylated liposomes elicit an anti-PEG IgM reponse in a T cell-independent manner. *J Control Release* 122: 349–355
53. Ishida T, Kashima S, Kiwada H (2008) The contribution of phagocytic activity of liver macrophages to the accelerated blood clearance (ABC) phenomenon of PEGylated liposomes in rats. *J Control Release* 126: 162–165
54. Ganson NJ, Kelly SJ, Scarlett E, Sundy JS, Hershfield MS (2006) Control of hyperuricemia in subjects with refractory gout, and induction of an antibody against poly(ethylene glycol) (PEG),

in a phase I trial of subcutaneous PEGylated urate oxidase. *Arthritis Res Ther* 8: R12
55. Sundy JS, Ganson NJ, Kelly SJ, Scarlett EL, Rehrig CD, Huang W, Hershfield MS (2007) Pharmacokinetics and pharmacodynamics of intravenous PEGylated recombinant mammalian urate oxidase in patients with refractory gout. *Arthritis Rheum* 56: 1021–1028 Erratum in: (2007) *Arthritis Rheum* 56: 1370
56. Armstrong JK, Hempel G, Koling S, Chan LS, Fisher T, Meiselman HJ, Garratty G (2007) Antibody against poly(ethylene glycol) adversely affects PEG-asparaginase therapy in acute lymphoblastic patients. *Cancer* 110: 103–111
57. Amgen, Annual Financial Report (2007) http://www.amgen.com/pdfs/Investors_2007_AnnualReport.pdf
58. Schering-Plough, Annual Financial Report (2007) http://thomson.mobular.net/thomson/7/2707/3250/
59. Roche Pharmaceuticals, Annual Financial Report (2007) http://www.roche.com/gb07e04.pdf
60. Enzon, Annual Financial Report (2007) http://investor.enzon.com/annuals.cfm
61. OSI Pharmaceuticals, Annual Financial Report (2007) http://media.corporate-ir.net/media_files/irol/70/70584/2007_OSIP_Annual_Report.pdf
62. Caliceti P, Schiavon O, Veronese FM (2001) Immunological properties of uricase conjugated to neutral soluble polymers. *Bionconj Chem* 12: 515–522
63. Roberts MJ, Green ME, Baker MR (2002) Antibodies specific for poly(ethylene glycol). *Patent Cooperation Treaty* WO/2002/094853
64. Sherman MR, Saifer MGP, Perez-Ruiz F (2008) PEG-uricase in the management of gout and hyperuricemia. *Adv Drug Deliv Rev* 60: 59–68
65. Li WM, Bally MB, Schutzde-Redelmeier MP (2002) Enhanced immune response to T-independent antigen using CpG oligodeoxynucleotides encapsulated in liposomes. *Vaccine* 20: 148–157
66. Boeckler C, Dautel D, Schelté P, Frisch B, Wachsmann D, Klein JP, Schuber F (1999) Design of highly immunogenic liposomal constructs combining structurally independent B cell and T helper cell peptide epitopes. *Eur J Immunol* 29: 2297–2308
67. de Jong S, Chikh G, Sekirov L, Raney S, Semple S, Klimuk S, Yuan N, Hope M, Cullis P, Tam Y (2007) Encapsulation in liposomal nanoparticles enhances the immunstimulatory, adjuvant and anti-tumor activity of subcutaneously administered CpG ODN. *Cancer Immunol Immunother* 56: 1251–1264
68. O'Hagan DT, Singh M (2003) Microparticles as vaccine adjuvants and delivery systems. *Exp Rev Vaccines* 2: 269–283
69. Ishida T, Ichihara M, Wang XY, Kiwada H (2006) Spleen plays an important role in the induction of accelerated blood clearance of PEGylated liposomes. *J Control Release* 115: 243–250
70. Armstrong JK, Leger R, Wenby RB, Meiselman HJ, Garratty G, Fisher TC (2003) Occurrence of an antibody to poly(ethylene glycol) in normal donors. *Blood* 102: 556A
71. General laboratory methods (1999) In: Vengelen-Tyler V (ed.): *AABB Technical Manual* (13th edition). American Association of Blood Banks, Bethesda MD, 646
72. Fisher TC, Armstrong JK, Wenby RW, Meiseman HJ, Leger R, Garratty G (2003) Isolation and identification of a human antibody to poly(ethylene glycol). *Blood* 102: 559A
73. Schick MJ (ed.) (1967) *Nonionic surfactants.* Marcel Dekker, New York
74. Smyth HF, Carpenter CP, Shaffer CB (1947) The toxicity of high molecular weight polyethylene glycols; chronic oral and parenteral administration. *J Am Pharm Assoc* 36: 157–160
75. Olser BL, Olser M (1957) Nutritional studies on rats on diets containing high levels of partial ester emulsifiers: IV Mortality and *post mortem* pathology; general conclusions. *J. Nutrition* 61: 235–252
76. Hooker E (2004) Final report on amended safety assessment of PEG-5, -10, -16, -25, -30 and -40 soy sterol. *Int J Toxicol* 23: 23–47
77. Shaffer CB, Critchfield FH, Carpenter CP (1948) Renal excretion and volume distribution of some polyethylene glycols in the dog. *Am J Physiol* 151: 93–99
78. Shaffer CB, Critchfield FH (1947) The absorption and excretion of the solid polyethylene glycols ("Carbowax" compounds). *J Am Pharm Assoc* 36: 152–157
79. Yamaoka T, Tabata Y, Ikada Y (1994) Distribution and tissue uptake of poly(ethylene glycol) with different molecular weights after intravenous administration in mice. *J Pharm Sci* 83: 601–606
80. Yamaoka T, Tabata Y, Ikada Y (1995) Fate of water-soluble polymers administered via different routes. *J Pharm Sci* 84: 349–354
81. Danielson GK, Dubilier LD, Bryant LR (1970) Use of Pluronic F-68 to diminish fat emboli and

hemolysis during cardiopulmonary bypass. *J Thorac Cardiovasc Surg* 59: 178–184
82. Schaer GL, Spaccavento LJ, Browne KF, Krueger KA, Krichbaum D, Phelan JM, Fletcher WO, Grines CL, Edwards S, Jolly MK et al. (1996) Beneficial effects of RheothRx injection in patients receiving thrombolytic therapy for acute myocardial infarction. Results of a randomized, double-blind, placebo-controlled trial. *Circulation* 94: 298–307
83. Grindel JM, Jaworski, T, Emanuele RM, Culbreth P (2002) Pharmacokinetics of a novel surface-active agent, purified poloxamer 188, in rat, rabbit, dog and man. *Biopharmaceut Drug Dispos* 23: 87–103
84. Willcox ML, Newman MM, Paton BC (1978) A study of labeled Pluronic F-68 after intravenous administration into the dog. *J Surg Res* 25: 349–356
85. Wang ZYJ, Stern IJ (1975) Disposition in rats of a polyoxypropylene-polyoxyethylene copolymer used in plasma fractionation. *Drug Metabol Dispos* 3: 536–542
86. Jewell RC, Khor SP, Kisor DF, LaCroix KAK, Wargin WA (1997) Pharmacokinetics of RheothRx injection in healthy male volunteers *J Pharm Sci* 86: 808–812
87. Yang BB, Lum PK, Hayashi MM, Roskos LK (2004) Polyethylene glycol modification of fil-grastim results in decreased renal clearance of the protein in rats. *J Pharm Sci* 93: 1367–1373
88. Asgeirsson D, Venturoli D, Fries D, Rippe B, Rippe C (2007) Glomerular sieving of three neutral polysaccharides, polyethylene oxide and bikunin in rat. Effects of molecular size and conformation. *Acta Physiol* 191: 237–246
89. Webster R, Didier E, Harris P, Siegel N, Stadler J, Tilbury L, Smith D (2007) PEGylated proteins: Evaluation of their safety in the absence of definitive metabolism studies. *Drug Metab Dispos* 35: 9–16
90. Fusco S, Borzacchiello A, Netti PA (2006) Perspectives on: PEO-PPO-PEO triblock copolymers and their biomedical applications. *J Bioact Compat Polymers* 21: 149–164
91. Landsteiner K (1924) Experiments on anaphylaxis to azoproteins. *J Exp Med* 39: 631–637
92. Tsai NM, Cheng TL, Roffler SR (2001) Sensitive measurement of polyethylene glycol modified proteins. *BioTecniques* 30: 396–402
93. Chan B, Wara D, Bastian J, Hershfield MS, Bohnsack J, Azen CG, Parkman R, Weinberg K, Kohn DB (2005) Long-term efficacy of enzyme replacement therapy for adenosine deaminase (ADA)-deficient severe combined immunodeficiency (SCID). *Clin Immunol* 17: 133–143
94. Apostolidou E, Swords R, Alvarado Y, Giles FJ (2007) Treatment of acute lymphoblastic leukaemia: A new era. *Drugs* 67: 2153–2171
95. Avramis VI, Tiwari PN (2006) Asparaginase (native ASNase or pegylated ASNase) in the treatment of acute lymphoblastic leukemia. *Int J Nanomed* 1: 241–254
96. Kemmer N, Neff GW (2007) Managing chronic hepatitis C in the difficult-to-treat patient. *Liver Int* 27: 1297–1310
97. Tydén G, Kumlien G, Efvergren M (2007) Present techniques for antibody removal. *Transplantation* 84: S27–S29

Pegfilgrastim – designing an improved form of rmetHuG-CSF

Graham Molineux

Amgen Inc., Mailstop 15-2-A, One Amgen Center Drive, Thousand Oaks, California 91320, USA

Abstract

rmetHuG-CSF is the recombinant version of natural granulocyte colony-stimulating factor, the dominant stimulator in the production of neutrophilic leukocytes (neutrophils). Neutrophils represent the first line of defense against invading pathogens and when neutrophil numbers are suppressed by cancer chemotherapy, patients become liable to life-threatening infections.
The clearance of rmetHuG-CSF is effected by a combination of neutrophil mediated degradation and renal filtration. Site-directed addition of a single, linear PEG molecule yielded a form of G-CSF (pegfilgrastim) that was shown to be resistant to renal elimination yet remained sensitive to neutrophil-mediated destruction. This semi-synthetic cytokine drug can persist in the plasma for extended periods in neutropenic conditions, yet is cleared rapidly when neutrophils recover. This lends a degree of automation to the therapeutic control of neutrophil numbers which has been exploited in clinical practice since its approval for human use in 2002.

Introduction

Natural Granulocyte Colony-Stimulating Factor (G-CSF) is a circulating glycoprotein that regulates neutrophil production and activity. Neutrophils normally comprise around 30% of leukocytes and are major effectors of innate immunity. Neutrophils remain in the blood only a matter of hours so they need to be replaced by rapid proliferation of their precursors in the bone marrow (Tab. 1). This high turnover rate makes neutrophil production susceptible to cancer chemotherapy and neutropenia (lack of neutrophils) following chemotherapy leaves many patients prone to infection and hospitalization. The use of G-CSF to stave off such complications has become widespread practice over the last 20 years. This was made possible by the cloning and characterization of recombinant human G-CSF (rHuG-CSF) in the early 1980s [1–3] culminating in the expression of r-metHuG-CSF in *E. coli*, trials in humans and eventually approval of Filgrastim® for administration to US patients in 1991 [4–7]. Alternative forms of G-CSF also exist, among them a form expressed in Chinese hamster ovary (CHO) cells (lenograstim; Granocyte®), which is glycosylated and lacks the N-terminal methionine required in the *E. coli*-derived version [8] another form with a deliberately mutated amino acid sequence [9] and several follow-on versions (Tevagrastim®, Ratiograstim®, ratiopharm Filgrastim® and Biograstim®).

Table 1. Production rate of human blood cells

Blood cell type	Concentration of cells in blood	Total cells in blood (5L blood volume)	Lifespan of cells	Production rate
Neutrophils	$3 \times 10^3/\mu L$	15×10^9	8 h	4.5×10^{10}/day
Erythrocytes	$5 \times 10^6/\mu L$	25×10^{12}	120 days	21×10^{10}/day

Blood cell numbers are stable in normal individuals. The lifespan of each of the cell types is quite different, but to maintain stable circulating numbers cell production must precisely balance cell loss. This process is under the control of hormone-like cytokines, in the case of neutrophils, granulocyte colony-stimulating factor, and in the case of erythrocytes, erythropoietin. Production rates can be increased dramatically or reduced to almost zero in response to physiological demand.

Initially, administration of G-CSF was for the treatment of cancer chemotherapy-induced neutropenia (CIN) and the prevention of associated infections, but over the years the indications have been broadened to include use in severe congenital neutropenia, AIDS, aplastic anemia and myelodysplastic syndromes. In addition, the serendipitous finding that G-CSF can 'mobilize' large numbers of transplantable 'stem cells' to the blood has been exploited in both cancer patients and normal donors [10, 11]. In this setting G-CSF can cause peripheral blood progenitor cells (PBPC) to move to the blood, where they may be harvested to allow remedial treatment of damaged or diseased bone marrow. To date over 4 million patients have received G-CSF for various indications, with the only major reported side effect being bone pain – perhaps part of the bone marrow's normal response to G-CSF.

rHuG-CSF has to be injected into the body because it is degraded in the stomach and is too large to pass unaided through the skin. Continuous infusion is the most effective way to administer G-CSF, followed by twice daily injection, and daily administration; from there effectiveness is yet further reduced as injections are more widely spaced. The requirement for frequent G-CSF injections stems from its rapid clearance from the body mainly via the kidney but also due to neutrophil-mediated processes. Clinical experience has shown that G-CSF can be administered to patients intravenously (IV) [5, 6], subcutaneously (SC) [12] or intramuscularly [13] and in all cases the neutrophil response is similar. After IV injection G-CSF levels increase within a matter of minutes [5, 6] though SC administration also shows very rapid absorption [7]. Clearance of G-CSF from the body is very fast as illustrated by a serum half-life of between 1–2 h in several tested species [14–16]. Normal humans clear G-CSF with a half-life of less than 2 h [17], but neutropenic patients take almost 5 h to clear half the drug concentration – *prima facie* evidence that G-CSF half-life is related inversely to absolute neutrophil count (ANC). This suggests that neutrophils themselves may play a role in the clearance of G-CSF – not an unprecedented suggestion since the elimination of hematopoietic cytokines by the products of those cytokines' action has been proposed for several important regulators [18–25]. In this model (Fig. 1) at least one cytokine

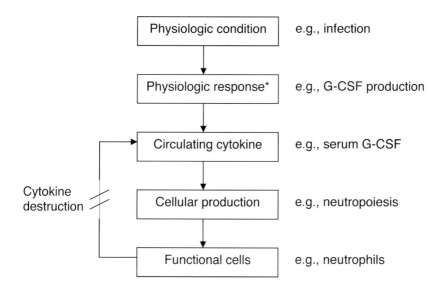

Figure 1. Autoregulation of lineage specific cytokine levels by terminally differentiated cells of the affected cellular lineage. * some cytokines are inducible (e.g., G-CSF), others are produced constitutively (e.g., thrombopoietin, the main stimulator of platelet production), the model is unaffected.

involved in the control of blood cell development would be a lineage specific regulator, i.e., one that shows high fidelity for that single cellular lineage, for example thrombopoietin for platelets or G-CSF for neutrophils. The terminal cell type of that lineage would then control the circulating levels of the regulator possibly by receptor-mediated internalization and degradation. In the specific case of G-CSF the reciprocal relationship between G-CSF levels and neutrophil counts has been reported numerous times [26–29], and the ability of neutrophils to destroy G-CSF *in vitro* [30] has also been documented. Neutrophil-mediated clearance of G-CSF is not the only contributor to its rapid removal, indeed an alternative route – renal excretion, may account for the majority of clearance in some situations [31, 32]. Renal loss is of sufficient magnitude that in order to maintain effective serum levels even in conditions of absolute neutropenia, daily injections are required.

To summarize the effect of these two modes of clearance consider the fate of the first of a series of filgrastim injections. If this injection were made to a normal individual, with normal ANC, then as injected dose and serum concentration increased, the drug would be absorbed by the neutrophil population (and lost from both measurements of serum G-CSF and prevented from influencing neutropoiesis). When that compartment was saturated, the excess would then be circulating in the serum where it could stimulate neutrophil production, but would be eliminated from the serum via the kidney with a half-life of about 5 h. Should the initial injection have been made into a neutropenic individual then none of the drug would be lost to the neutrophil pool and all would be subject to renal elimination with a 5 h half-life, stimulating neutrophil production in the meantime. In considering the fate of subsequent injections account must be taken of the accumulating response to G-CSF, i.e., ascending ANC. As neutrophil numbers increase over time the persistence of each injection would be progressively shorter as more drug was absorbed and destroyed by the ANC pool; indeed, this was what was actually observed in early clinical studies [25]. Thus the predominant clearance pathway for G-CSF will depend on the saturable (and therefore non-linear) ANC-related route and the linear renal pathway which will in turn depend on the response to the drug (ANC increases as G-CSF has its effects). So unmodified G-CSF displays a degree of 'self regulation' – it induces the means of its own destruction – but this process is of relatively minor importance because most of the drug is lost through the relentless renal process.

These considerations suggested a strategy that might be employed to design a successor to filgrastim – it was reasoned that the contribution of the two routes of clearance could be manipulated independently to engineer a derivative of G-CSF that resisted renal clearance yet retained sensitivity to neutrophil-mediated destruction. In considering the properties of a next generation therapeutic as many of the desirable properties listed below should be built in to the molecule:

1) The safety record of the parent drug must not change, i.e., no non-G-CSFR mediated effects and no increase in antigenicity would be tolerated
2) The formulation properties of the drug (stability, solubility, etc.) should be at least maintained
3) The persistence of the drug in the body must be increased to cover, if possible, a complete cycle of chemotherapy from a single injection
4) The 'exposure profile' should be optimized, i.e., more drug should be provided at the time it is most needed, less when less is needed.

To deliver as many of these properties as possible filgrastim was retained at the core of the molecule as data to date would suggest that this protein has a single cellular receptor, and that in turn that receptor has but a single ligand – retaining this core would reduce the potential for introduction of new off-target activities. The stability of filgrastim is good as formulated, but G-CSF is inherently unstable at physiological pH, temperature and salt concentration – this was considered a relatively easy profile upon which improvements could be made. An increase in size presented a simple and proven method to evade renal clearance, but absent detailed knowledge of the optimum exposure profile for G-CSF was not known whether this change would in itself be sufficient to satisfy the longevity requirements. In considering what was known about the optimum exposure profile, inferences could be drawn from the literature; it is established 1) that continuous infusion offers superior efficacy [25, 33, 34], 2) that increased serum levels provoke a greater response [4] and 3) that early provision of G-CSF is important to maximize benefit [26].

Strategies for the improvement of rmetHuG-CSF

Two broad strategies were considered to improve G-CSF; sustained release and sustained duration. The physicochemical properties of G-CSF do not lend themselves well to the fabrication of a depot formulation. Neither is it an easy protein to deliver through skin or via the gastrointestinal tract. These delivery approaches are feasible to a degree but are less readily controlled than modification of the residence time of G-CSF in the circulation.

A sustained duration form of G-CSF that could sustain its effects for four weeks would need to be administered at relatively high doses (Filgrastim is administered at 5 μg/kg/day, so a 70 kg patient would require 9.8 mg for 28 days treatment assuming the introduction of zero inefficiency). Such a large amount of drug would unavoidably offer front-loading when administered as a single injection. Also, if this injection were administered shortly after chemotherapy then the resulting high serum concentration would coincide with the time at which the marrow required maximum impetus to launch recovery.

Protein therapeutics have been modified in various ways to extend their persistence in the body. Recently notable has been the success of glycoengineering an analog of erythropoietin to prolong its half-life [35]. In this exercise understanding the role carbohydrate played in controlling the elimination of erythropoietin was extended to engineer a hyperglycosylated variant with three times the residence time in the body [36, 37]. Though natural G-CSF is a glycoprotein (with a single O-linked carbohydrate on threonine 133) a comparison of two forms; one of which is produced in eukaroytic cells and glycosylated while filgrastim is made in prokaryotic cells and has no attached sugars, revealed their pharmacokinetics to be identical. This would suggest that the carbohydrate component is entirely optional for activity and further suggests that this may not be the most rewarding pathway to an extended duration derivative. Having thus eliminated glycoengineering from our considerations what other strategies are likely to work? A serum albumin conjugate of G-CSF has been discussed [38] and poly[ethylene glycol] (PEG) derivatives have been known for some years [16, 39–41].

As discussed above, a detailed consideration of the dual routes of elimination of G-CSF suggested that separation of these processes might lead to a new and particularly useful form. G-CSF is normally administered to patients who are neutropenic following cancer chemotherapy. In neutropenia only the kidney effects G-CSF clearance. The development of a form of G-CSF that could resist renal clearance, yet retain neutrophil-mediated clearance held the promise of enhancing the self-regulation which was already a feature of the parent molecule.

PEGylation

Several drugs; enzymes [42, 43], interferons ([44, 45], see elsewhere this volume) and cytokines [46–49], have been developed by covalently attaching

PEG because it confers benefits such as reduced immunogenicity, prolonged residence time or improved formulation properties [42, 43, 50–52]. In general the advantages gained by the protein conjugation are the properties of PEG itself, especially the unique ability of PEG to occupy a disproportionately large volume in aqueous solution. As an increase in size was one of the main targets of filgrastim derivitization, PEG was thought to be uniquely suitable for our purposes.

Several different ways of attaching PEG to proteins have been reported [47, 53], and most are similar in that they rely on the nucleophilic attack of amino groups (or other active protein components) on the terminal ethylene glycol group of PEG. In most cases such a reaction can be shown to yield reproducible forms with consistent location and number of attached PEGs [54]. Though the actual sites of attachment cannot be realistically determined in advance and are in effect controlled by the chemistry, precisely this approach has proven useful for two FDA approved PEG-enzyme conjugates [42, 43]. Considering a similar approach for a cytokine presents a different set of constraints. Cytokines tend to be large molecules with complex three-dimensional structure, and their receptors also tend to be similarly large and complex. Employing a non-selective PEGylation strategy in such a circumstance is likely to yield a suboptimal product. This is because of the contrasting effects of increasing serum residence time on one hand, but on the other hand lowering the affinity of the ligand/receptor interaction, probably by steric hindrance. The interplay of these factors will determine the usefulness of any derivative and selection of the final candidate tends to be a semi-empirical process. To avoid the vagaries of non-directed PEGylation several site-directed approaches have been developed. These techniques have been used for, e.g., topographical mapping of attachment sites [51]. In these cases, PEGylation can be targeted to, say, specific lysines in the amino acid sequence, but general sites of attachment can also be targeted such as the N- or C-terminus.

Targeting specific amino acids, most commonly lysine, can be useful especially if a limited number of such lysines are in desirable locations within the protein. If the sites at which such amino acids are found are not deemed desirable then lysines may be substituted for less reactive arginines or new lysines may be inserted at the appropriate site. However, a large protein may have several potential attachment sites that would require extensive re-engineering to remove from the molecule. These numerous sites would tend to produce a multi-PEGylated protein though some have been shown to retain substantial biological activity. However, in the case of G-CSF, the four lysines found throughout the molecule tend to be in regions that are not good target areas. A second site-directed approach mediated via specific amino acids is targeting the thiols of cysteine residues. Again cysteines can be introduced or removed to lend a protein to this type of chemistry. However, the three dimensional configuration of G-CSF is stabilized by several disulfide bonds between cysteines and derivitization of any of them may upset the structure leading to potential affects on activity and immunogenicity (Fig. 2).

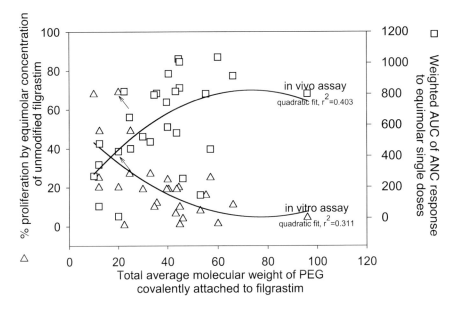

Figure 2. The activity of various PEG-G-CSFs assessed both *in vivo* and *in vitro* and related to the amount of PEG added per molecule. Many of the candidates comprised defined blends of non-PEGylated, mono-, di-, tri- and tetraPEGylated G-CSFs. The amount of PEG per molecule represents the product of molecular weight of the PEG moiety and the average number of additions per molecule. $\rightarrow$ – indicates the data obtained with the final pegfilgrastim selected for further development. $\triangle$ – the proliferation of 32D clone 3 cells (a murine G-CSF dependent cell line) as measured by reduction of Alomar blue. $\square$ – weighted AUC was obtained from daily average ANC from mice (5 per timepoint) weighted by multiplying by the number of days after injection, then summed. This weights selection in favor of longer acting forms.

Targeting the N-terminal residue of proteins is an attractive option, offering the benefits of a single, defined site, a known relationship to the receptor binding domain of the cytokine and relative simplicity in the required chemistry. Several methods have been developed to target the N-terminus for PEGylation including chemical activation and enzyme ligation; however, few have been exploited to develop viable product entities. The method employed for the fabrication of pegfilgrastim was based on a reductive alkylation process to direct conjugation of PEG to the N-terminal methionine of filgrastim (r-metHuG-CSF). This was achieved by taking advantage of the different pKa of the α-amino group of the N-terminal methionine (pKa 7.6–8.0) in contrast to the ε-amino group (pKa 10–10.2 [55]) found on the lysines throughout the molecule.

Mono-N-terminal poly(ethylene glycol) conjugates of filgrastim

Numerous PEG-G-CSF conjugates were prepared for activity screening. Linear mono-functional monomethoxy PEG aldehydes of various molecular weights (between 12 and 30 kD) were used to prepare derivatives. More complex, branched PEG forms were also assessed. The method included stirring a cold buffered (pH 5) solution of rmetHuG-CSF in the presence of a five fold molar excess of mPEG aldehyde in 20 mM sodium cyanoborohydride. The degree of PEGlyation was tracked with HPLC until after around 10 h 92% of the protein was shown to be mono-PEG conjugate. The site of PEG attachment was determined by endoproteinase mapping and confirmed to be single site of PEG conjugation at the N-terminus of the protein [55, 56].

Preclinical and clinical development of Pegfilgrastim

Screening activity

The screening process for PEGylated derivatives was designed to select a candidate with prolonged action *in vivo* and retention of the maximum *in vitro* activity. As mentioned above, the engineering of darbepoetin is in some ways analogous to the development of pegfilgrastim and some lessons from that program are salutory. The literature published on the development of darbepoetin illustrated that the derivatives that were most potent *in vivo* were among the least active *in vitro*. In the case of erythropoietin analogs the form with the highest affinity for the receptor and the highest *in vitro* activity (a deglycosylated form) paradoxically had no detectable activity *in vivo*; presumably because it was cleared from the body in a matter of only minutes. At the other extreme, highly modified forms (with high sialic acid content) had lower affinity for the receptor, were also several fold less active *in vitro* yet were the most spectacularly effective when injected *in vivo*. This led us to conclude that in assessing the activity of derivitized cytokines, assays carried out *in vitro* where affinity is a dominant determinant, may be misleading in candidate selection. The aim in developing a pharmacokinetically advantaged derivative is not to increase affinity (indeed the opposite would appear to be true) but to obtain the optimum blend between longevity and the [likely?] reduction in affinity detected by somewhat artificial *in vitro* assays. On a practical level this meant that though assessment of the *in vitro* activity of various PEGylated G-CSFs was performed, it was considered with little weight against the *in vivo* assessments of activity.

A relationship was defined between the molecular weight of PEG added to various PEG-G-CSF derivatives and the performance of the conjugates both *in vitro* and *in vivo*. There existed weak relationships between average MW of added PEG and activity *in vivo* (a positive relationship) and *in vitro* (an inverse relationship) – see Figure 2. The final selection was made based upon several parameters including retention of around 70% of the *in vitro* activity of the parent molecule in combination with substantial improvement in weighted (in favor of longer acting forms) AUC of ANC response. Other *in vivo* parameters were also considered as part of the selection process – mobilization of PBPC and reversal of 5-fluorouracil induced myelosuppression in mice in addition to factors such as consistency/robustness of the production process, availability of raw materials and formulation properties.

Having made the selection based on data from mice that indicated a prolonged mode of action, several pieces of information were collected to assess whether other design parameters had been met. In a study in groups of normal or bilaterally nephrectomized rats, an intravenous dose of pegfilgrastim was

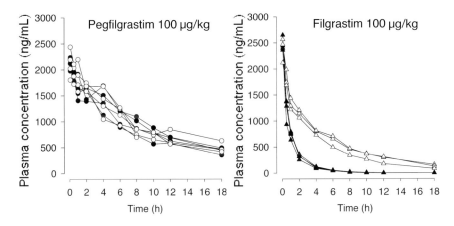

Figure 3. Pegfilgrastim clearance from the plasma of treated rats is independent of renal function. Plasma levels after a single intravenous injection of 100 µg/kg of pegfilgrastim (left) and filgrastim (right) in normal (closed symbols) or bilaterally nephrectomized (open symbols) rats (n = 3 or 4, individual data shown). Note on the left the presence or absence of kidneys makes little difference to the clearance of pegfilgrastim, but the clearance of filgrastim (right) is significantly affected by the existence of a functional kidney. Adapted from Yang et al. [61].

cleared with identical kinetics in both groups (see Fig. 3). Filgrastim, in contrast, was eliminated much more rapidly in normal animals than in those lacking kidney function. This suggested that the new form was resistant to renal clearance as was hoped from the design process. The second feature of the molecule that was considered essential in the design stage was that it should remain sensitive to neutrophil-mediated destruction. Figure 4 illustrates that both filgrastim and pegfilgrastim are removed from culture supernatent by neutrophils isolated from the blood of normal volunteers. Pegfilgrastim is relatively protected, but since both PEG- and non-PEG-G-CSF were removed both could possibly be cleared from the body by this process.

The stage was therefore set to initiate more advanced testing. Toxicology studies had revealed no new safety concerns – the only observations made were associated with exaggerated pharmacology, as would be expected with a more active derivative of G-CSF.

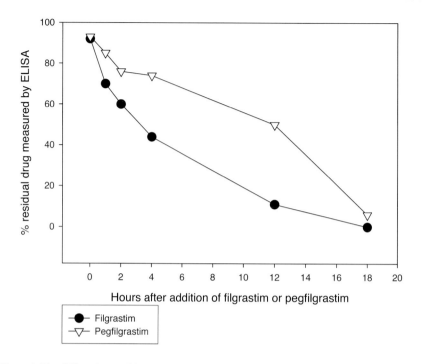

Figure 4. The ability of normal human neutrophils to remove filgrastim and pegfilgrastim from culture supernatent. Adapted from Briddell et al. [30].

Clinical development

Daily dosing with filgrastim is required for clinical efficacy and experiments in animals have illustrated that no matter how far the dose of filgrastim is escalated, the requirement for frequent administration cannot be avoided [14, 57].

Among the early clinical experiments was a simple dose escalation study in normal volunteers (see Fig. 5). Neutrophil counts increased in a dose dependent manner. Other Phase I trials were uneventful and a Phase II trial in patients with non small-cell lung cancer [58] confirmed that many of the initial design objectives had been fulfilled for pegfilgrastim, including an extended duration of action. This study employed an interesting cycle 0/cycle 1 design in which patients intended for treatment received pegfilgrastim prior to chemotherapy (cycle zero) then again immediately after chemotherapy (cycle 1). This allowed each patient to act as their own control and made possible analysis of the effects of chemotherapy induced neutropenia on pegfilgrastim. In cycle zero there was a dose dependent neutrophilia not dissimilar to the data reported from the earlier Phase I trial. The chemotherapy, as expected, caused a significant neutropenia in cycle 1 (Fig. 6A), but the critical analysis from this paper, from a mechanism perspective, is the variation in pharmacokinetics from cycle 0 to cycle 1 (Fig. 6B). Peak serum levels attained in response to

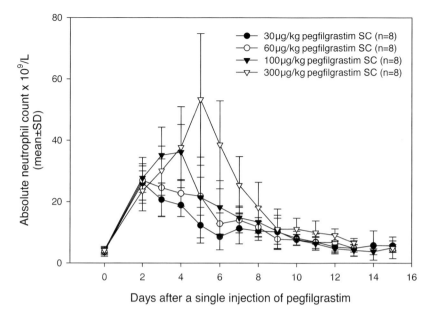

Figure 5. Neutrophil response in normal volunteers injection subcutaneously with a single escalating dose of pegfilgrastim (adapted from Molineux et al. [57]).

100 µg/kg either before or after chemotherapy were similar – at around 100 ng/mL compared with 10–20 ng/mL peak levels in the group receiving the recommended dose of filgrastim. Obviously, the pegfilgrastim recipients received 20 times more drug that accounts for the higher maximum concentration attained, though the rate of loss, once underway, was broadly comparable between pegfilgrastim and filgrastim. The main difference between cycle 0 and cycle 1 was the time at which that clearance began. For several days post chemotherapy no pegfilgrastim is lost from the serum–serum concentration remained constant for several days – a phenomenon that had not been seen in the pre-chemotherapy cycle. However, starting around nine days after chemotherapy pegfilgrastim was lost from the serum – and lost at a precipitous rate. This turning point coincides with the recovery of neutrophils after chemotherapy. This observation is compatible with the concept of self-regulation where pegfilgrastim levels remain broadly stable accelerating neutrophil recovery and when that neutrophil recovery begins the new neutrophils then clear the drug.

Various Phase II trials uncovered no untoward activities of pegfilgrastim and two randomized double blind Phase III trials were initiated with slightly different designs. Both were conducted in breast cancer patients receiving doxorubicin and docetaxel chemotherapy but in one trial later to be reported by Green et al. [59] patients received pegfilgrastim at a fixed dose of 6 mg irrespective of body weight, the complementary trial reported by Holmes et al. [60]

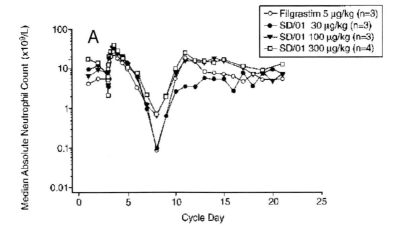

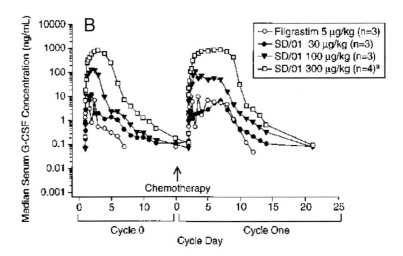

Figure 6. Phase II data with pegfilgrastim (SD/01) in lung cancer patients treated with escalating doses in cycle 0 (pre-chemotherapy) or cycle 1 (post-chemotherapy). Panel A. Neutrophil counts after chemotherapy in cycle 1. Panel B. Pharmacokinetics of pegfilgrastim (SD/01) in cycle 0 (pre-chemotherapy, normal ANC) and in cycle 1 (post-chemotherapy and neutropenic – see Panel A). Note the prolonged exposure in cycle 1 *versus* cycle 0 and the precipitous clearance of SD/01 in parallel with neutrophil recovery. Johnston et al. [58].

used conventional dosing by body weight at 100 μg/kg. The somewhat unusual step of using a fixed dose of a biological was taken based upon analysis of the various Phase II trials in terms of the total dose received by individual patients (of body weights ranging from 46–125 kg) and the duration of their severe neutropenia (see Tab. 2). It is apparent that irrespective of body weight, the days of severe neutropenia (DSN) were similar, perhaps even shorter at the higher body weights. Both Phase III trials focused on DSN as the primary end-

Table 2. The duration of severe neutropenia in the 2 phase 3 trials of pegfilgrastim in which the drug was dosed based upon patient body weight or administered as a single fixed dose.

		Days of severe neutropenia			
		Cycle 1	Cycle 2	Cycle 3	Cycle 4
Per weight dosing (100µg/kg)[59]	pegfilgrastim	1.1	0–1 (in 98% of patients)		
	filgrastim	1.6	0–1 (in 96% of patients)		
Fixed dose (6mg)[60]	pegfilgrastim	1.8	1.1	1.1	1.0
	filgrastim	1.6	0.9	0.9	1.0

point and in both cases DSN was shown to be non-inferior to filgrastim (an unusual endpoint useful in making a statistical comparison to an active control). In the trial where patients received a 6 mg fixed dose, the 77 patients who received pegfilgrastim and the 75 who received filgrastim had 1.8 and 1.6 days of severe neutropenia, and in the by-weight trial 1.1 and 1.6 respectively.

The first warning sign that a neutropenic cancer patient may be developing an infection is becoming febrile (having an elevated temperature). Febrile neutropenia is defined as a temperature of greater than 38.2 °C when accompanied by neutropenia and often prompts the use of anti-infectives even though in many cases an infection cannot be confirmed. Combined data from both the by-weight and fixed dose trials showed that pegfilgrastim reduced significantly the occurrence of febrile neutropenia even compared to filgrastim (11% *versus* 19% – no placebo control group was reported). It is unknown to date why this may be the case. It is tempting to speculate that the front-loading, high dose, or lack of daily fluctuations in drug or ANC levels in the pegfilgrastim recipients may play a role, but dissecting out each of these components has not proven feasible to date.

Bone pain, which is the major side-effect reported for filgrastim, remained the only significant pegfilgrastim-related event that could be teased out of the complex symptoms reported by cancer patients undergoing chemotherapy.

Conclusions

Pegfilgrastim is a rationally designed cytokine derivative engineered specifically to enhance its properties as a therapeutic. The design evolved from understanding the limitations placed on the parent drug by its brief residence time in the body. Of the two routes of filgrastim clearance that contribute to its rapid loss, one – the neutrophil-mediated pathway, is related to the product of the drug's effects, the other is a relentless, linear process based on loss through the kidney. Pegfilgrastim, for the first time, separated these effects and removed renal loss as a significant phenomenon. This left neutrophil-mediated destruction as the only significant route of drug elimination. Since stimulation of neutrophil production is the reason why G-CSF is administered to patients, this novel drug eliminates the requirement for dosing based on the patient's individual characteristics or response.

Though pegfilgrastim is the latest of a new generation of 'designer cytokines' it is unlikely to be the last. The evolution of protein therapeutics from natural materials purified from animal or human sources, through the fabrication of recombinant equivalents to semi-synthetic hybrid molecules (like pegfilgrastim) and eventually to fully synthetic drugs will continue to optimize the utility of this class of drugs improving patient convenience, compliance and response rates but hopefully retaining the exquisite specificity and side effect profile of the parent hormones.

References

1. Nicola NA, Begley CG, Metcalf D (1985) Identification of the human analog of a regulator that induces differentiation in murine leukemic cells. *Nature* 314: 625–628
2. Souza LM, Boone TC, Gabrilove J, Lai PH, Zsebo KM, Murdock DC, Chazin VR, Bruszewski J, Lu H, Chen KK (1986) Recombinant human granulocyte colony-stimulating factor: effects on normal and leukemic myeloid cells. *Science* 232: 61–65
3. Nagata S, Tsuchiya M, Asano S, Kaziro Y, Yamazaki T, Yamamoto O, Hirata Y, Kubota N, Oheda M, Nomura H (1986) Molecular cloning and expression of cDNA for human granulocyte colony-stimulating factor. *Nature* 319: 415–418
4. Bronchud MH, Scarffe JH, Thatcher N, Crowther D, Souza LM, Alton NK, Testa NG, Dexter TM (1987) Phase I/II study of recombinant human granulocyte colony-stimulating factor in patients receiving intensive chemotherapy for small cell lung cancer. *British Journal of Cancer* 56: 809–813
5. Gabrilove JL, Jakubowski A, Fain K, Grous J, Scher H, Sternberg C, Yagoda A, Clarkson B, Bonilla MA, Oettgen HF (1988) Phase I study of granulocyte colony-stimulating factor in patients with transitional cell carcinoma of the urothelium. *Journal of Clinical Investigation* 82: 1454–1461
6. Gabrilove JL, Jakubowski A, Scher H, Sternberg C, Wong G, Grous J, Yagoda A, Fain K, Moore MA, Clarkson B (1988) Effect of granulocyte colony-stimulating factor on neutropenia and associated morbidity due to chemotherapy for transitional-cell carcinoma of the urothelium. *New England Journal of Medicine* 318: 1414–1422
7. Morstyn G, Campbell L, Souza LM, Alton NK, Keech J, Green M, Sheridan W, Metcalf D, Fox R (1988) Effect of granulocyte colony stimulating factor on neutropenia induced by cytotoxic chemotherapy. *Lancet* 1: 667–672
8. Ono M (1994) Physicochemical and biochemical characteristics of glycosylated recombinant human granulocyte colony stimulating factor (lenograstim). *European Journal of Cancer* 30A: Suppl-11
9. Uzumaki H, Okabe T, Sasaki N, Hagiwara K, Takaku F, Tobita M, Yasukawa K, Ito S, Umezawa Y (1989) Identification and characterization of receptors for granulocyte colony-stimulating factor on human placenta and trophoblastic cells. *Proceedings of the National Academy of Sciences of the United States of America* 86: 9323–9326
10. Molineux G, Pojda Z, Hampson IN, Lord BI, Dexter TM (1990) Transplantation potential of peripheral blood stem cells induced by granulocyte colony-stimulating factor. *Blood* 76: 2153–2158
11. Sheridan WP, Begley CG, Juttner CA, Szer J, To LB, Maher D, McGrath KM, Morstyn G, Fox RM (1992) Effect of peripheral-blood progenitor cells mobilised by filgrastim G-CSF on platelet recovery after high-dose chemotherapy. *Lancet* 339: 640–644

12. Morstyn G, Campbell L, Lieschke G, Layton JE, Maher D, O'Connor M, Green M, Sheridan W, Vincent M, Alton K (1989) Treatment of chemotherapy-induced neutropenia by subcutaneously administered granulocyte colony-stimulating factor with optimization of dose and duration of therapy. *Journal of Clinical Oncology* 7: 1554–1562

13. Di Leo A, Bajetta E, Nole F, Biganzoli L, Ferrari L, Oriana S, Riboldi G, Bohm S, Spatti G, Raspagliesi F (1994) The intramuscular administration of granulocyte colony-stimulating factor as an adjunct to chemotherapy in pretreated ovarian cancer patients: an Italian Trials in Medical Oncology (ITMO) Group pilot study. *British Journal of Cancer* 69: 961–966

14. Tanaka H, Tokiwa T (1990) Pharmacokinetics of recombinant human granulocyte colony-stimulating factor studied in the rat by a sandwich enzyme-linked immunosorbent assay. *Journal of Pharmacology & Experimental Therapeutics* 255: 724–729

15. Tanaka H, Kaneko T (1992) Pharmacokinetic and pharmacodynamic comparisons between human granulocyte colony-stimulating factor purified from human bladder carcinoma cell line 5637 culture medium and recombinant human granulocyte colony-stimulating factor produced in Escherichia coli. *Journal of Pharmacology & Experimental Therapeutics* 262: 439–444

16. Eliason JF, Greway A, Tare N, Inoue T, Bowen S, Dar M, Yamasaki M, Okabe M, Horii I (2000) Extended activity in cynomolgus monkeys of a granulocyte colony-stimulating factor mutein conjugated with high molecular weight polyethylene glycol. *Stem Cells (Miamisburg)* 18: 40–45

17. Kearns CM, Wang WC, Stute N, Ihle JN, Evans WE (1993) Disposition of recombinant human granulocyte colony-stimulating factor in children with severe chronic neutropenia. *Journal of Pediatrics* 123: 471–479

18. Nichol JL, Hokom MM, Hornkohl A, Sheridan WP, Ohashi H, Kato T, Li YS, Bartley TD, Choi E, Bogenberger J (1995) Megakaryocyte growth and development factor. Analyses of *in vitro* effects on human megakaryopoiesis and endogenous serum levels during chemotherapy-induced thrombocytopenia. *Journal of Clinical Investigation* 95: 2973–2978

19. Marsh JC, Gibson FM, Prue RL, Bowen A, Dunn VT, Hornkohl AC, Nichol JL, Gordon-Smith EC (1996) Serum thrombopoietin levels in patients with aplastic anaemia. *British Journal of Haematology* 95: 605–610

20. Emmons RV, Reid DM, Cohen RL, Meng G, Young NS, Dunbar CE, Shulman NR (1996) Human thrombopoietin levels are high when thrombocytopenia is due to megakaryocyte deficiency and low when due to increased platelet destruction. *Blood* 87: 4068–4071

21. Fielder PJ, Hass P, Nagel M, Stefanich E, Widmer R, Bennett GL, Keller GA, de Sauvage FJ, Eaton D (1997) Human platelets as a model for the binding and degradation of thrombopoietin. *Blood* 89: 2782–2788

22. Kato M, Kamiyama H, Okazaki A, Kumaki K, Kato Y, Sugiyama Y (1997) Mechanism for the nonlinear pharmacokinetics of erythropoietin in rats. *Journal of Pharmacology & Experimental Therapeutics* 283: 520–527

23. Redman BG, Flaherty L, Chou TH, Kraut M, Martino S, Simon M, Valdivieso M, Groves E (1992) Phase I trial of recombinant macrophage colony-stimulating factor by rapid intravenous infusion in patients with cancer. *Journal of Immunotherapy* 12: 50–54

24. Bauer RJ, Gibbons JA, Bell DP, Luo ZP, Young JD (1994) Nonlinear pharmacokinetics of recombinant human macrophage colony-stimulating factor (M-CSF) in rats. *Journal of Pharmacology & Experimental Therapeutics* 268: 152–158

25. Layton JE, Hockman H, Sheridan WP, Morstyn G (1989) Evidence for a novel *in vivo* control mechanism of granulopoiesis: mature cell-related control of a regulatory growth factor. *Blood* 74: 1303–1307

26. Shimazaki C, Uchiyama H, Fujita N, Araki S, Sudo Y, Yamagata N, Ashihara E, Goto H, Inaba T, Haruyama H (1995) Serum levels of endogenous and exogenous granulocyte colony-stimulating factor after autologous blood stem cell transplantation. *Experimental Hematology* 23: 1497–1502

27. Takatani H, Soda H, Fukuda M, Watanabe M, Kinoshita A, Nakamura T, Oka M (1996) Levels of recombinant human granulocyte colony-stimulating factor in serum are inversely correlated with circulating neutrophil counts. *Antimicrobial Agents & Chemotherapy* 40: 988–991

28. Stute N, Santana VM, Rodman JH, Schell MJ, Ihle JN, Evans WE (1992) Pharmacokinetics of subcutaneous recombinant human granulocyte colony-stimulating factor in children. *Blood* 79: 2849–2854

29. Sturgill MG, Huhn RD, Drachtman RA, Ettinger AG, Ettinger LJ (1997) Pharmacokinetics of intravenous recombinant human granulocyte colony-stimulating factor (rhG-CSF) in children receiving myelosuppressive cancer chemotherapy: clearance increases in relation to absolute neu-

trophil count with repeated dosing. *American Journal of Hematology* 54: 124–130

30. Briddell R, Stonev G, Molineux G (2001) Investigation of the mode of clearance of filgrastim and filgrastim SD/01 by human peripheral blood neutrophils *in vitro*. *Experimental Hematology* 29

31. Tanaka H, Tokiwa T (1990) Influence of renal and hepatic failure on the pharmacokinetics of recombinant human granulocyte colony-stimulating factor krn-8601 in the rat. *Cancer Research* 50: 6615–6619

32. Kuwabara T, Ishikawa Y, Kobayashi H, Kobayashi S, Sugiyama Y (1995) Renal clearance of a recombinant granulocyte colony-stimulating factor, nartograstim, in rats. *Pharmaceutical Research* 12: 1466–1469

33. Tanaka J, Miyake T, Shimizu T, Wakayama T, Tsumori M, Koshimura K, Murakami Y, Kato Y (2002) Effect of continuous subcutaneous administration of a low dose of G-CSF on stem cell mobilization in healthy donors: A feasibility study. *International Journal of Hematology* 75: 489–492

34. Furuya H, Wakayama T, Ohguni S, Yamauchi K, Tanaka J, Hatazoe T, Kato Y (1995) Effect of continuous subcutaneous administration of a small dose of granulocyte colony stimulating factor (G-CSF) by the use of a portable infusion pump in patients with non-Hodgkin's lymphoma receiving chemotherapy. *International Journal of Hematology* 61: 123–129

35. Egrie JC, Dwyer E, Browne JK, Hitz A, Lykos MA (2003) Darbepoetin alfa has a longer serum half-life and greater *in vivo* potency than recombinant human erythropoietin. *Experimental Hematology* 31: 290–299

36. Elliott S, Egrie J, Browne J, Lorenzini T, Busse L, Rogers N, Ponting I (2004) Control of rHuEPO biological activity: The role of carbohydrate. *Experimental Hematology* 32: 1146–1155

37. Heatherington AC, Schuller J, Mercer AJ (2001) Pharmacokinetics of novel erythropoiesis stimulating protein (NESP) in cancer patients: preliminary report. *British Journal of Cancer* 84: Suppl-6

38. Halpern W, Riccobene TA, Agostini H, Baker K, Stolow D, Gu M-L, Hirsch J, Mahoney A, Carrell J, Boyd E et al. (2002) Albugranin™, a recombinant human granulocyte colony stimulating factor (G-CSF) genetically fused to recombinant human albumin induces prolonged myelopoietic effects in mice and monkeys. *Pharmaceutical Research* 19: 1720–1729

39. Jensen-Pippo KE, Whitcomb KL, Deprince RB, Ralph L, Habberfield AD (1996) Enternal bioavailability of human granulocyte colony stimulating factor conjugated with poly(ethylene glycol). *Pharmaceutical Research (New York)* 13: 102–107

40. Choi SH, Lee H, Park TG (2003) PEGylation of G-CSF using cleavable oligo-lactic acid linkage. *Journal of Controlled Release* 89: 271–284

41. DeFrees S, Wang Z-G, Xing R, Scott AE, Wang J, Zopf D, Gouty DL, Sjoberg ER, Panneerselvam K, Brinkman-Van der Linden ECM et al. (2006) GlycoPEGylation of recombinant therapeutic proteins produced in Escherichia coli. *Glycobiology* 16: 833–843

42. Mueller H-J, Loening L, Horn A, Schwabe D, Gunkel M, Schrappe M, von Schuetz V, Henze G, da Palma JC, Ritter J et al. (2000) Pegylated asparaginase (Oncaspar™) in children with ALL: Drug monitoring in reinduction according to the ALL/NHL-BFM 95 protocols. *British Journal of Haematology* 110: 379–384

43. Melton A, Hershfield MS, Greenberg ML, Hatem C, Markert ML, Kurtzberg J, Abuchowski A, Buckley RH (1987) Treatment of adenosine deaminase-deficient severe combined immune deficiency ADA-SCID with polyethylene glycol-modified bovine adenosine deaminase PEG-ADA. *Journal of Allergy & Clinical Immunology* 79: 252

44. Karasiewicz R, Nalin C, Rosen P (1995) Peg-interferon conjugates. *Official Gazette of the United States Patent & Trademark Office Patents*: 17, 1995

45. Bailon P, Palleroni A, Schaffer CA, Spence CL, Fung W-J, Porter JE, Ehrlich GK, Pan W, Xu Z-X, Modi MW et al. (2001) Rational design of a potent, long-lasting form of interferon: A 40 kDa branched polyethylene glycol-conjugated interferon alpha-2a for the treatment of hepatitis C. *Bioconjugate Chemistry* 12: 195–202

46. Malik F, Brew J, Maidment SA, Delgado C, Francis GE (2000) PEG-modified erythropoietin with improved efficacy. *Experimental Hematology* 28: 106

47. Sato H, Yamamoto K, Hayashi E, Takahara Y (2000) Transglutaminase-mediated dual and site-specific incorporation poly(ethylene glycol) derivatives into a chimeric interleukin-2. *Bioconjugate Chemistry* 11: 502–509

48. Hokom MM, Lacey D, Kinstler OB, Choi E, Kaufman S, Faust J, Rowan C, Dwyer E, Nichol JL, Grasel T et al. (1995) Pegylated megakaryocyte growth and development factor abrogates the lethal thrombocytopenia associated with carboplatin and irradiation in mice. *Blood* 86:

4486–4492
49. Jolling K, Ruixo JJP, Hemeryck A, Piotrovskij V, Greway T (2004) Population pharmacokinetic analysis of pegylated human erythropoietin in rats. *Journal of Pharmaceutical Sciences* 93: 3027–3038
50. Kuan C-T, Wang Q-C, Pastan I (1994) Pseudomonas exotoxin A mutants: Replacement of surface exposed residues in domain II with cysteine residues that can be modified with polyethylene glycol in a site-specific manner. *Journal of Biological Chemistry* 269: 7610–7616
51. Tsutsumi Y, Onda M, Nagata S, Lee B, Kreitman RJ, Pastan I (2000) Site-specific chemical modification with polyethylene glycol of recombinant immunotoxin anti-Tac(Fv)-PE38 (LMB-2) improves antitumor activity and reduces animal toxicity and immunogenicity. *Proceedings of the National Academy of Sciences of the United States of America* 97: 8548–8553
52. Benhar I, Wang Q-C, Fitzgerald D, Pastan I (1994) Pseudomonas exotoxin A mutants: Replacement of surface-exposed residues in domain III with cysteine residues that can be modified with polyethylene glycol in a site-specific manner. *Journal of Biological Chemistry* 269: 13398–13404
53. Gaertner HF, Offord RE (1996) Site-specific attachment of functionalized poly(ethylene glycol) to the amino terminus of proteins. *Bioconjugate Chemistry* 7: 38–44
54. Wang M, Lee LS, Nepomich A, Yang J-D, Conover C, Whitlow M, Filpula D (1277) Single-chain Fv with manifold N-glycans as bifunctional scaffolds for immunomolecules. *Protein Engineering* 11: 1277–1283
55. Kinstler OB, Brems DN, Lauren SL, Paige AG, Hamburger JB, Treuheit MJ (1996) Characterization and stability of N-terminally PEGylated rhG-CSF. *Pharmaceutical Research* 13: 996–1002
56. Kinstler O, Molineux G, Treuheit M, Ladd D, Gegg C (2002) Mono-N-terminal poly(ethylene glycol)-protein conjugates. *Advanced Drug Delivery Reviews 54(4):* 477–485
57. Molineux G, Kinstler O, Briddell B, Hartley C, McElroy P, Kerzic P, Sutherland W, Stoney G, Kern B, Fletcher FA et al. (1999) A new form of Filgrastim with sustained duration *in vivo* and enhanced ability to mobilize PBPC in both mice and humans. *Experimental Hematology* 27: 1724–1734
58. Johnston E, Crawford J, Blackwell S, Bjurstrom T, Lockbaum P, Roskos L, Yang B-B, Gardner S, Miller-Messana MA, Shoemaker D et al. (2000) Randomized, dose-escalation study of SD/01 compared with daily filgrastim in patients receiving chemotherapy. *Journal of Clinical Oncology* 18: 2522–2528
59. Green MD, Koelbl H, Baselga J, Galid A, Guillem V, Gascon P, Siena S, Lalisang RI, Samonigg H, Clemens MR et al. (2003) A randomized double-blind multicenter phase III study of fixed-dose single-administration pegfilgrastim *versus* daily filgrastim in patients receiving myelosuppressive chemotherapy. *Annals of Oncology* 14: 29–35
60. Holmes FA, O'Shaughnessy JA, Vukelja S, Jones SE, Shogan J, Savin M, Glaspy J, Moore M, Meza L, Wiznitzer I et al. (2002) Blinded, randomized, multicenter study to evaluate single administration pegfilgrastim once per cycle *versus* daily filgrastim as an adjunct to chemotherapy in patients with high-risk stage II or stage III/IV breast cancer. *Journal of Clinical Oncology* 20: 727–731
61. Yang B-B, Lum PK, Hayashi MM, Roskos LK (2004) Polyethylene glycol modification of filgrastim results in decreased renal clearance of the protein in rats. *Journal of Pharmaceutical Sciences* 93: 1367–1373

PEGylated Protein Drugs: Basic Science and Clinical Applications
Edited by F.M. Veronese
© 2009 Birkhäuser Verlag/Switzerland

PEGylation of human growth hormone: strategies and properties

Rory F. Finn

Pfizer Inc, 700 Chesterfield Parkway West, Chesterfield, USA

Abstract

Recombinant human growth hormone (hGH) is a well characterized molecule with broad acceptance as a treatment for growth hormone deficiencies (GHD). However, treatment with hGH requires daily injections due to the drug's short duration of action. Many groups have focused on PEGylation of hGH as a means to extend its half-life and generate less frequent dosage forms. This chapter provides a review of the preclinical and clinical results obtained from the many approaches directed towards modification of hGH with PEG. The chapter will describe a historical progression of PEGylation strategies and results. The first half of the chapter will discuss initial studies that utilized multiple 5 kDa PEG attachments for extension of hGH half-life and the subsequent development of a PEGylated hGH receptor antagonist, pegvisomant, a successful therapy for acromegaly. The latter half of the chapter will summarize more recent and current work focusing on site selective mono-PEGylation of hGH.

Introduction

Human growth hormone (hGH, human somatotropin) is secreted by the pituitary gland in a pulsatile and episodic manner. It acts both directly and indirectly on tissues to elicit a wide range of growth enhancing effects (Fig. 1). Pharmacological responses to hGH are mediated through the GH receptor (GHR). These include immediate responses such as the production of the insulin-like growth factors I and II (IGF-1 and IGF-2), the stimulation of triglyceride hydrolysis in adipose tissues, and the stimulation of hepatic glucose output. Human growth hormone also mediates multiple intermediate responses such as anabolism and growth promotion in the form of chondrogenesis, skeletal and soft tissue growth that are mediated by insulin-like growth factor-1 (IGF-1). IGF-1 is produced primarily by the liver following hGH-induced increases in IGF-1 mRNA transcription [1, 2]. The IGF-1 is disseminated in the plasma to the assorted sites of growth promotion in complex with a binding protein (IGFBP-3) and the acid-labile subunit protein (ALS) and hGH plays a critical role in the genesis and regulation of this insulin-like growth factor-1 binding protein complex [3]. Recombinant hGH is prescribed worldwide for both pediatric and adult Growth Hormone Deficiency (GHD), as well as, Chronic Renal Insufficiency, and treatment of short stature in patients with Turner's Syndrome and children born small for gestational age

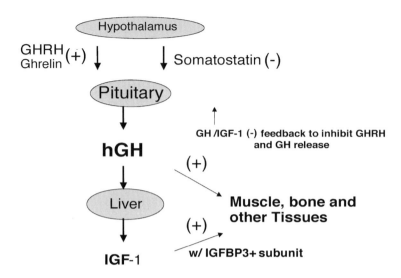

Figure 1. Schematic of growth hormone action and regulation. hGH is secreted by the pituitary under direct regulation by the hypothalamus and can stimulate growth directly through GH receptors in tissues and muscle and indirectly through induction of IGF-1 production in the liver.

(SGA). Six major pharmaceutical companies currently market hGH with annual worldwide sales exceeding 2.7 B (US$).

Due to the rapid clearance rate of hGH in the blood, replacement therapy typically involves daily subcutaneous injections over the duration of treatment which generally extends through puberty in pediatric GHD. Therefore, development of an effective long duration form of hGH would be desirable as a means to decrease the number and frequency of injections. Toward this end, many approaches are being pursued in order to extend the duration of action of hGH. These include specific slow subcutaneous release formulations [4–8] and hGH fused to proteins such as albumin [9]. Alternatively, conjugation of polyethylene glycol (PEG), described as PEGylation, affords a proven means to enhance the duration of action, stability, solubility and/or safety of a given therapeutic polypeptide [10–16], and many groups are studying the effects of PEGylation on hGH pharmacology as an approach to reducing parenteral injection frequency.

hGH structure and receptor binding

Human growth hormone is a 22,000 Dalton (Da) protein comprised of a single chain of 191 amino acids with a tertiary structure containing four anti-parallel α-helices and two disulfide bonds. Human growth hormone interacts with growth hormone receptor (GHR) expressed in target tissues with a stoichiometry of one hGH molecule to two GHR suggesting dimerization of the recep-

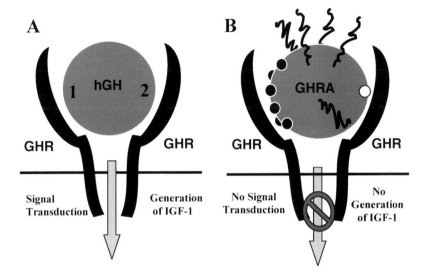

Figure 2. Interaction of hGH and GHRA with GH receptor. A) hGH binding to GHR dimer through binding sites 1 and 2 induces conformational change for signal transduction and IGF-1 production. B) Representation of GHRA (Pegvisomant), with site one directed amino acid mutations (●) and PEGylation ( ). GHRA binds to GHR at binding site 1 with lower affinity presumably due to steric hindrance by PEG. Binding site two mutation, G120K (○) inhibits conformational change needed for signal transduction and IGF-generation.

tor. hGH binds through two separate binding sites, respectively, to each monomer in the GHR dimer, forming an active ligand-receptor trimer, that is necessary for initial signal transduction and pathway activation (Fig. 2A). Originally, a model was proposed whereby hGH bound sequentially via hGH binding site one to a GHR monomer followed by subsequent binding to a second GHR through hGH binding site two to effect receptor dimerization and signal transduction [17]. Recent studies suggest a pre-formed GHR dimer is present on cell surfaces and hGH binding induces a conformational change in the receptor for activation [18, 19]. In either case, binding of both hGH binding sites are crucial for activity, and interference with either site is a concern when developing long acting forms of the molecule.

Multi-PEGylated hGH

Initial studies supporting the use of PEGylation to extend hGH action were reported by Clark and co-workers [20]. As larger molecular weight PEG reagents (>5 kDa) had not yet become readily available, the authors utilized first generation PEG technology in which multiple PEG molecules were conjugated to available amine residues in a relatively nonselective fashion to afford extended pharmacokinetics (PK). The primary structure of hGH contains ten available primary amines (nine lysine residues and the N-terminus) [21]. The authors prepared a series of PEGylated conjugates with an increasing number of PEG polymers attached by reaction with the N-hydroxysuccinimide (NHS) ester [22] of 5 kDa PEG (PEG 5000 NHS). Since this reagent will react readily with all available amines, the reaction produced a heterogeneous mixture of PEG–hGH species. The PEGylation products were fractionated and subsequently characterized as to the number and the locations of the modified amines. Conjugates with 2–7 PEG moieties attached per hGH molecule were identified. Preclinical studies were then carried out to determine the effect of the various extents of PEGylation on receptor binding affinity, hGH circulating half-life and *in vivo* potency. Although the most reactive amines did not lie in close proximity to the two receptor binding sites, increasing numbers of PEG bound to hGH did result in a respective decrease in affinity for the GH receptor up to 1,500-fold. *In vivo* efficacy, however, increased with increasing levels of PEG modification in conjunction with increased circulation half lives suggesting that decreased affinity can be overcome by extension of plasma residency time. Comparisons of the effective sizes of the PEG-GH conjugates, as determined by size exclusion chromatography, with rat PK data showed that an average of 4–5 PEG 5000 polymers was required to achieve an effective molecular weight that exceeded the 70 kDa cut-off for rapid renal clearance. Clearance significantly decreased as greater than 5 PEG moieties were attached. Subsequent *in vivo* weight gain studies in hypophysectomized rats determined that hGH modified with five PEG 5000 groups yielded the most effective long-acting molecule. Moreover, the 5×5 kDa PEG-hGH conjugate was as effective as daily hGH injections in its growth promoting activity when administered on a once per week basis. Clearly, these studies demonstrated that improving hormone clearance properties could offset reduced binding affinity due to PEG interference and that a PEGylated hGH could be dosed less frequently and, therefore, could be viable as an alternative to daily injections of hGH. This approach was used successfully in developing pegvisomant, a PEGylated growth hormone antagonist for treatment of human disease.

Pegvisomant

B2036 discovery

Kopchick and co-workers hypothesized that stabilization of the third helix in bovine growth hormone (bGH) would enhance receptor binding affinity and increase bGH potency. Interestingly, transgenic mice, expressing a (bGH) variant with three amino acids substitutions for increased amphiphilicity in helix three, demonstrated lower circulating levels of IGF-1 and a dwarf phenotype [23]. Continued studies determined that the glycine at position 120 in hGH (binding site 2) was crucial for hGH activity and substitution of any amino acid except alanine led to conversion of hGH to a potent GHR antagonist [24].

The mutant molecule, hGH G120K, binds to the GH receptor dimer, is internalized, yet does not generate the receptor conformational change required for signaling [19, 25]. In designing this molecule as a possible therapeutic GHR antagonist molecule, eight other residues in hGH G120K were mutated in order to increase the affinity at binding site 1 generating the GHR antagonist molecule known as B2036 [26].

PEGylation of B2036

The GHR antagonist molecule, B2036, was determined to have a very short half-life similar to that of hGH (15–20 min). Therefore, B2036 was PEGylated to extend its duration of action. 4–6 5 kDa PEG molecules were attached to free amino groups on B2036, and this molecule is known as pegvisomant (Figs 2B and 3) [27, 28]. As with GH, the PEGylation greatly reduces affinity for the GH receptor (20-fold reduction) [19, 28], however, as expected, the half-life is prolonged to 72–75 h [29, 30]. Subsequent studies on rhesus monkeys dosed with pegvisomant by intravenous or subcutaneous injection demonstrated dose-dependent reductions of IGF-1 and IGF binding protein-3 (IGFBP-3) with IGF-1 levels remaining suppressed for seven days after a single 1 mg/kg subcutaneous dose [31, 32].

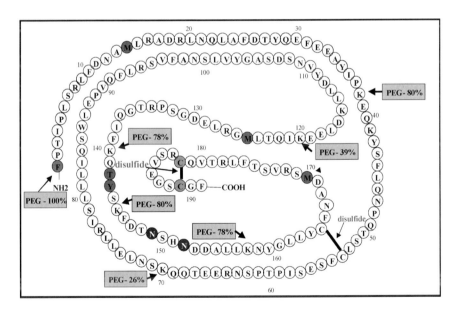

Figure 3. Ball and chain representation of primary structure of pegvisomant. The B2036 amino acid sequence with disulfide bridges and potential PEGylation sites. Pegvisomant is a mixture of 4–6 × 5 kDa PEG molecules/B2036 distributed as depicted by the relative reactivity of each (compared to that of the N-terminus indicated by %). Originally published in [50]. Reproduced with kind permission of Taylor and Francis Group LLC.

GH receptor antagonists as a therapy for acromegaly

Acromegaly is a chronic disorder in which a benign pituitary adenoma effects hyper-secretion of growth hormone leading to elevated levels of insulin growth factor-I (IGF-I). Acromegalic patients have abnormally high basal (non-pulsatile) hGH production, hGH pulsatile secretion and burst frequency. The disorder results in increased body weight and height as well as a wide range of secondary symptoms such as headaches, cerebral nerve disfunction, visual field defects, impotence, infertility, and amenhorrhea. Acromegalic patients also have increased incidence of cardiovascular and respiratory diseases, and increased risk of malignancies [33, 34]. Historically, treatments for the disorder included surgical removal of the adenoma [35] and radiation therapy [36] as well as medical treatment with dopamine [37] and somatostatin agonists [38]. In each case these treatments have limitations with successful treatment in only 50–60% of patients.

The above treatments target decreased levels of hGH and IGF-1 through inhibition of hGH secretion from the pituitary. It was speculated that a similar effect might be achieved by specifically blocking hGH signaling at the receptor level. Toward this end, pegvisomant was pursued as the first GHR antagonist for treatment of acromegaly generating an alternate approach for treating previously non-responsive patients.

PEGvisomant; a clinical success

Somavert® (pegvisomant for injection) is currently marketed for the treatment of acromegaly (GH excess) in patients who have had an inadequate response to surgery and/or radiation therapy and/or other medical therapies, or for whom these therapies are not appropriate. A number of reviews and clinical study compilations have been published in recent years on pegvisomant in the treatment of acromegaly [32, 39, 40]. Early Phase I and II clinical studies with weekly 1 mg/kg single subcutaneous dosing demonstrated that the half-life of pegvisomant was approximately 72 h with peak plasma levels at 36 h and maximal suppression of IGF-1 around five days [41]. Plasma IGF-1 levels decreased as much as 30%, however, IGF-1 levels failed to normalize in the majority of these subjects. Further pharmacokinetic analysis suggested that this weekly dosing schedule was not sustaining the pegvisomant plasma levels needed for normalization of IGF-1 levels. Subsequent studies showed that daily dosing with the same total amount of pegvisomant could raise trough levels to that required for the normalization of IGF-1 levels [42].

The first Phase III clinical study reporting the efficacy of pegvisomant as a therapy for acromegaly was published in 2000. In this study, 112 patients, who had previously unsuccessfully undergone treatment by surgery, radiation, dopamine, somatostatin agonists or combinations of each, demonstrated that daily dosing (up to 20 mg for 12 weeks) normalized IGF-1 levels in 82% of the subjects [43]. A larger 160 patient study, all of whom had undergone previous treatments, resulted in 97% normalization of IGF-1 levels following 12 month 40 mg/day dosing [44]. In all studies, decreased IGF-1 levels yielded corresponding decreases in the signs and symptoms related to acromegaly such as fatigue, soft tissue swelling, and excess perspiration. Multiple studies have shown that pegvisomant improves insulin sensitivity and glucose tolerance as well [44–47]. A recent longer term (24 month) observational study of pegvisomant safety and efficacy under field conditions in clinical practice concluded that the drug showed a well tolerated safety profile similar to that of the clinical trials [40, 43, 48] and effectively reduced IGF-1 levels in patients previously shown to be non-responsive to other therapies [49]. Although more studies are needed to determine long-term efficacy and safety (either as a mono or combination therapy), currently pegvisomant appears to be well accepted as an effective choice of therapy for patients resistant to somatostatin analogs [32, 40]. The success of pegvisomant, which is essentially a PEGylated GH variant molecule, has given promise for PEGylated hGH as a treatment for GHD.

Mono-PEGylation of hGH

Development of pegvisomant was challenging due to the heterogeneous nature of the PEGylation. Regulatory approval required a product with a reproducible, uniform mixture of PEG isomers. PEGylation reaction kinetic models and refined purification and analytical methodologies were needed in order to generate a pegvisomant product with a well-defined PEG-profile with consistent mixtures of four, five, or six molecules of PEG per molecule of protein [50]. In recent years, regulatory guidelines for approval of PEGylated products have become more stringent with a need for characterization of all isomers (whenever possible), and therefore, groups are focusing on mono-PEGylation of proteins [15, 51, 52]. This trend has been influenced by improvements in PEGylation technologies, notably the development of low polydispersity PEG reagents of increasing size and selectivity. Typically, PEG sizes of 12–20 kDa or greater are chosen to maximize hydrodynamic volume and exceed renal exclusion cut-off criteria (>70 kDa effective molecular weight). Generally, single modification of therapeutic proteins can have pronounced advantages over multiple attachments. In the case of PEG hGH, a single PEG attachment not only would reduce heterogeneity and streamline process development, but also might minimize PEG interference with receptor binding.

One approach toward selective mono-PEGylation of proteins is to attach a PEG moiety to the N-terminus via reductive alkylation using the method of Kinstler et al. [52]. PEG aldehydes will preferentially react with the amino terminus of proteins under mildly acidic conditions due to the relative differences in pKa values between the epsilon amino groups of lysine and the alpha amino group at the N-terminus. In this manner, systematic studies of the effect of mono N-terminal PEG additions have been carried out on hGH using larger

PEG polymers (up to 40,000 Da), with the concomitant analysis of the products with respect to synthetic and analytical properties as well as biological function [53, 54]. The intent was to determine whether a single larger molecular weight PEGylation at a specific site, could simplify production and characterization methods, while yielding similar biological responses to the previously described multi-PEGylated $4-6 \times 5,000$ Da PEG-hGH [20]. PEG-hGH conjugates were evaluated for their pharmacodynamic and pharmacokinetic profile, as well as for other factors critical in the development of a protein therapeutic, such as synthetic yields, homogeneity, and identification of conjugation sites. A series of mono-PEGylated hGH variants were synthesized utilizing assorted activated polyethylene glycol (PEG) reagents (Nektar Therapeutics) of various lengths and geometries together with amine coupling chemistry (NHS and aldehyde). Typically mono-PEGylation yields were inversely proportional to PEG size and directly proportional to selectivity (Tab. 1). Biochemical characterization of reaction products demonstrated that aldehyde PEG-hGH reactions generally yielded products with less heterogeneity. The purified mono-PEG aldehyde-hGH reaction products possessed a single N-terminal site of PEG attachment as determined by proteolytic mapping experiments. Proteolytic mapping and RP–HPLC demonstrated that the purified mono-NHS (succinimidyl ester)-hGH reaction products were heterogeneous with multiple PEG binding isomers [54].

Compounds were tested in the hypophysectomized rat model of growth hormone deficiency (GHD) to evaluate the effect of number, polymer size, and conjugation site upon biological response. Once weekly dosing of linear 5, 20, 30 and branched 40 kDa PEG-hGH conjugates yielded increases in IGF-1 levels (Fig. 4) with subsequent weight gains of 36%, 79%, 92 and 102% respectively, when compared to hGH administered by daily injection [53] (Tab. 2).

Table 1. Comparison of product yields for synthesis and purification of a panel of monoPEGylated hGH conjugates. Final product yields are reported as a % relative to the starting hGH protein concentration. Conjugates were 90–95% pure by SEC-HPLC and SDS-PAGE and shown to be mono-PEGylated by Maldi-TOF MS. Purified mono-PEG aldehyde-hGH conjugates had a single N-terminal site of attachment as determined by proteolytic mapping experiments Proteolytic mapping and RP–HPLC suggested that the purified mono-NHS (succinimidyl ester)-hGH conjugates were heterogeneous with multiple PEG binding isomers

PEG hGH conjugate	Yield %
NHS chemistry	
$4-6 \times 5$ kDa PEG Hgh	25%
1×20 kDa PEG hGH	28%
1×30 kDa PEG hGH	28%
Aldehyde Chemistry	
(1×20 kDa PEG ALD) hGH	50%
(1×30 kDa PEG ALD) hGH	45%
(1×40 kDa PEG ALD) hGH	30%

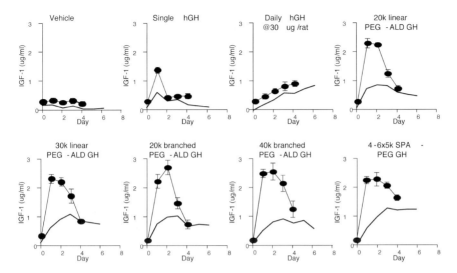

Figure 4. Assessment of hGH stimulation of IGF-1 release. hGH or PEG-GH conjugates were administered to hypophysectomized rats (growth hormone deficiency model) on day 0 and IGF-1 levels were monitored on a daily basis by immunoassay. A subcutaneous (SC) dose of 1.8 mpk PEG-hGH was compared to an equivalent accumulative daily SC dose of GH (0.3 mpk per day). The IGF-1 levels are plotted *versus* time (solid circles) and compared in each plot to the growth effects in a companion study (lines no circles). Mono PEGylated GH conjugates can induce IGF-1 release to levels similar to daily GH and 4–6 × 5,000 Da PEG- hGH.

Enhancement of growth was additionally confirmed by increases in tibia bone length. Pharmacokinetic (PK) studies on the series showed an extension of plasma half-life directly proportional to size (hydrodynamic volume) for the N-terminally PEGylated hGH (Fig. 5) [53]. Interestingly, size (Tab. 3) and PK/PD (Figs 4 and 5) did not correlate well when comparing the N-terminal

Table 2. Hypophysectomized Rat Growth Model. The efficacy of once weekly PEG-hGH was compared to daily hGH in an 11 day growth study using female hypophysectomized (pituitary removed) rats. Total body weight gain was assessed at day 11 following dosing at day 0 and day 6 with 1.8 mpk GH equivalence and compared to a daily GH dose of 0.3 mpk (mg/kg) and reported as % relative to daily hGH dosing. *Lin* denotes linear PEG, *BR* denotes branched PEG. *ALD* denotes aldehyde chemistry (Nektar Therapeutics). *SPA* denotes succinimidyl propionate chemistry (Nektar Therapeutics)

Conjugate tested at 1.8 mpk on days 0 and 6	Weight gain (% relative to daily hGH)
Daily hGH (0.3 mpk)	*100% (n = 24)*
hGH	39% (n = 8)
5 kDa Lin PEG-ALD hGH	36% (n = 8)
20 kDa Lin PEG-ALD hGH	73% (n = 24)
20 kDa Br PEG-ALD hGH	87% (n = 3)
30 kDa Lin PEG-ALD hGH	93% (n = 24)
40 kDa Br PEG-ALD hGH	102% (n = 24)
4–6 × 5 kDa PEG-SPA hGH	132% (n = 16)

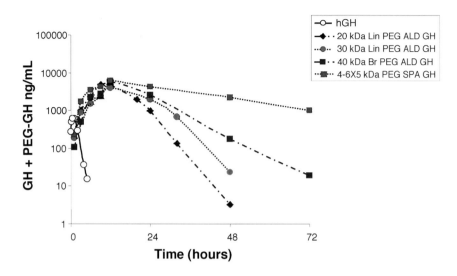

Figure 5. Pharmacokinetics of N-terminally Mono-PEGylated hGH. hGH or PEG-GH conjugates were administered as single subcutaneous bolus injections (1.8 mpk) to hypophysectomized female Sprague Dawley rats and GH/PEG hGH levels in the blood were monitored by immunoassay.

mono-PEGylated conjugates to the multiply PEGylated 4–6 × 5 kDa hGH molecule. The latter had a molecular size similar to the 20 kDa PEG-Aldehyde hGH, yet had the longest observed half-life. These data suggest clearance for PEG hGH is not only a function of PEG size but also related to the number and/or site(s) of attachment.

Table 3. PEG hGH conjugates were assessed for effective molecular size by size exclusion chromatography (SEC). Due to the large hydrodynamic volume contributed by PEG, each molecule elutes at a much larger molecular weight as compared to a protein standard curve (2–3 times theoretical MW)

Size Exclusion Chromatography

	MW (Theoretical)	Size (SEC)
GH	22,000	21,000
4–6 × 5K PEG GH	47,000	128,000
20K Linear PEG GH	42,000	120,000
20K Branched PEG GH	42,000	114,000
40K Branched PEG GH	62,000	205,000

Clinical studies with N-terminal PEG-hGH

The aforementioned studies identified a hGH candidate with a 40 kDa branched PEG N-terminally attached (PHA-794428) that had an approximately 20-fold decrease in binding affinity [55] yet induced weight gains similar to daily dosing of hGH when dosed weekly in hypophysectomized rat growth models. This molecule was chosen for clinical studies in both normal volunteers and adult hGH deficiency (AGHD) patients [56]. 56 normal volunteers were given seven escalating doses (range 3–500 μg/kg; eight subjects per group) and monitored for GH and IGF-1 levels [57, 58]. Plasma concentrations of hGH increased greatly with a maximum at 48 h and maintenance of supra-physiological concentrations up to 246 h. Correspondingly, IGF-1 levels also increased. PHA-794428 was well tolerated up to 300 μg/kg which was defined as the maximum tolerated dose. PEG hGH dosed at 60 μg/kg resulted in an increase in IGF-1 from a 160 ng/mL basal level to 400 ng/mL at 48 h and concentrations remained elevated over basal levels for a period up to 240 h. These results suggest once weekly administration may be feasible. Two doses (20 μg/kg and 60 μg/kg) were given to patients with AGHD (eight per group) [59]. hGH deficient patients receiving 60 μg/kg doses demonstrated hGH serum concentration curves similar to those seen in normal volunteers. IGF-1 levels increased by 220 ng/mL which peaked after 48 h and remained above baseline for 336 h. Although both hGH and IGF-1 level increases were significantly smaller with the 20 μg/kg dose, detectable levels remained for 96 h and 240 h, respectively. PHA-794428 was well tolerated in these studies with two adverse effects in the patient group (mild diarrhea and mild fatigue) both which resolved. No serious adverse events were seen, with no injection site reactions, and no specific anti-PHA-794428 or anti-hGH IgG's observed. However, injection site lipoatrophy was observed after single injections in late Phase II trials, and subsequently, clinical studies on PHA-794428 were terminated [www.clinicaltrial.gov].

Insertion of specific PEG attachments sites into hGH

Modification of native hGH as described above is limited to 'natural' fixed sites of attachment and maintenance or loss of activity is dependent on the number and proximity of those sites to receptor binding sites on the protein. In order to circumvent these losses in activity due to PEG interference and minimize heterogeneity due to PEGylation of multiple residues on the protein, several groups have studied the effects of engineering specific PEG attachment sites into the hGH molecule.

Free sulfhydryl PEG attachment

In a study by Cox et al., an unpaired free cysteine residue was incorporated into hGH via site-directed mutagenesis [60]. Over 40 mutants with cysteine insertions at various sites were expressed in E. coli and screened for biological activity. A molecular variant with cysteine substituted for threonine at residue 3 (hGH T3C) was identified that could be correctly refolded with the free thiol intact and readily available for PEG coupling. Subsequent PEGylation of hGH T3C using a specific sulfhydryl reactive 20 kDa PEG (vinyl-sulfone PEG) yielded a homogeneous mono-PEGylated product with only a slight (3–4-fold) attenuation in hGH *in vitro* biological activity. The 20 kDa PEG hGH TC3 demonstrated an extended half-life (eight-fold) over nonPEGylated hGH in rat pharmacokinetic models and an increased duration of action in growth studies carried out in hypophysectomized rats. Rat weight gain and tibial bone growth resulting from administration of 20k Da PEG hGH T3C every third day was comparable to the weight gain and tibial bone growth observed with hGH administered daily.

Chemical/enzymatic modification of hGH for PEG attachment

Site-selective PEGylation of hGH has also been demonstrated through the enzymatic insertion of reactive handles at the C-terminus via Carboxypeptidase Y (CPY) transpeptidation [61]. Both ketone and azide moieties were incorporated at the C-terminus for PEG coupling using oxime ligation and copper(I) catalyzed [2 + 3] catalyzed cycloaddition reactions, respectively. Comparisons of 30 kDa linear and two 40 kDa branched species revealed that the size and geometry of the PEG coupled at the C-terminus can influence the biological activity. *In vitro* potency assessment using an hGH receptor transfected BAF cell assay showed reduced activities from 30–180-fold with activity decreasing with increased PEG size and branching.

In another case of reactive handle generation, Dorwald describes regioselective PEGylation of hGH via periodate oxidation of an engineered amino terminal serine residue forming an aldehyde which is then reacted with amino benzoyl PEG. An example of hGH PEGylated in this manner with a 40 kDa branched PEG was shown to be active in an *in vitro* hGH receptor transfected BAF cell assay with a 10-fold attenuation in potency relative to unmodified hGH [62]. This group also describes regioselective PEGylation of hGH by utilizing transglutaminase to selectively transaminate glutamine 141 with a nitrogen containing nucleophile such as 1,3-diamino-2-propanol, yielding a handle that can be targeted for PEG conjugation. Examples of hGH conjugated in this manner with linear and branched PEGs of various sizes are shown to retain activity in a hGH receptor transfected BAF cell assay with slight attenuations in potency (6–20-fold) relative to unmodified hGH [62, 63].

Genetic incorporation of a reactive moiety into hGH

A recent strategy for site specific PEGylation in hGH utilizes a novel technology developed by Peter Schultz and co-workers in which unnatural amino acids are genetically incorporated into proteins expressed by *E. coli* [64]. The group is able to alter the cell's translational workings by incorporating a new tRNA/tRNA synthetase pair specific to charge the desired non-native amino acid onto a tRNA evolved to recognize an amber stop codon. In this manner, unnatural amino acids with functional groups for site-selective modification may be introduced [65]. Specifically, a keto amino acid, p-acetyl-phenylalanine, was incorporated into hGH as a chemical handle for the specific linkage of PEG molecules. 20 different PEG hGH conjugates were made each with a single 5, 20 or 30 kDa PEG attached at different locations. Of these variants, several were identified that retained *in vitro* binding activity presumably due to PEG coupling in locations distal to hGH-hGHR interactions. These conjugates were shown to have extended pharmacokinetics in rats. One molecule (ARX201) generated the same weight gain for a single weekly dose as daily hGH in the hypophysectomized rat model [66]. Phase I/II clinical trials were initiated in 2007 on ARX201 to determine the safety, tolerability, PK/PD profile following single escalation and repeated dosing (Ambrx press release, 2-12-2007). No results have been reported to date.

Reversible PEGylation

Reversible PEGylation has been investigated as a means to circumvent PEG related inactivation through slow release of the active protein moiety in a prodrug manner. In one hGH study, 40 kDa PEG release and subsequent reactivation of free hGH was demonstrated *in vitro* with a half-life of 11 h in phosphate buffered saline (PBS) [67]. A second study described a novel PEG-hGH linkage which released over a span of 3–5 days in PBS [68]. These reagents offer the possibility of dosing a long acting hGH molecule which would 'activate' over period of time. However, future studies are needed to fully characterize the *in vivo* pharmacokinetics and pharmacodynamics of hGH conjugates with releasable PEGs.

Conclusions

Recombinant human growth hormone (hGH) is a well characterized molecule with broad acceptance as a treatment for growth hormone deficiencies (GHD). Patients, however, must endure daily injections due to drug's short duration of action. PEGylation, as one of many approaches to PK modulation of therapeutic proteins, has been a very important and successful means of increasing half-life and decreasing dosing frequency. Therefore, PEGylation of hGH as a means to develop a long lasting, more conveniently dosed therapy for GHD was a natural progression.

PEGylation of growth hormone has spanned the evolution of PEG technologies from non-specific multi-PEG decoration to single specific attachment to reversible couplings for delayed release. Studies have been carried out with a wide range of PEG sizes both linear and branched in design. Early studies characterizing binding sites and bioactivity of hGH modified with multiple smaller (5 kDa) PEGs provided a set of convincing preclinical data for extending the duration of action of hGH and additional work lead to the successful development of a new therapy for acromegaly (Somavert®).

More recently, several groups have succeeded in the generation of mono-PEGylated forms of hGH which demonstrate extended action in animal models. However, it has yet to be determined whether any of these molecules will be successful clinically as once weekly or less frequently dosed therapies for hGH deficiencies.

Acknowledgements
The author would like to thank Christine E. Smith and John J. Finnessy for critical review of the manuscript, Graeme Bainbridge, Rob Webster, John J. Buckley and Mark E. Gustafson for helpful discussions, and Serdar Aykent for illustration of Figure 3.

References

1 Ascoli M, Segaloff S (1996) Adenohypophyseal hormones and their hypothalamic releasing factors. In: al He (ed.): *Goodman and Gilman's The Pharmacological Basis of Therapeutics* McGraw-Hill, New York, 1363–1382

2 Kupfer SR, Underwood L, Baxter RC, Clemmons DR (1993) Enhancement of the anabolic effects of growth hormone and insulin-like growth factor I by use of both agents simultaneously. *Journal of Clinical Investigation* 91:391–396

3 Matthews LS, Norsted G, Palmiter RD (1986) Regulation of insulin-like growth factor gene expression by growth hormone. *Proceedings of the National Academy of Sciences of the United States of America* 83:9343–9347

4 Lee HJ, Riley G, Johnson O, Cleland JL, Kim N, Charnis M, Bailey L et al. (1997) *In vivo* characterization of sustained-release formulations of human growth hormone. *Growth Hormone and IGF Research* 11:41–48

5 Reslow M, Lonsson M, Laakso T (2003) Sustained-release of human growth hormone from PLG-coated starch microspheres. *DDS&S* 2:103–109

6 Reiter EO, Attie KM, Moshang T Jr, Silverman BL, Kemp SF, Neuwirth RB, Ford KM et al. (2001) A multicenter study of the efficacy and safety of sustained release GH in the treatment of naive pediatric patients with GH deficiency. *J Clin Endocrinol MeTab*. 86:4700–4706

7 Govardhan CK, Jung CW, Simeone B, Higbie A, Qu S, Chemmalil L, Pechenov S et al. (2005) Novel long-acting crystal formulation of human growth hormone. *Pharmaceutical Research* 22:1461–1470

8 Cleland JL, Duenas E, Daugherty A, Marian M, Yang J, Wilson M, Celniker AC et al. (1997) Recombinant human growth hormone poly(lactic-co-glycolic acid) (PLGA) microspheres provide a long lasting effect. *Journal of Controlled Release* 49:193–205

9 Osborn BL, Sekut L, Corcoran M, Poortman C, Sturm B, Chen G, Mather D et al. (2002) Albutropin: a growth hormone-albumin fusion with improved pharmacokinetics and pharmacodynamics in rats and monkeys. *European Journal of Pharmacology* 456:149–158

10 Veronese FM, Pasut G (2005) PEGylation, successful approach to drug delivery. *Drug Discovery Today* 10:1451–1458

11 Zalipsky S, Lee C (1992) Use of functionalized poly(ethylene glycol)s for modification of polypeptides. *Poly(Ethylene Glycol) Chem* 347–370

12 Fu CH, Sakamoto KM (2007) PEG-asparaginase. *Expert Opinion on Pharmacotherapy* 8:1977–1984

13 Chan B, Wara D, Bastian J, Hershfield MS, Bohnsack J, Azen CG, Parkman R et al. (2005) Long-term efficacy of enzyme replacement therapy for adenosine deaminase (ADA)-deficient severe combined immunodeficiency (SCID). *Clinical Immunology* 117:133–143

14 Bailon PP, Schaffer CA, Spence CL, Fung WJ, Porter JE, Ehrlich GK, Pan W et al. (2001)

Rational design of a potent, long-lasting form of interferon: A 40 kDa branched polyethylene glycol-conjugated interferon-2a for the treatment of hepatitis C. *Bioconjugate Chemistry* 12:195–202

15 Wang Y-S, Youngster S, Grace M, Bausch J, Bordens R, Wyss DF (2002) Structural and biological characterization of pegylated recombinant interferon alpha-2b and its therapeutic implications. *Advanced Drug Delivery Reviews* 54:547–570

16 Molineux G (2004) The design and development of Pegfilgrastim (PEG-rmetHuG-CSF, Neulasta®. *Current Pharmaceutical Design* 10:1235–1244

17 Behncken SN, Waters MJ (1999) Molecular recognition events involved in the activation of the growth hormone receptor. *J Molecular Recognition* 12:255–362

18 Waters MJ, Hoang HN, Fairlie DP, Pelekanos RA, Wan Y, McKinstry WJ, Paleothorpe K et al. (2006) New insights into growth hormone action. *J Mol Endocrinology* 36:1–7

19 Ross RJ, Leung KC, Maamra M, Bennett W, Doyle N, Waters MJ, Ho KKY (2001) Binding and functional studies with the growth hormone receptor antagonist, B2036-PEG (pegvisomant), reveal effects of PEGylation and evidence that it binds to a receptor dimer. *Journal of Clinical Endocrinology and Metabolism* 86:1716–1723

20 Clark R, Olson K, Fuh G, Marian M, Mortensen D, Teshima G, Chang S et al. (1996) Long-acting growth hormones produced by conjugation with polyethylene glycol. *The Journal of Biological Chemistry* 271:21969–21977

21 Wells JA, Cunnigham BC, Fuh G, Lowman HB, Bass SH, Mulkerrin MG, Ultsch M et al. (1993) The molecular basis for growth hormone-receptor interactions. *Recent Progress in Hormone Research* 48:253–275

22 Morpurgo M, Veronese F (2004) Conjugates of peptides and proteins to polyethylene glycols. In: Niemeyer CM (ed.): *Methods in Molecular Biology*, vol Bioconjugation Protocols. Humana Press, Totowa, New Jersey, United States, 45–69

23 Chen WY, Wight DC, Wagner TE, Kopchick JJ (1990) Expression of a mutated bovine growth hormone gene suppresses growth of transgenic mice. *Proceedings of the National Academy of Sciences of the United States of America* 87:5061–5065

24 Chen WY, Chen NY, Yun J, Wagner TE, Kopchick JJ (1994) *In vitro* and *in vivo* studies of antagonistic effects of human growth hormone analogs [published erratum appears in *J Biol Chem* (1994) 269(32):20806] *J Biol Chem* 269:15892–15897

25 Maamra MK, Kopchick JJ, Strasburger CJ, Ross, RJM (2004) Pegvisomant, a growth hormone-specific antagonist, undergoes cellular internalization. *Journal of Clinical Endocrinology and Metabolism* 89:4532–4537

26 Fuh GC, Brian C, Fukunaga R, Nagata S, Goeddel DV, Wells JA (1992) Rational design of potent antagonists to the human growth hormone receptor. *Science (Washington, DC, United States)* 256:1677–1680

27 Olson K, Gehant R, Mukku V, O'Connell K, Tomlinson B, Totpal K, Winkler M (1997) Preparation and characterization of poly(ethylene glycol)ylated human growth hormone antagonist. In: Harris MJ, Zalipski, S (ed.): *Poly(ethylene glycol) Chemistry and Biological Applications* Oxford University Press, New York, United States 170–181

28 Pradhananga S, Wilkinson I, Ross RJM (2002) Pegvisomant: structure and function. *Journal of Molecular Endocrinology* 29:11–14

29 Rodvold KA, Bennet WF, Zib KA (1997) Single-dose safety and pharmacokinetics of B2036-PEG (Somavert) after subcutaneous administration in healthy volunteers. *J Clinical Pharmacology* 37:869

30 Rodvold KA, van der Lely AJ (1999) Pharmacokinetics and pharmacodynamics of B2036-PEG, a novel growth hormone receptor antagonist, in acromegalic subjects. *Proceedings 81st Annual Meeting of the Endocrine Society* 1–49

31 Wilson ME (1998) Effects of estradiol and exogenous insulin-like growth factor I (IGF-I) on the IGF-I axis during growth hormone inhibition and antagonism. *Journal of Clinical Endocrinology and Metabolism* 83:4013–4021

32 Ben-Shlomo A, Melmed S (2008) Acromegaly. *Endocrinology & Metabolism Clinics of North America* 37:101–122

33 Roelfsema F, Biermasz NR, Pereira AM, Romijn J (2006) Nanomedicines in the treatment of acromegaly: focus on pegvisomant. *International Journal of Nanomedicine* 1:385–398

34 Colao A, Ferone D, Marzullo P (2004) Systemic complications of acromegaly: epidemiology, pathogenesis, and management. *Endocrine Rev* 25:102–152

35 Freda PU, Wardlaw SL, Post KD (1998) Long-term endocrinological follow-up evaluation in 115

patients who underwent transsphenoidal surgery for acromegaly. *Journal of Neurosurgery* 89:353–358

36 Biermasz NR, Dulken HV, Roelfsema F (2000) Postoperative radiotherapy in acromegaly is effective in reducing GH concentration to safe levels. *Clinical Endocrinology (Oxford, United Kingdom)* 53:321–327

37 Abs R, Verhelst J, Maiter D, Acker KV, Nobles F, Coolens JL, Mahler C, Beckers A (1998) Cabergoline in the treatment of acromegaly. *J Clinical Endocrinology* 83:374–378

38 Lancranjan I, Atkinson AB (1999) The Sandostatin LAR group results of a European multicentre study with sandostatin LAR in acromegalic patients. *Pituitary* 1:105–114

39 van der Lely AJ, Kopchick JJ (2006) Growth hormone receptor antagonists. *Neuroendocrinology* 83:264–268

40 Roelfsema F, Biermasz NR, Pereira AM, Romijn JA (2008) The role of pegvisomant in the treatment of acromegaly. *Expert Opinion on Biological Therapy* 8:691–704

41 Thorner MO, Strasburger CJ, Wu Z, Straume M, Bidlingmaier M, Pezzoli S, Zib K et al. (1999) Growth hormone (GH) receptor blockade with a PEG-modified GH (B2036-PEG) lowers serum insulin-like growth factor-I but does not acutely stimulate serum GH. *Journal of Clinical Endocrinology and Metabolism* 84:2098–2103

42 Parkinson C, Trainer PJ (2001) The place of pegvisomant in the management of acromegaly. *Expert Opinion on Investigational Drugs* 10:1725–1735

43 Trainer PJ, Drake WM, Katznelson L, Freda PU, Herman-Bonert V, van der Lely AJ, Dimaraki V et al. (2000) Treatment of acromegaly with the growth hormone-receptor antagonist pegvisomant. *New England Journal of Medicine* 342:1171–1177

44 van der Lely AJ, Hutson RK, Trainer PJ, Besser GM, Barkan AL, Katznelson L, Klibanski A et al. (2001) Long-term treatment of acromegaly with pegvisomant, a growth hormone receptor antagonist. *Lancet* 358:1754–1759

45 Lindberg-Larsen R, Moeller N, Schmitz O, Nielsen S, Andersen M, Oerskov H, Joergensen JOL (2007) The impact of pegvisomant treatment on substrate metabolism and insulin sensitivity in patients with acromegaly. *Journal of Clinical Endocrinology and Metabolism* 92:1724–1728

46 Parkinson C, Drake WM, Roberts ME, Meeran K, Besser GM, Trainer PJ (2002) A comparison of the effects of pegvisomant and octreotide on glucose, insulin, gastrin, cholecystokinin, and pancreatic polypeptide responses to oral glucose and a standard mixed meal. *Journal of Clinical Endocrinology and Metabolism* 87:1797–1804

47 Colao A, Pivonello R, Auriemma RS, De Martino MC, Bidlingmaier M, Briganti F, Tortora F et al. (2006) Efficacy of 12-month treatment with the GH receptor antagonist pegvisomant in patients with acromegaly resistant to long-term, high-dose somatostatin analog treatment: effect on IGF-I levels, tumor mass, hypertension and glucose tolerance. *European Journal of Endocrinology* 154:467–477

48 Feenstra J, deHerder WW, ten Have SMTH, van den Beld AW, Feelders R, Janssen J, van der Lely AJ (2005) Combined therapy with somatostatin analogues and weekly pegvisomant in active acromegaly. *Lancet* 365:1644–1646

49 Schreiber I, Buchfelder M, Droste M, Forssmann K, Mann K, Saller B, Strasburger CJ (2007) Treatment of acromegaly with the GH receptor antagonist pegvisomant in clinical practice: safety and efficacy evaluation from the German pegvisomant observational stud. *European Journal of Endocrinology* 156:75–82

50 Buckley JJ, Finn RF, Mo J, Bass LA, Ho SV (2008) PEGylation of biological macromolecules. In: Gadamasetti KB, Tamim (ed.): *Process Chemistry in the Pharmaceutical Industry*, vol Challenges in an Ever Changing Climate. CRC Press Taylor & Francis Group, Boca Raton, FL, United States 383–402

51 Rajender Reddy K, Modi MW, Pedder S (2002) Use of peginterferon alfa-2a (40 KD) (Pegasys®) for the treatment of hepatitis C. *Advanced Drug Delivery Reviews* 54:571–586

52 Kinstler O, Mollineux G, Treuheit M, Ladd D, Gegg C (2002) Mono-N-terminal poly(ethylene glycol)-protein conjugates. *Advanced Drug Delivery Reviews* 54:477–485

53 Finnessy J, Girard T, Finn R, Zheng J, Kindle J, Siegel N (2004) PEGylated human growth hormone: effect of PEG size on the biology of PEG-hGH. In: 31st Annual Meeting and Exposition of the Controlled Release Society, vol 2004 Transactions. Controlled Release Society, Honolulu, Hawaii, United States p. 94

54 Finn R, Thiele B, Finnessy J, Liao W, Hall T, Nicastro P, Martin S et al. (2004) PEGylated growth hormone: biochemical properties of a series of conjugates. In: 31st Annual Meeting and

Exposition of the Controlled Release Society, vol 2004 Transactions Controlled Release Society, Honolulu, Hawaii, United States p. 468

55 Webster R, Xie R, Didier E, Finn R, Finnessy J, Edgington A, Walker D (2008) PEGylation of somatropin (recombinant human growth hormone): impact on the clearance in humans. *Xenobiotica* 38(10):1340–1351

56 Clemmons DR (2007) Long-acting forms of growth hormone-releasing hormone and growth hormone: effects in normal volunteers and adults with growth hormone deficiency. *Hormone Research* 68:178–181

57 Harris PE, Didier E, Kantaridis C, Boonen A, Weissberger G (2006) First in human study of PEGylated human growth hormone. *Hormone Res* 65:30

58 Xie R, Didier E, Harris PE, Milligan PA, Karlsson MO (2006) Population pharmacokinetic/pharmacodynamic analysis for PEGylated recombinant human growth hormone (PHA 794428) in healthy male volunteers. *Annual Meeting of the Endocrine Society Program and Astracts*

59 Abs R, Didier EA, Boonen A, Kantaridis CG, Weissberger G, Harris PE (2006) The pharmacokinetics, pharmacodynamics, and safety of PEGylated recombinant human growth hormone after single subcutaneous injections in adult male patients with growth hormone deficiency. *Hormone Research* 2006:Suppl 4

60 Cox GN, Rosendahl MS, Chlipala EA, Smith DJ, Carlson SJ, Doherty DH (2007) A long-acting, mono-PEGylated human growth hormone analog is a potent stimulator of weight gain and bone growth in hypophysectomized rats. *Endocrinology* 148:1590–1597

61 Peschke B, Zundel M, Bak S, Clausen TR, Blume N, Pedersen A, Zaragoza F et al. (2007) C-terminally PEGylated hGH-derivatives. *Bioorganic & Medicinal Chemistry* 15:4382–4395

62 Dorwald FZ (2007) New protein conjugates and methods for their preparation. International Patent Application PCT/EP2007/056819, filed 5 July 2007

63 Dorwald FZ, Johansen NL, Iversen LF (2006) Transglutaminase mediated conjugation of growth hormone. International Patent Application PCT/EP2006/063246, filed 15 June 2006

64 Wang L, Brock A, Herberich B, Schultz PG (2001) Expanding the genetic code of Escherichia coli. *Science* 292:498–500

65 Wang L, Zhang Z, Brock A, Schultz PG (2003) Addition of the keto functional group to the genetic code of Escherichia coli. *Proceedings of the National Academy of Sciences* 100:56–61

66 Cho HS, Buechler Y, Bussell S, Djavahishvili T, Hays A-M, Kraynov V, Litzinger D et al. (2005) Engineering the next generation of therapeutic proteins. *Abstracts of Papers, 229th ACS National Meeting, San Diego, CA, United States*

67 Tsubery H, Mironchik M, Fridkin M, Shechter Y (2004) Prolonging the action of protein and peptide drugs by a novel approach of reversible polyethylene glycol modification. *Journal of Biological Chemistry* 279:38118–38124

68 Pasut G, Caboi F, Schrepfer R, Tonin G, Schiavon O, Veronese FM (2007) New active poly (ethylene glycol) derivative for amino coupling. *Reactive and Functional polymers* 67:529–539

PEGylated α interferons: two different strategies to achieve increased efficacy

Gianfranco Pasut

Department of Pharmaceutical Sciences, University of Padua, Via F. Marzolo 5, 35131 Padua, Italy

Abstract

Conjugation of poly(ethylene glycol) to proteins is a well known technique used to prolong half-life and reduce immunogenicity. In the case of interferon α, improving pharmacokinetics to reduce dosing frequency was the driving force in the development of two long-acting derivatives: PEG-Intron® by Schering-Plough and Pegasys® by Roche Pharmaceuticals. These conjugates, even though developed with a similar approach and for the same clinical use, present several differences that offer the possibility for an interesting and unique comparison. The basic PEGylation chemistry and characterization of the conjugates will be described together with an analysis of the pharmacokinetic/pharmacodynamic behaviors.

Introduction

The hepatitis C virus (HCV) is the leading cause of chronic liver diseases and the major cause of cirrhosis and hepatocellular carcinoma [1]. It is also the leading cause for liver transplants in both Europe and the United States [2]. It is estimated that about 170 million people are chronically infected with HCV. The virus is not directly cytopathic and liver lesions are mainly related to immune-mediated mechanisms. HCV high genetic variability forms the basis of its ability to escape the host immunoresponse, a characteristic that also complicates the treatments directed to eradicate the virus. This variability is dependent upon the RNA polymerase properties that, as for other RNA viruses, is highly error prone and lacks proofreading capabilities. There are eleven major viral subtypes numbered 1–11. There are also several subtypes, 1a, 1b, etc., and approximately 100 different strains of the virus [3]. Genotypes 1–3 are found worldwide. Types 1a and 1b account for ~60% of global infections. Type 1a is common in North America and Northern Europe. Southern Europe, Eastern Europe and Japan most frequently have type 1b present. Unfortunately genotype 1 has a poor response to interferon alone. The sustained response rates are 30% in genotype 1 and nearly 60% in genotype non-1. Genotype 2 is found globally but less frequently than type 1. Type 3 is endemic to south-east Asia but distribution varies.

Treatment of HCV with interferon has recently been reviewed [4–6]. Interferon alfa (INF-α) as mono-therapy for HCV has been marketed under the trade names Intron® A (interferon alfa-2b, recombinant, Schering-Plough) and Roferon® (interferon α-2a, recombinant, Roche Pharmaceuticals).

Only about half of the patients treated with standard INF-α 2b three times a week for 6–12 months responded to the standard therapy. Even those who initially responded had a 50–80% relapse rate [7]. The poor response by patients with subtype 1 to interferon mono therapy led to the creation of consensus interferon [8]. Consensus interferon was launched in 1997 and marketed as Infergen (interferon alfacon-1, initially by Amgen, later by InterMune and currently by Valeant) as a recombinant non-naturally occurring protein. Consensus interferon (CIFN) was created by determining the most conserved amino acids from several interferon-alfa subtypes to arrive at a 'consensus' sequence with four additional amino acid changes. Early clinical trials with CIFN demonstrated equivalent efficacy to INF-α 2b [9]. Later trials demonstrated CIFN could be used for patients previously treated with interferon therapy who were relapsers and non-responders [10, 11]. The changes to the protein sequence alone provided some improvements in therapeutic benefit but IFN-α was shown to be better in reducing the viral load if given in combination with ribavirin, a guanoside-like nucleoside analog with a broad spectrum of antiviral activity [12]. One study suggested that ribavarin acts more as immunomodulator than as an antiviral because it has small or no effect on serum HCV RNA levels [13], and when administered alone showed an insignificant effect on liver histology [14, 15]. The combination therapy of ribavarin and interferon for 24 or 48 weeks gives overall sustained virological response rates of 33% and 41%, respectively *versus* 6% and 16% with interferon alone [16–18]. Even in combination with ribavirin the interferon has to be administered three times a week like the interferon alone treatment. Besides lacking broad activity against all types of HCV this frequent *s.c.* administration schedule is the major problem of interferons. Due to the short circulation time the three times weekly dosing created extended periods of time in which the hepatitis C virus is not exposed to high concentrations of drug. New information on HCV viral kinetics showed that HCV replicated rapidly and that the antiviral effect of alfa interferon occurred within the first day and it was dose-dependent. Furthermore, 48 h after the injection, the viral load is increased again [19]. Therefore, the standard schedule of administration of 3 million units (MU) three times a week may not be the most appropriate approach.

PEGylation, the chemical link of poly(ethylene glycol), to α-interferons was pursued to extend the half-life of these protein preventing their fast kidney clearance. All three interferon drugs were PEGylated and studied.

General overviews of the chemistry of PEGylation, pharmacokinetic properties and biodistribution can be found in the introductory chapter of this book and in other reviews [20, 21]. Although the effects of PEGylation on a given protein vary, several conjugates are currently marketed and others are in various phases of clinical trials. However, all of the positive effects of PEGylation are not observed in every modified protein. This chapter is focused on the application of PEG to the various interferons in use or under potential development for hepatitis C.

PEGylation of α interferons

Any PEGylation study of a protein for human use must take into account several factors to reach a successful final outcome. In fact, the main aim of this strategy is to prolong the *in vivo* half-life of the conjugated protein by slowing its kidney clearance, preventing its eventual immunogenicity and reducing its enzymatic degradation. Altogether, these improvements lead to improved pharmacodynamic profiles (PD) and prolonged therapeutic effects in comparison to non-modified protein. However, any coupling of a polymer to a protein, and especially to those that act through receptors, will face a balance between the increase of protein half-life and the reduction of native protein biological activity. In fact, either a single PEG chain or a large number of short PEGs (both cases yielding a high mass of coupled PEG) can ensure sufficient hydrodynamic volume increase to reduce the kidney clearance or shielding against enzymatic degradation. On the other hand, this high amount of coupled polymer might hamper the protein-receptor recognition thus affecting the activity. Therefore, the coupling strategy needs to be tailored to the specific requirements of the investigated proteins. As reported later in this chapter for Pegasys®, the reduction in native protein *in vitro* activity is well counterbalanced by the prolongation in half-life and steady drug concentrations in the blood.

Several types of PEGylation reagents, also known as 'activated PEGs' or 'PEGylating agents', are commercially available, allowing investigation of different options for better conjugation results. The basics of PEGylation chemistry are discussed in the chapter by Bonora of this book.

The two marketed INF-α drugs are made with acylating reagents which differ both in PEG size and coupling chemistry. Acylating agents primarily target lysine amino groups forming stable bonds, but depending upon the pH and reactivity of the specific reagent, other amino acids can also be modified, such as the amino group at the N-terminus or the imidazole ring of histidine.

PEG-Intron®

Early PEGylation of INF-α-2b was done by scientists at Enzon Inc. while the development and commercialization as PEG-Intron® was done by Schering-Plough Corporation. The product entered the market in the 2000.

Initial screening of conjugates was done with PEGs of 5,000 Da, but later with the aim to reach a desired pharmacokinetic prolongation, allowing once per week dosing, PEGs of 12,000 Da were used. The investigated PEGylating agents were either mono-methoxypoly(ethylene glycol)-succinimidyl carbonate (mPEG-SC) or mono-methoxypoly(ethylene glycol)-2-pyridyl carbonate reagents (mPEG-PC) [22].

Active carbonates and esters react with Nε of lysines or the *N*-terminal amine on proteins forming stable urethane or amide linkages, respectively, as shown below in Scheme 1 [23–28] but are also capable of reacting with histidine and tyrosine residues [29].

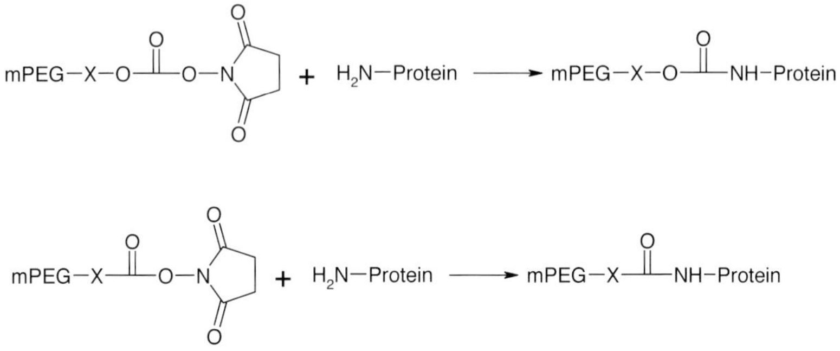

Scheme 1. Coupling of active PEG carbonates or esters to protein's amine. X is a linker between the PEG and activated group

Typical conjugation reactions are carried out at basic pHs to achieve deprotonated, and therefore nucleophilic, ε amino group of lysine. Zalipsky found the optimal pH for succidimyl carbonate, SC-PEG [23] conjugation to be around pH 9.3. This pH however, is not always suitable for many proteins, from a stability point of view.

INF-α-2b conjugation was investigated at several pH values ranging from acid to basic. In general conjugation at slightly acidic pH preferably leads to histidine conjugation (Sch. 2), which yielded conjugates with better retention of activity than conjugates made at higher pH [30].

The selected reaction conditions, sodium phosphate buffer pH 6.5, yielded a conjugate mixture composed of 14 positional isomers, among which the conjugate at histidine-34 represented approximately the 47% [31], with an overall retention of *in vitro* activity in the commercial product of 28%. Structural characterization of the purified conjugate mixture by MALDI-MS indicated that it

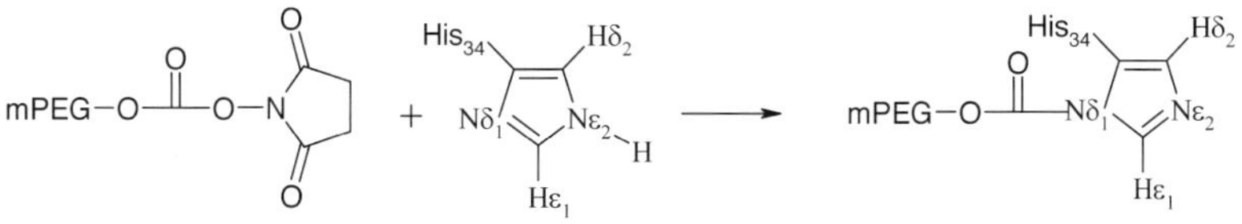

Scheme 2. Conjugation of PEG-SC to the ring nitrogen of imidazole group

was composed of mono-PEGylated isomers, with traces of diPEGylated adducts and native interferon. Circular dichroism spectra of the conjugate mixture and of native protein were nearly superimposable in the near to far UV suggesting that there was no significant secondary or tertiary structural impact caused by the PEGylation [32]. Although His[34] de-PEGylation could be accelerated at basic pH or with hydroxylamine, the monoPEGylated interferon conjugate, as a whole, was stable at pH 6.8 for 30 days at room temperature (only 4% of the conjugate dissociated). Basic pH however, accelerated the conjugate degradation yielding 22% dissociation in 24 h at pH 8.6 [30]. Due to the instability of the PEG-His[34] bond upon long-term storage in solution, PEG-Intron® is marketed as a lyophilized powder that has to be reconstituted before use. The excess of reconstituted solution should be discarded and not stored for future use [33].

^{1}H-PEG is very hydrophilic and the ethylene oxide monomer can coordinate several water molecules, making the polymer looks bigger with respect to a globular protein of the same molecular weight when analyzed by sizing methods such as SDS PAGE or size exclusion chromatography. Interferon alfa-2b has a molecular weight of 19,271 Da. The coupling with a 12,000 Da PEG yields a calculated molecular weight of ~31,000 Da but it gives the appearance of a conjugate that is almost 5 times larger than its true weight.

^{1}H-NMR investigation, aimed to identify and characterize the major positional isomer, suggested that PEGylation occurred at the 1 position in the imidazole ring of His[34] [34]. Interferon-α2b has two additional histidines, His[7] and His[57] but carboxyalkylation at these two positions was not detected. All positional isomers were isolated and characterized in terms of *in vitro* biological activity. The conjugate at the His[34] is the most active, maintaining 37% of the native INF *in vitro* activity [31].

The higher activity of this interferon preparation during the *in vitro* assays was initially thought to be related to the ability to release free and fully active interferon by slow hydrolysis of the His-PEG bond. Argument against this correlation is the fact that the level of free interferon did not increase during the course of the assay. Furthermore, if PEG release was playing an important role in the PEG-Intron® activity the increasing of PEG size would not have attenuated the activity as much. Therefore, it can be concluded that the higher activity of the conjugate at the His[34] might be due to location of this amino acid in the protein sequence, which is not involved in the receptor recognition as demonstrated by X-ray crystallography [35].

As stated above, PEG Intron® *in vitro* activity is 28% of the native IFN-α2b. When dosed on a per-unit basis relative to Intron® A as a standard, the dose response curves are almost identical [32]. Thus although specific activity and therefore potency were reduced, the potential biotherapeutic potency of the antiviral response was unchanged by the PEGylation. PEG Intron® and Intron A were shown to be mechanistically indistinguishable [36] and were not different in their immunotherapeutic profiles.

Pegasys®

Preliminary PEGylation studies with INF-α-2a were conducted using mPEG 5,000 Da conjugated to ε-amine of lysines via an urea linkage [37]. The sites of PEGylation were determined by peptide mapping on each positional isomer purified by cation exchange HPLC. Interestingly all eleven lysines were modified but no PEGylation was found at the N-terminus of the protein. The residual activity of these isomers was in the range of 6–40% of the antiviral specific activity of the native protein. Furthermore, the isoforms also retained a good level of the antiproliferative activity, encouraging clinical development. However, in Phase II clinical studies the INF-α-2a-PEG 5,000 Da conjugate failed to demonstrate sufficient superiority with respect to the native protein (Roferon-A). Modeling studies of the pharmacodynamics and clinical trial data suggested that to overcome the still short pharmacokinetic profile of INF-α-2b-PEG 5,000 Da, a higher molecular weight PEG had to be used [38–40]. Thorough studies were conducted comparing INFα-2a conjugates obtained with PEG varying chain length (5,000–20,000 Da) and architecture (linear *versus* branched). Interestingly, the conjugate with two linear PEG 10,000 Da chains possessed a similar pharmacokinetic profile to that obtained using a single 20,000 Da branched chain. This resulted in favoring the branched derivative because its single point attachment simplified the analytical analyses and investigations of the monoPEGylated isomers with respect to the diPEGylated isomers. Furthermore, the 'umbrella-like' structure of branched PEG better shield the protein surface from approaching antibodies and proteolytic enzymes. A high molecular weight branched PEG of 40,000 Da was chosen for the development of a monoPEGylated INF-α-2b conjugate [39, 40].

The 40,000 Da branched succinimidyl PEG (PEG₂-NHS) was conjugated using a 3: 1 PEG/protein molar ratio in 50 mM sodium borate buffer pH 9 (Sch. 3) [39].

PEGylation under these conditions led to a mixture containing 45–50% mono-substituted protein, 5–10% poly-substituted (essentially diPEGylated IFN) and 40–50% unmodified interferon. The purified monoPEGylated conjugate fraction was thoroughly characterized by a combination of high performance cation exchange chromatography, peptide mapping, amino acid

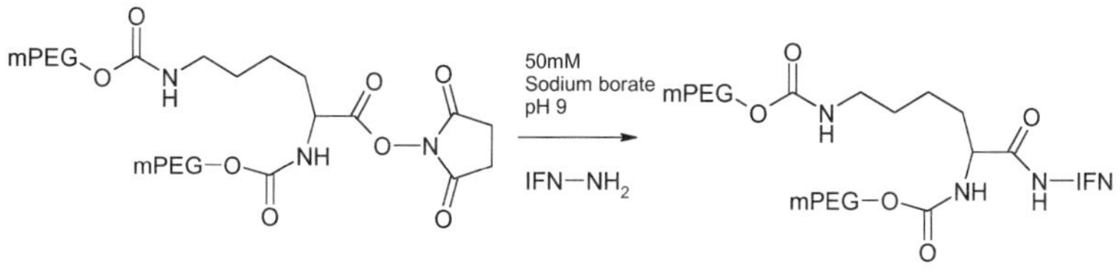

Scheme 3. Conjugation of branched PEG 40,000 Da to INF through the coupling of lysines ε-amino group

sequencing and mass spectroscopy analysis. The 94% of the all isoforms was represented by conjugation at Lys^{31}, Lys^{121}, Lys^{131} or Lys^{134}, with minor sites of PEGylation at Lys^{70} and Lys^{83} [39]. It is worth noting that the use of a larger PEG reduced the number of isoforms from eleven to four going from the 5,000 Da to the 40,000 Da PEG. Unfortunately, the bigger PEG also created greater steric hindrance for the INF recognition by its receptor, markedly reducing the native residual activity to the low value of 7%, as tested *in vitro*. On the other hand, the *in vivo* activity, measured as the ability to reduce the size of various human tumors, was higher than that of free IFN [39]. The positive result could be related to the extended blood residence time of the conjugated form (20 h) with respect to the native INF (1 h) measured in rats [40]. Clinical studies demonstrating improved sustained virological responses relative to native protein supported entry into the market a long lasting interferon conjugate, Pegasys® that was marketed in 2002 [40]. The stability of amide linkages between PEG and lysines in solutions allowed the conjugate to be dispensed in liquid form [41].

Comparisons of PEG-Intron® and Pegasys®

There are no multi-center head-to head clinical trials in which the efficacies of therapy with PEG-IFN-α-2a and PEG-IFN-α-2b are compared directly but only several comparative studies with different designs. Therefore the following discussion is firstly based on data obtained and reported in clinical trials or studies that investigated the conjugates separately. Comments on data from comparative studies have also been included. Note that differences, such as drug dose, duration of therapy and protocols for combination with ribavarin, might influence some parameters but the evaluation of the best IFN therapy is beyond the scope of this chapter.

The main aim that fuelled the INF-α PEGylation was to achieve a high blood residence time of the drug at a stable concentration, thus ensuring a prolonged HCV exposure to the drug. This goal was reached by both products since they achieved the once a week labeling in comparison to the native INF that is administered thrice weekly (Tab. 1).

Absorption

Both PEGylated products are administered subcutaneously creating a 'depot' effect because the conjugates must diffuse from the injection site to reach the blood flow, and this process is slowed by the PEG chain. The diffusion is significantly different between the two preparations and it reflects the different conjugates' molecular weights. Pegasys® being bigger than PEG-Intron®, it's time to achievement of maximum serum concentration (t_{max}) is about 80 h [42] with a mean absorption half-life ($t_{1/2\ abs}$) of 50 h [43] compared to a t_{max} of

Table 1. Comparison between Pegasys® and PEG-Intron®

Characteristic	PEGylated-INF-α-2b (PEG-Intron®)	PEGylated-INF-α-2a (Pegasys®)
PEG	PEG-SC, linear, 12,000 Da	PEG-NHS, branched, 40,000 Da
Positional isomers (relative abundance)	His[34] (47.8%), Cys[1] (13.2%), Lys[121] (7.3%), Lys[131] and Lys[164] (6.9%) Lys[31] (5.7%), Lys[49] (4.5%), Lys[83] (3.6%), Lys[112], Lys[134], His[7], Tyr[129], Ser[163], Lys[133], traces of diPEGylated conjugates and native INF-α-2b	Lys[31], Lys[121], Lys[131] or Lys[134] (altogether 94%), traces of conjugates at Lys[70] and Lys[83]
Residual antiviral activity	28% for the mixture, 37% for His[34]	7% for the mixture
How supplied	Lyophilized powder. 74, 118.4, 177.6 and 222 µg/vials	Injectable solution, prefilled syringe. 180 µg/1 ml or 180 µg/0.5 ml vial
Dose	1.5 µg/kg/week	180 µg/week
t_{max} (h)	15–44	80
$t_{1/2\ abs}$ (h)	4.6	50
V_d (l)	Similar to IFN (31–73)	6–14
$t_{1/2\beta}$ (h)	27.2–39.3	61–110
Sustained virological response	23% (alone), 54% (+ ribavarin) [56]	36–39% (alone) [57, 58], 56% (+ ribavarin) [59, 60]

15–44 h of the last with a $t_{1/2\ abs}$ of 4.6 h [44] and to a t_{max} of 7.3–12 h for standard INF α [42, 45] with $t_{1/2\ abs}$ of 2.3 h [44].

Distribution

The PEG size is also controlling the distribution of the conjugates. In fact, the mean apparent volume of distribution for Pegasys® is reduced to the range of ~6–14 L compared to the 31–73 L of native INF [43, 46], whereas PEG-Intron® has a volume of distribution comparable to that of native INF [44]. The limited distribution of Pegasys®, to mainly blood and liver [47], is the reason why it can be administered as a standard dose (180 µg/week) rather than on a dose-per-weight basis, as is done for PEG-Intron® (1.5 µg/kg/week).

Elimination

The half-life of standard INF ranges from 4–16 h. Therefore the protein is rapidly cleared from the body, and its metabolism is mainly due to proteolytic degradation during the reabsorption process in the proximal tubules after the filtration of the protein in the glomeruli. As consequence small amounts of native IFN are found in the urine [45]. PEGylation significantly modified the half-life of IFN α. In particular, PEG-Intron®, in an escalation dose study, showed mean half-life values ranging between 27.2 and 39.3 h [44]. As expected, the 40,000 Da PEG of Pegasys® further increased the elimination half-life to the range of 61–110 h, in young and elderly patients, respectively [48]. It must be highlighted that an accurate half-life determination for Pegasys® might be affected by its prolonged absorption time [48]. In contrast with native IFN-α, intact Pegasys® is primarily metabolized by the liver, as shown in animal studies, and then the degraded products, sequence fragments of IFN attached to the whole branched PEG or to a PEG moiety, are cleared by the kidney [47].

Therapeutic effectiveness

A study directly comparing pharmacokinetics and viral dynamics of PEG-Intron® (1 µg/kg weekly dose, 12 patients) and Pegasys® (180 µg/wk dose, 10 patients) reported that the former has a t_{max} of 24 h after which the IFN serum levels decreased rapidly, being undetectable at 168 h in 11 out of 12 patients. In contrast, Pegasys® showed a t_{max} ranging between 48–168 h and the IFN was detectable in all 10 patients at 168 h [49]. These differences were claimed by the authors as probably responsible for the significantly lower serum HCV RNA levels at 12 weeks (a parameter used to monitor the effectiveness of anti-HCV therapy) in patients treated with Pegasys® with respect to those treated with PEG-Intron®. The undetectable drug level at 168 h (1 week) in almost all patients taking PEG-Intron® (1.0 µg/kg/week) might suggest that this preparation is not ensuring a continuous drug exposure for HCV, a fact, supported by another study reporting the concomitant HCV RNA increase with the decline of PEG-Intron® blood concentration [50] (note however, that the approved dosing for PEG-Intron® is 1.5 µg/kg/week). In conclusion, both studies are suggesting that a single week administration of PEG-Intron® might not be sufficient for continuous IFN exposure, which should be reachable by twice-weekly dosing. On the other hand, it must be noted that there are other several studies evaluating the effectiveness of PEG-Intron® and Pegasys®, taking into consideration different HCV genotypes and analyzing the HCV infection by many markers [49, 51–55]. Interestingly, most of these studies reported that there are no statistically significant differences in efficacy and safety between Pegasys® and PEG-Intron®, even though the former leads to slightly better percentages of early viral response and sustained viral response. Since the main limit of these studies resides often in the number of patients enrolled, a larger-scale multi-center trial would be needed to elucidate these observations in greater detail. However, the gold standard of care for HCV treatment now includes co-administration with Ribavirin® and both drugs perform well when administered as combination therapy.

Conclusion

Both PEGylated products demonstrated the potential of PEGylation in creating drugs with increased therapeutic value. This not only from the point of view of effectiveness, being the treatment with Pegasys® or PEG-Intron® was more successful than that with native interferon in eradicating the viral infection, but also for the patient compliance. In fact, PEGylation allowed shifting the administration schedule from the thrice weekly injection of interferons to the once weekly for the conjugates. This goal has been reached thanks to the reduced clearance and increased half life of Pegasys® and PEG-Intron®, which at the same time ensured a more sustained therapeutic serum level of the protein.

It is worth noting how two quite different approaches of PEGylation, namely exploiting a PEG of 12,000 Da or 40,000 Da for PEG-Intron® and Pegasys®, respectively, finally lead to similar pharmacodynamic outcomes starting from significantly diverse pharmacokinetic profiles (summarized in Tab. 1). These two examples of PEG research, together with the numerous related studies, demonstrated how flexible and useful the PEGylation technique can be.

References

1. Cohen J. (1999) The scientific challenge of hepatitis C. *Science* 285: 26–30.
2. Liang TJ, Rehermann B, Seeff LB, and Hoofnagle JH. (2000) Pathogenesis, natural history, treatment, and prevention of hepatitis C. *Annals of Internal Medicine* 132: 296–305.
3. Simmonds P. (1999) Viral heterogeneity of the hepatitis C virus. *Journal of Hepatology* 31 (Suppl 1): 54–60.
4. Souvignet C, Lejeune O, and Trepo C. (2007) Interferon-based treatment of chronic hepatitis C. *Biochimie* 89: 894–898.
5. Palumbo E. (2007) PEG-interferon alpha-2b for acute hepatitis C: a review. *Mini Rev Med Chem* 7: 839–843.
6. Zeuzem S. (2008) Interferon-based therapy for chronic hepatitis C: current and future perspectives. *Nat Clin Pract Gastroenterol Hepatol*. 5: 610–622.
7. Heathcote EJ, Keeffe EB, Lee SS, Feinman SV, Tong MJ, and Reddy KR et al. (1998) Re-treatment of chronic hepatitis C with consensus interferon. *Hepatology* 27: 1136–1143.
8. Blatt LM, Davis JM, Klein SB, and Taylor MW. (1996) The biologic activity and molecular characterization of a novel synthetic interferon-alpha species, consensus interferon. *Journal of Interferon and Cytokine Research* 16: 489–499.
9. Melian EB and Plosker GL. (2001) Interferon alfacon-1: a review of its pharmacology and therapeutic efficacy in the treatment of chronic hepatitis C. *Drugs* 61: 1661–1691.
10. Keeffe EB and Hollinger FB. (1997) Therapy of hepatitis C: consensus interferon trials. Consensus Interferon Study Group. *Hepatology* 26: 101S–107S.
11. Heathcote EJ, James S, Mullen KD, Hauser SC, Rosenblate H, and Albert DG Jr (1999) Chronic hepatitis C virus patients with breakthroughs during interferon treatment can successfully be retreated with consensus interferon. The Consensus Interferon Study Group. *Hepatology* 30: 562–566.
12. Sjogren MH, Sjogren R, Holtzmuller K, Winston B, Butterfield B, and Drake S et al. (2005) Interferon alfacon-1 and ribavirin *versus* interferon alpha-2b and ribavirin in the treatment of chronic hepatitis C. *Digestive Diseases and Sciences* 50: 727–732.
13. Thomas HC, Torok ME, Forton DM, Taylor-Robinson SD. (1999) Possible mechanisms of action and reasons for failure of antiviral therapy in chronic hepatitis C. *J. Hepatol*. 31 (Suppl. 1): 15229.
14. Dusheiko G, Main J, Thomas H, Reichard O, Lee C, and Dhillon A et al. (1996) Ribavirin treatment for patients with chronic hepatitis C: results of a placebo-controlled study. *J. Hepatol*. 25:

591–598.

15. Di Bisceglie AM, Conjeevaram HS, Fried MW, Sallie R, Park Y, and Yurdaydin C et al. (1995) Rivabirin as therapy for chronic hepatitis C. A randomized, double-blind, placebo-controlled trial. *Ann Intern Med* 123: 897–903.

16. Weiland O. (1999) Treatment of naive patients with chronic hepatitis. C. *J. Hepatol.* 31 (Suppl. 1): 168–173.

17. Poynard T, Marcellin P, Lee SS, Niederau C, Minuk GS, and Ideo G et al. (1998) Randomised trial of interferon alpha2b plus ribavirin for 48 weeks or for 24 weeks *versus* interferon alpha2b plus placebo for 48 weeks for treatment of chronic infection with hepatitis C virus. International Hepatitis Interventional Therapy Group (IHIT). *Lancet* 352: 1426–1432.

18. McHutchison JG, Gordon SC, Schiff ER, Shiffman ML, Lee WM, and Rustgi VK et al. (1998) Interferon alfa-2b alone or in combination with ribavirin as initial treatment for chronic hepatitis C. Hepatitis Interventional Therapy Group. *New England Journal of Medicine* 339: 1485–1492.

19. Lam NP, Neumann AU, Gretch DR, Wiley TE, Perelson AS, and Layden TJ. (1997) Dose-dependent acute clearance of hepatitis C genotype 1 virus with interferon alpha. *Hepatology* 26: 226–231.

20. Veronese FM, and Harris JM. (2002) Introduction and overview of peptide and protein pegylation. *Advanced Drug Delivery Reviews* 54: 453–456.

21. Veronese FM, and Pasut G. (2005) PEGylation, successful approach to drug delivery. *Drug Discovery Today* 10: 1451–1457.

22. Gilbert CW, and Park-Cho M-O. (1999) *Interferon Polymer Conjugates* US 5951974.

23. Zalipsky S, Seltzer R, and Menon-Rudolph S. (1992) Evaluation of a new reagent for covalent attachment of polyethylene glycol to proteins. *Biotechnology and Applied Biochemistry* 15: 100–114.

24. Miron T, and Wilchek M. (1993) A simplified method for the preparation of succinimidyl carbonate polyethylene glycol for coupling to proteins. *Bioconjugate Chemistry* 4: 568–569.

25. Pasut G, Guiotto A, and Veronese FM. (2004) Protein, peptide and non-peptide drug PEGylation for therapeutic application. *Expert Opinion on Therapeutic Patents* 14: 859–894.

26. Kozlowski A, Charles SA, and Harris JM. (2001) Development of pegylated interferons for the treatment of chronic hepatitis C. *BioDrugs* 15: 419–429.

27. Kozlowski A, and Harris JM. (2001) Improvements in protein PEGylation: pegylated interferons for treatment of hepatitis C. *J Control Release* 72: 217–224.

28. Zalipsky S. (1995) Chemistry of polyethylene glycol conjugates with biologically active molecules. *Advanced Drug Delivery Reviews* 16: 157–182.

29. Roberts MJ, Bentley MD, and Harris JM. (2002) Chemistry for peptide and protein PEGylation. *Advanced Drug Delivery Reviews* 54: 459–476.

30. Wylie DC, Voloch M, Lee S, Liu YH, Cannon-Carlson S, and Cutler C et al. (2001) Carboxyalkylated histidine is a pH-dependent product of pegylation with SC-PEG. *Pharmaceutical Research* 18: 1354–1360.

31. Wang Y-S, Youngster S, Grace M, Bausch J, Bordens R, and Wyss DF. (2002) Structural and biological characterization of pegylated recombinant interferon alpha-2b and its therapeutic implications. *Advanced Drug Delivery Reviews* 54: 547–570.

32. Grace M, Youngster S, Gitlin G, Sydor W, Xie L, and Westreich L et al. (2001) Structural and biologic characterization of pegylated recombinant IFN-alpha2b. *Journal of Interferon and Cytokine Research* 21: 1103–1115.

33. Peginterferon alfa-2b Product Information. Kenilworth, NJ, USA: Schering Corporation.

34. Wang YS, Youngster S, Bausch J, Zhang R, McNemar C, and Wyss DF. (2000) Identification of the major positional isomer of pegylated interferon alpha-2b. *Biochemistry* 39: 10634–10640.

35. Nagabhushan TL, Reichert P, Walter MR, and Murgolo NJ. (2002) Type I interferon structures: Possible scaffolds for the interferon-alpha receptor complex. *Canadian Journal of Chemistry* 80: 1166–1173.

36. Chou CP, Yu CC, Lin WJ, Kuo BY, andWang WC. (1999) Novel strategy for efficient screening and construction of host/vector systems to overproduce penicillin acylase in *Escherichia coli.* *Biotechnology and Bioengineering* 65: 219–226.

37. Monkarsh SP, Ma Y, Aglione A, Bailon P, Ciolek D, and DeBarbieri B et al. (1997) Positional isomers of monopegylated interferon alpha-2a: isolation, characterization, and biological activity. *Analytical Biochemistry* 247: 434–440.

38. Nieforth KA, Nadeau R, Patel IH, and Mould D. (1996) Use of an indirect pharmacodynamic

stimulation model of MX protein induction to compare *in vivo* activity of interferon alfa-2a and a polyethylene glycol-modified derivative in healthy subjects. *Clinical Pharmacology and Therapeutics* 59: 636–646.

39. Bailon P, Palleroni A, Schaffer CA, Spence CL, Fung WJ, and Porter JE et al. (2001) Rational design of a potent, long-lasting form of interferon: a 40 kDa branched polyethylene glycol-conjugated interferon alpha-2a for the treatment of hepatitis C. *Bioconjugate Chemistry* 12: 195–202.

40. Reddy RK, Modi MW, and Pedder S. (2002) Use of peginterferon alfa-2a (40 KD) (Pegasys) for the treatment of hepatitis C. *Advanced Drug Delivery Reviews* 54: 571–586.

41. Peginterferon alfa-2a Complete Product Information. Nutley, NJ, USA: Roche Pharmaceuticals.

42. Algranati NE, Sy S, and Modi M. (1999) A branched methoxy 40 KDA polyethylene glycol (PEG) moiety optimizes the pharmacokinetics (PK) of peginterferon alpha-2a (peginterferon) and may explain its enhanced efficacy in chronic hepatitis C (CHC). *Hepatology* 30:190A. Abstract 120.

43. Harris JM, Martin NE, and Modi M. (2001) Pegylation: A novel process for modifying pharmacokinetics. *Clin. Pharmacokinet.* 40: 539–551.

44. Glue P, Fang JW, Rouzier-Panis R, Raffanel C, Sabo R, and Gupta SK et al. (2000) Pegylated interferon-alpha2b: pharmacokinetics, pharmacodynamics, safety, and preliminary efficacy data. Hepatitis C Intervention Therapy Group. *Clinical Pharmacology and Therapeutics* 68: 556–567.

45. Wills RJ. (1990) Clinical Pharmacokinetics of interferons. *Clin. Pharmacokinetic.* 19: 390–399.

46. Lamb MW, and Martin NE. (2002) Weight-based *versus* fixed dosing of peginterferon (40 kDa) alfa-2a. *Ann Pharmacothe* 36: 933–935.

47. Modi M, Fulton JS, Buckman DK, Wright TL, and Moore D. (2000) Clearance of PEGylated (40 kDa) Interferon Alfa-2a (PEGASYS) is Primarily Hepatic. *Hepatology* 32: 371A.

48. Martin NE, and Modi MW. (2000) Characterization of pegylated (40KDA) interferon alfa-2a (PEGASYS) in the elderly. *Hepatology* 32: 348A. Abstract 755.

49. Bruno R, Sacchi P, Ciappina V, Zochetti C, Patruno S, and Maiocchi L et al. (2004) Viral dynamics and pharmacokinetics of peginterferon alpha-2a and peginterferon alpha-2b in naïve patients with chronic hepatitis C: a randomized, controlled study. *Antiviral Therapy* 9: 491–497.

50. Formann E, Jessner W, Bennett L, and Ferenci P. (2003) Twice-weekly administration of peginterferon-a-2b improves viral kinetics in patients with chronic hepatitis C genotype 1. *J. Vir. Hep.* 10: 271–276.

51. Sporea I, Danila M, Sirli R, Popescu A, Laza A, and Baditoiu L. (2006) Comparative study concerning the efficacy of Peg-IFN-2a *versus* Peg-IFN alpha-2b on the early virological response (EVR) in patients with chronic viral C hepatitis. *J. Gastrointestin. Liver Dis.* 15: 125–130.

52. Yenice N, Mehtap O, Gümrah M, and Arican N. (2006) The efficacy of pegylated alpha 2a or 2b plus ribavarin in chronic hepatitis C patients. *Turk. J. Gastroenterol.* 17: 94–98.

53. Escudero A, Rodriguez F, Serra MA, Del Olmo JA, Montes F, and Rodrigo JM. (2008) Pegylated alpha-interferon-2a plus ribavarin compared with pegylated alpha-interferon-2b plus ribavarin for initial treatment of chronic hepatitis C virus: prospective, non-randomized study. *J. Gastroenterol. Hepatol.* 23: 861–866.

54. Laguno M, Cifuentes C, Murillas J, Veloso S, Larrousse M, Payeras A et al. (2009) Randomized trial comparing pegylated interferon α-2b *versus* pegylated interferon α-2a, both plus ribavarin, to treat chronic hepatitis C in human immunodeficiency virus patients. *Hepatology* 49: 22–31.

55. Chou R, Carson S, and Chan BKS. (2008) Pegylated interferons for chronic hepatitis C virus infection: an indirect analysis of randomized trials. *J. Viral Hep.* 15: 551–570.

56. Manns MP, McHutchinson J, Gordon S, Rustgi VK, Shiffman M, Reindollar R et al. (2001) Peginterferon alfa-2b plus ribavarin for initial treatment of chronic hepatitis C: a randomized trial. *Lancet* 358: 958–965.

57. Reddy KR, Wright TL, Pockros PJ, Shiffman M, Everson G, Reindollar R et al. (2001) Efficacy and safety of pegylated (40-KD) interferon alpha-2a compared with interferon alpha-2a in patients with in noncirrhotic patients with chronic hepatitis C. *Hepatology* 33: 433–438.

58. Zeuzem S, Feinman SV, Rasenack EJ, Heathcote EJ, La MJ, Gane E et al. (2000) Peginterferon alfa-2a in patients with chronic hepatitis C. *New Eng. J. Med.* 343: 1666–1672.

59. Fried MW, Shiffman ML, Reddy KR, Smith C, Marino G, Goncales F et al. (2001) Pegylated interferon alfa-2a (Pegasys) in combination with ribavarin: efficacy and safety results from a phase III, randomized, actively controlled, multicenter study. *Gastroenterology* 120: A55.

60. Fried MW, Shiffman ML, Reddy KR, Smith C, Marinos G, Goncales FL Jr et al. (2002) Peginterferon alfa-2a plus ribavirin for chronic hepatitis C virus infection. *N Engl J Med.* 347: 975–982.

Development of PEGylated mammalian urate oxidase as a therapy for patients with refractory gout

Michael S. Hershfield, John S. Sundy, Nancy J. Ganson and Susan J. Kelly

Duke University Medical Center, Durham, NC 27710, USA

Abstract

Gout is a form of arthritis caused by inflammatory crystals of monosodium urate, which deposit in joints when the plasma concentration of uric acid chronically exceeds the limit of solubility, ~7 mg/dL (0.42 mM). The human species is predisposed to hyperuricemia and gout by mutation of the urate oxidase gene during evolution. Urate oxidases from various sources have been used as a model to investigate the effects of PEGylation in animals. More than 15 years ago we initiated a project to develop a PEGylated recombinant mammalian urate oxidase as an Orphan Drug for treating patients with refractory gout. Clinical testing of this PEG-uricase, now called pegloticase, began in 2001. Pegloticase was found to have a half-life in plasma of about two weeks, and when infused at 2–4 week intervals to rapidly correct hyperuricemia. PEGylation was effective in limiting immune recognition of the recombinant uricase protein, but antibodies to PEG develop in some patients, resulting in the rapid clearance of pegloticase and loss of efficacy. However, in many patients with refractory gout, treatment with pegloticase maintains plasma urate at well below saturating concentrations, leading to elimination of tissue urate deposits and control of disease.

Introduction

Urate oxidase (uricase) and adenosine deaminase (ADA) were among the first enzymes used as models to examine the effects of PEGylation on pharmacokinetics and immune reactivity [1, 2]. An interest in diseases of purine metabolism led to our early, and continuing, involvement in the clinical development of PEGylated derivatives of both of these enzymes, beginning with a trial of PEGylated bovine ADA to treat a child with a rare, fatal immunodeficiency disease caused by inherited deficiency of ADA [3]. Based on findings in this and subsequent patients, PEG-ADA (Adagen®, manufactured by Enzon Pharmaceuticals), became the first PEGylated therapeutic, and first enzyme replacement therapy, to receive approval from the US Food and Drug Administration; Adagen remains an important treatment for ADA deficiency.

A 1981 study in volunteer cancer patients showed that infusing a PEGylated fungal urate oxidase could lower serum uric acid concentration, suggesting its potential use for controlling elevated urate levels associated with gout or with renal failure during cancer chemotherapy [4]. Shortly after starting the PEG-

ADA trial, we obtained permission for the compassionate use of a PEGylated bacterial urate oxidase, provided by Enzon, to treat the latter condition in a lymphoma patient who was unable to take other uric acid lowering medications [5, 6]. In spite of these early pilot clinical studies, full-scale clinical testing of a PEG-uricase did not begin until 2002. This chapter will review the biochemistry of uricase and its potential clinical applications, the extensive animal testing of PEGylated uricases during the 1980s and 1990s, and the preclinical and clinical development of a PEGylated recombinant mammalian uricase for the treatment of refractory gout.

Uricase structure, function and evolution

The biochemistry and phylogeny of urate oxidase have been of interest for more than a century. Early studies showed that isolated uricase transforms urate into allantoin, and that most mammals have high uricase activity in the liver and excrete allantoin in urine, whereas humans and some primates lack uricase and excrete uric acid [7, 8]. Molecular studies later showed that uricase gene mutations had been acquired during evolution of the latter species [9, 10].

Based on these observations, it was believed until recently that uricase is solely responsible for the metabolism of urate to allantoin. Elegant research during the last decade has revealed that uricase catalyzes only the initial step in this pathway, in which urate reacts with O_2 and water to form 5-hydroxy-isourate (HIU) with release of H_2O_2 [11]. The HIU product is unstable and undergoes non-enzymatic hydrolysis to 2-oxo-4-hydroxy-4-carboxy-ureidoimidazoline (OHCU), which then decarboxylates spontaneously to form racemic allantoin [12]. While these non enzymatic steps can occur *in vivo*, recent investigation has identified two enzymes, HIU hydrolase and OHCU decarboxylase, that catalyze the much more rapid sequential conversion of HIU to OHCU and then to (S)-allantoin [13]. Both HIU hydrolase and OHCU decarboxylase are expressed in species that possess active uricase, but not in several species that lack uricase.

In mammals, urate oxidase is found at highest levels in peroxisomes of liver, where it is associated with the central 'crystalloid core', along with xanthine oxidoreductase, the enzyme responsible for urate synthesis [14]. Isolated mammalian uricases are sparingly soluble in buffers under physiologic conditions [15]. Like uricase, HIU hydrolase is localized to the peroxisome (the intracellular localization of OHCU decarboxylase is uncertain) [13]. Purified uricases from several plant, microbial, and animal sources are homotetrameric with subunit molecular weight of about 30–35,000. Crystallographic analysis of recombinant *Aspergillus flavus* urate oxidase shows a globular structure traversed by a central tunnel ('T-fold'), with four active sites located at the subunit interfaces; that enzyme, and other purified urate oxidases, contain no metal or other cofactors [16].

Uricase therapy for management of hyperuricemia

The concentration of urate in plasma and extracellular fluid is significantly higher in humans than in species that express uricase. In a portion of the human population, plasma urate can exceed the limit of solubility (~7 mg/dL or 0.42 mM). This condition, *hyperuricemia*, results over time in the deposition of crystalline monosodium urate (MSU) in tissues. Chronically hyperuricemic individuals are at risk for developing gout, a relatively common disorder in which MSU crystals can trigger a very painful, and potentially destructive inflammatory arthritis [17, 18]. Patients with longstanding gout may also develop uric acid kidney stones, as well as tophi – nodular deposits of MSU crystals surrounded by inflammatory cells – within bone, joints, and other soft tissues, which can in various ways result in local damage and loss of function. In patients with leukemia and lymphoma, rapid turnover of malignant cells can lead to a marked increase in renal uric acid excretion, particularly during chemotherapy. This can result in bilateral obstruction of kidney tubules, causing acute renal failure (tumor lysis syndrome) [19].

A primary goal of treatment for both gout and uric acid nephropathy is to reduce the concentration of urate in plasma to below the limit of solubility; a widely accepted target is a plasma urate of <6 mg/dL (0.36 mM). When hyperuricemia is mild, this may be achieved by weight reduction or change in diet, but it often requires treatment with drugs that either promote renal urate excretion (probenecid, sulfinpyrazone), or that block uric acid synthesis by inhibiting xanithine oxidoreductase. Allopurinol, currently the only FDA-approved xanthine oxidase inhibitor, is the most widely used uric acid lowering agent (it may soon be joined by febuxostat, a non-purine xanthine oxidase inhibitor). These 'conventional' drugs are usually well tolerated, and at adequate doses are effective in controlling hyperuricemia in the majority of patients. However, allopurinol may cause serious hypersensitivity, and for various reasons, it is relatively ineffective in a subset of patients with gout. If hyperuricemia is not well controlled and urate levels remain elevated, gout can progress to a chronic stage that can be very difficult to manage [20].

There has long been interest in parenteral uricase as an alternative therapy for controlling hyperuricemia. Purified uricase from the fungus *Aspergillus neoformans* (Uricozyme) has been used for more than 40 years in Europe, primarily for preventing acute uric acid nephropathy during chemotherapy for hematologic malignancies. A recombinant preparation of this enzyme (rasburicase, Elitek) is now used, and has been approved for this indication in the US [21]. Because of a relatively short circulating life (about 18 h), this unmodified uricase is infused daily or every other day. Because of its potential immunogenicity, it has only been approved for a single treatment course of 1–2 weeks duration in patients undergoing chemotherapy. Rasburicase has been used to treat several patients with gout, but its safety and efficacy for this indication is uncertain [22, 23]. PEGylation has been explored as a means of prolonging circulating life and reducing immunogenicity of purified uricase to permit more extensive and chronic clinical use.

Investigation of PEG-uricase in animals

Chen et al were among the first to examine the effects of PEGylation on the physicochemical and pharmacologic properties of uricase [24]. Uricases from hog liver and *Candida utilits* were coupled with monomethoxyPEG (mPEG) of average molecular weight 5,000 (5 kDa mPEG), using cyanuric chloride activation. Modification of about 60% of reactive amino groups significantly increased enzyme solubility, but eliminated about 75% of catalytic activity and had minor effects on pH and temperature optima, and on Km for uric acid. PEGylation resulted in a large change in antigenicity, immunogenicity, and circulating life: the 60%-modified enzymes did not react with antibodies raised in mice against the native uricases, and did not elicit precipitating anti-uricase antibodies in mice or rabbits. After repeated intravenous dosing of mice with the 60%-PEGylated uricases, blood circulating life was well maintained, in contrast with markedly enhanced clearance of both native and less extensively (37–47%) PEGylated enzymes after a few injections. The same laboratory reported [25] that both 35% and 70% PEGylated uricase reduced the IgE antibody response elicited by a subsequent exposure to the native uricase. The 35% modified uricase was more 'toleragenic' than the 70% modified enzyme. A tolerizing effect of PEGylated allergens had previously been reported [26, 27].

Tsuji et al. [28] found that modifying 45% of *Candida* uricase amino groups with cyanuric chloride activated mPEGs of 5, 7.5, and 10 kDa eliminated 91–94% of catalytic activity, and as little as 4–9% modification with 10 kDa mPEG eliminated ≥70% of activity. The antibody response in rabbits and guinea pigs immunized with these preparations in the presence of Freund's adjuvant diminished with increasing degree of modification. IgG antibody from the immunized animals reacted with the PEGylated, but not with the native uricases. Thus, PEGylation appeared to mask epitopes present in native uricase, but generated novel eptiopes not present in the native enzyme. The ability of antisera induced by PEGylated uricase to bind PEGylated superoxide dismutase (PEG-SOD) suggested that the novel epitopes were partially due to the structure of the coupling agent and the PEG-protein juncture, or to the PEG itself [28].

Caliceti et al. [29, 30] modified *Candida* uricase with succinimidyl ester-activated linear (5 kDa) or branched (10 kDa) PEGs (other polymers used have not yet been employed clinically and will be not discussed). PEGylation was performed in the presence of urate to protect the active site and limit the loss of catalytic activity, which nevertheless ranged from 20–80% of the activity of the native enzyme. For pharmacokinetic analysis, the enzymes were labeled with [3H]-proprionate. PEGylation greatly extended circulating life and diminished uptake of labeled enzyme by tissues [29]. Uptake by the liver and spleen was greater with enzyme modified with branched than linear PEG, but both conjugates were retained only transiently in tissues, apparently after phagocytosis by resident reticuloendothelial cells, and were ultimately eliminated by the kidney. In other studies, Caliceti et al. [30] found that antibodies raised against native *Candida* uricase showed minimal binding of uricase conjugated with either linear or branched PEG. Both PEG conjugates were less immunogenic than the native enzyme after repeated intraperitoneal administration to mice in the presence of Freund's adjuvant. Uricase conjugated with the branched PEG was less immunogenic than enzyme conjugated with linear PEG, and it induced lower levels of IgM and IgG antibodies to PEG. The effect of anti-polymer antibody on pharmacokinetics of PEGylated *Candida* uricase was not reported [30].

The laboratory species used in the studies described above possessed endogenous uricase, and the PEGylated enzymes administered to them had lost significant catalytic activity during modification. However, one study was conducted in leghorn chickens, a species that lacks uricase and has uric acid levels similar to humans [31]. Infusion of a large amount (10–20 U) of native or 61% PEGylated *Candida* uricase rapidly lowered plasma uric acid to undetectable levels, although levels returned to the half normal range within 48 h. With repeated dosing, an antibody response to the native enzyme resulted in rapid clearance and loss of its effect on plasma urate, whereas the PEGylated enzyme remained effective after five infusions.

A PEGylated mammalian uricase (pegloticase) for treatment of refractory gout: Pre-clinical development

More than five years after reporting our experience with a PEGylated bacterial uricase [5, 6], neither this or any other PEGylated uricase had become available for further clinical investigation. By then, the safety and efficacy of PEG-ADA were evident, and we were optimistic that a PEG-uricase might benefit patients with severe gout whose other therapeutic options were limited. So in 1993 we applied for, and were awarded, a technology transfer (STTR) grant from the US National Institute of Health to develop a PEGylated recombinant mammalian uricase as an Orphan Drug for the treatment of patients with refractory gout.

We initially compared the expression in *E. coli* of recombinant pig and baboon uricase (based on gene exon sequences, both mammalian enzymes were much more closely related than were microbial uricases to ancestral human uricase). We proceeded with the significantly more active porcine uricase, but replaced an Arg codon in the porcine cDNA with a Lys codon found at the same position in baboon uricase, a strategy for enhancing the ability of PEGylation to mask epitopes and reduce immunogenicity [32]. After purification, the Lys-mutated porcine uricase had the same specific activity as wild type native porcine uricase; PEGylation significantly increased its solubility with minimal loss of activity. Optimized conditions for PEGylating the recombinant uricase were then developed at Mountainview Pharmaceuticals, which had joined the project when the STTR grant was renewed in 1996 (reviewed elsewhere [33]). For testing the efficacy and immunogenicity of the resulting PEG-uricase, we used uricase knockout mice, which lacked endogenous uricase protein and developed an often fatal uric acid nephropathy [34]. We found that weekly intraperitoneal injections of PEG-uricase normalized urate levels in plasma and urine, and prevented nephropathy, in these mice [35]. PEG-uricase was well tolerated, and induced neither anti-uricase antibodies nor rapid enzyme clearance, whereas the unmodified recombinant uricase was both ineffective and highly immunogenic.

In 1998, Duke University and Mountainview Pharmaceuticals jointly licensed this PEGylated recombinant mammalian (porcine) uricase to Savient Pharmaceuticals for manufacture and clinical development. The material produced by Savient for toxicology studies and human clinical trials, which has received the generic name 'pegloticase', contains ~9 strands of 10 kDa mPEG per subunit (total mass ~500 kDa). The specific uricase activity of pegloticase is ~18 U/mg protein when assayed at 37 °C in our laboratory by a validated radiochemical-HPLC method [6].

Clinical investigation of pegloticase

In 2001, Puricase (pegloticase) received Orphan Drug status from the FDA Office of Orphan Products Development for the treatment of patients with refractory gout. This category includes hyperuricemic patients with symptomatic gout for whom currently available uric acid-lowering therapy is not tolerated or is contraindicated, or has been ineffective ('treatment failure gout' is another term applied to this patient population). Chronic refractory gout is a painful, disabling condition. Most of these patients have visible, often bulky tophi that can interfere significantly with normal activities and function. Comorbidities associated with gout, such as hypertension, hyperlipidemia, cardiovascular disease, diabetes, and renal insufficiency often further complicate their medical management [20].

Pharmcokinetics, pharmacodynamics and immunoreactivity in Phase I testing

Savient Pharmaceuticals sponsored two open-label, single escalating dose Phase I trials of pegloticase in hyperuricemic patients with refractory gout [36, 37]. The 37 participants had mean baseline pUA > 11 mg/dL, and more than 75% had tophi.

In the first trial [36] groups of four subjects received single subcutaneous (SC) injections of 4, 8, or 12 mg pegloticase; one subject received 24 mg.

Absorption of pegloticase was variable and generally slow. Maximum pUox occurred at about day 7 post-injection, coinciding with an average decline in pUA of almost 8 mg/dL from baseline; pUA normalized in 11/13 subjects (mean, 2.8 mg/dL). At doses ≥8 mg, mean pUA remained ≤6 mg/dL at 21 days post-injection. In eight subjects pUox was still measurable at 21 days post-injection, but in five others pUox was undetectable beyond 10 days. Rapid clearing of pUox coincided with appearance of IgM and then IgG antibodies to pegloticase, which reacted with the PEG rather than the protein moiety. Three subjects, all with anti-PEG antibody, had allergic skin reactions at 8–10 days after dosing, which resolved within a few days, in one case after treatment with corticosteroids.

The second Phase I trial [37] examined single intravenous (IV) infusions of 0.5, 1, 2, 4, 8, or 12 mg of pegloticase. Maximum pUox was proportional to dose, and AUC was linear up to a dose of 8 mg. The half-life of pUox was 6.4–13.8 days (much longer than that of rasburicase). After doses of 4–12 mg, mean pUA fell to 1.0 ± 0.5 mg/dL within 24–72 h, and the AUC for pUA was equivalent to maintaining pUA at a constant level of 1.2–4.7 mg/dL for the entire 21-day post-infusion observation period. The ratio of uric acid to creatinine (UA/Cr) in urine declined in parallel with pUA. IgG antibodies to pegloticase, mostly of the IgG2 subclass and specific for PEG, developed in nine of 24 subjects, but there were no allergic reactions.

Together these Phase I trials demonstrated that bioavailability, efficacy, and tolerability were superior with IV compared with SC pegloticase; and that single infusions of 4–12 mg of pegloticase could rapidly normalize pUA and greatly lower urinary uric acid for up to three weeks in markedly hyperuricemic gout patients who had failed prior uric acid lowering therapy.

Phase II and Phase III testing of intravenous pegloticase

A Phase II open-label trial evaluated the ability of repeated IV infusions of pegloticase to achieve and maintain a pUA of <6 mg/dL in 41 subjects with treatment failure gout (mean serum UA, 10.3 mg/dL; mean duration of gout, 14 years; 71% with tophi) [38]. Subjects were randomized to receive every two-week infusions of 4 mg or 8 mg pegloticase, or every four-week infusions of 8 mg or 12 mg, over 12–14 weeks. Within 6 h of the first dose, pUA fell to <6 mg/dL in all dose groups. A mean pUA of <6 mg/dL was maintained during the entire study period (through 28 days after the last dose of pegloticase) in the 8 mg and 12 mg dose groups, with 8 mg every two weeks being most effective (mean pUA, 1.42 ± 2.06 mg/dL). The mean plasma half-life of pegloticase was 289 h (12.0 days) after the first infusion and 268 h (11.1 days) after the last infusion. Gout flares occurred in 88% of subjects (flares are expected in patients with poorly controlled gout, and they are also a known consequence of initiating urate-lowering therapy). Most adverse events were mild or moderate in severity and were unrelated to treatment.

There were no anaphylactic reactions, but infusion day events, including muscle spasm, shortness of breath, and allergic reactions, accounted for 12/15 withdrawals from the study. Antibodies to pegloticase were detected in about ³/₄ of subjects, and were associated with reduced enzyme circulating life in some cases.

Two identical randomized, double-blind, placebo-controlled Phase III trials of pegloticase (GOUT1 and GOUT2) were completed in the fall of 2007. A total of 212 subjects with treatment failure gout received placebo (N = 43) or 8 mg of pegloticase, which was infused either every two weeks (N = 85) or every four weeks (N = 84) for six months. Data are still being evaluated, but abstracts reporting the following findings have been submitted: 1) Approximately 40% of pegloticase-treated subjects (combining the two- and four-week infusion groups), but none of the placebo subjects, achieved the primary trial goal, a reduction of pUA to <6 mg/dL in 80% of measurements made during both months three and six [39]. 2) At baseline, 73% of subjects had tophi. By six months, a significant reduction in tophus size had occurred in 40% of subjects treated with pegloticase every two weeks, with 20% showing complete resolution of at least one tophus by 13 weeks of treatment [40]. 3) Compared with placebo, pegloticase treatment was associated with a significant reduction in the number of tender joints, improvement in health-related quality of life measurements, and decreased disability [39, 41]. 4) Anti-PEG antibodies correlated strongly with the failure to meet the pUA reduction goal, and with infusion reactions [42]. Gout flares and infusion reactions were the most frequent reason for withdrawal from the study. About 24% of subjects treated with pegloticase, *versus* 12% for placebo, experienced serious adverse events.

In summary, clinical testing in patients with treatment failure gout has shown that infusions of 8 mg of pegloticase at 2–4 week intervals can rapidly correct hyperuricemia and maintain low levels of plasma and urinary uric acid, conditions that favor resolution of tophi and uric acid kidney stones. These benefits were achieved in some cases within 3–6 months, but longer treatment may be required in the most severely affected patients. Although PEGylation was effective in limiting immune recognition of the recombinant uricase protein, the induction of antibodies to PEG has led to rapid clearance of pegloticase and loss of efficacy in a proportion of treated patients. Understanding the basis for this anti-polymer response will be important for predicting susceptibility, and for developing strategies to suppress or prevent their development.

An open-label extension allowing Phase III trial participants to receive an additional 18 months of pegloticase therapy is currently in progress. This trial may provide a better picture of the ability of pegloticase therapy to modify the course of refractory/treatment failure gout, and of its long-term safety and tolerability. We are also conducting an open-label Phase II trial of pegloticase, sponsored by a grant from the FDA Office of Orphan Product Development, in which organ transplant recipients with refractory gout may participate (they were excluded from the other clinical trials). Results of this ongoing study may indicate whether immunosuppressive therapy, taken by these patients to prevent transplant rejection, can reduce the frequency of anti-PEG antibodies.

Acknowledgement
Drs Hershfield and Kelly are among inventors named on patents for PEGylated mammalian urate oxidase, and along with Duke University, they may receive royalties under the terms of a licensing agreement with Savient Pharmaceuticals. The authors have received grant support from Savient Pharmaceuticals. Dr. Hershfield has served on a few occasions as a paid consultant to Savient Pharmaceuticals.

References

1 Davis S, Abuchowski A, Park YK, Davis FF. Alteration of the circulating life and antigenic properties of bovine adenosine deaminase in mice by attachment of polyethylene glycol. Clin Exp Immunol. 1981;46:649–652.
2 Chen RH-L, Abuchowski A, van Es T, Palczuk NC, Davis FF. Properties of two urate oxidases modified by the covalent attachment of poly(ethylene glycol). Biochim Biophys Acta. 1981;660:293–298.
3 Hershfield MS, Buckley RH, Greenberg ML, Melton AL, Schiff R, Hatem C, Kurtzberg J, Markert ML, Kobayashi RH, Kobayashi AL, Abuchowski A. Treatment of adenosine deaminase deficiency with polyethylene glycol-modified adenosine deaminase. N Engl J Med. 1987;316:589–596.
4 Davis S, Park YK, Abuchowski A, Davis FF. Hypouricaemic effect of polyethylene glycol modified urate oxidase. Lancet. 1981;2:281–283.
5 Chua CC, Greenberg ML, Viau AT, Nucci M, Brenckman WD Jr, Hershfield MS. Use of polyethylene glycol-modified uricase (PEG-uricase) to treat hyperuricemia in a patient with non-Hodgkin lymphoma. Ann Int Med. 1988;109:114–117.
6 Greenberg ML, Hershfield MS. A radiochemical-high-performance liquid chromatographic assay for urate oxidase in human plasma. Anal Biochem. 1989;176:290–293.
7 Keilin J. The biologic significance of uric acid and guanine excretion. Biol Rev Cambridge Phil Soc. 1959;34:265–296.
8 Christen P, Peacock WC, Christen AE, Wacker WEC. Urate oxidase in primate phylogenesis. Eur J Biochem. 1970;12:3–5.
9 Wu X, Muzny DM, Lee CC, Caskey CT. Two independent mutational events in the loss of urate oxidase. J Mol Evol. 1992;34:78–84.
10 Oda M, Satta Y, Takenaka O, Takahata N. Loss of urate oxidase activity in hominoids and its evolutionary implications. Mol Biol Evol. 2002;19:640–653.
11 Kahn K, Serfozo P, Tipton PA. Identification of the true product of the urate oxidase reaction. J Am Chem Soc. 1997;119:5435–5442.
12 Kahn K, Tipton PA. Spectroscopic characterization of intermediates in the urate oxidase reaction. B. 1998;37:11651–11659.
13 Ramazzina I, Folli C, Secchi A, Berni R, Percudani R. Completing the uric acid degradation pathway through phylogenetic comparison of whole genomes. Nat Chem Biol. 2006;2:144–148.
14 Fahimi HD, Reich D, Volkl A, Baumgart E. Contributions of the immunogold technique to investigation of the biology of peroxisomes. Histochem Cell Biol. 1996;106:105–114.
15 Conley TG, Priest DG. Purification of uricase from mammalian tissue. Preparative Biochemistry. 1979;9:197–203.
16 Colloc'h N, El Haji M, Bachet B, L'Hermite G, Schiltz M, Prange T, Castro B, Mornon J-P. Crystal structure of the protein drug urate oxidase-inhibitor complex at 2.05 Å resolution. Nat Struct Biol. 1997;4:947–952.
17 Becker MA. Hyperuricemia and gout. In: Scriver CR, Beaudet al., Sly WS, Valle D, eds. The Metabolic and Molecular Bases of Inherited Disease (ed 8th). New York: McGraw-Hill; 2001:2513–2535.
18 Terkeltaub RA. Clinical practice. Gout. N Engl J Med. 2003;349:1647–1655.
19 Jones DP, Mahmoud H, Chesney RW. Tumor lysis syndrome: pathogenesis and management.

Pediatr Nephrol. 1995;9:206–212.

20 Sundy JS, Hershfield MS. Uricase and other novel agents for the management of patients with treatment-failure gout. Curr Rheumatol Rep. 2007;9:258–264.

21 Navolanic PM, Pui CH, Larson RA, Bishop MR, Pearce TE, Cairo MS, Goldman SC, Jeha SC, Shanholtz CB, Leonard JP, McCubrey JA. Elitek-rasburicase: an effective means to prevent and treat hyperuricemia associated with tumor lysis syndrome, a Meeting Report, Dallas, Texas, January 2002. Leukemia. 2003;17:499–514.

22 Vogt B. Urate oxidase (rasburicase) for treatment of severe tophaceous gout. Nephrol Dial Transplant. 2005;20:431–433.

23 Richette P, Briere C, Hoenen-Clavert V, Loeuille D, Bardin T. Rasburicase for tophaceous gout not treatable with allopurinol: an exploratory study. J Rheumatol. 2007;34:2093–2098.

24 Chen RH-L, Abuchowski A, van Es T, Palczuk NC, Davis FF. Properties of two urate oxidases modified by the covalent attachment of poly(ethylene glycol). Biochim Biophys Acta. 1981;660:293–298.

25 Savoca KV, Davis FF, Palczuk NC. Induction of tolerance in mice by uricase and monomethoxy-polyethylene glycol-modified uricase. Int Arch Allergy Appl Immunol. 1984;75:58–67.

26 Lee WY, Sehon AH. Suppression of reaginic antibodies with modified allergens. I. Reduction in allergenicity of protein allergens by conjugation to polyethylene glycol. Int Arch Allergy Appl Immunol. 1978;56:159–170.

27 Lee WY, Sehon AH, Akerblom E. Suppression of reaginic antibodies with modified allergens IV. Induction of suppressor T cells by conjugates of polyethylene glycol (PEG) and monomethoxy PEG with ovalbumin. Int Archs Allergy Appl Immun. 1981;64:100–114.

28 Tsuji J, Hirose K, Kasahara E, Naitoh M, Yamamoto I. Studies on the antigenicity of the polyethylene glycol-modified uricase. Int J Immunopharmacol. 1985;7:725–730.

29 Caliceti P, Schiavon O, Veronese FM. Biopharmaceutical properties of uricase conjugated to neutral and amphiphilic polymers. Bioconjugate Chem. 1999;10:638–646.

30 Caliceti P, Schiavon O, Veronese FM. Immunological properties of uricase conjugated to neutral soluble polymers. Bioconjug Chem. 2001;12:515–522.

31 Abuchowski A, Karp D, Davis FF. Reduction of plasma urate levels in the cockerel with polyethylene glycol-uricase. J Pharmacol Exp Ther. 1981;219:352–354.

32 Hershfield MS, Chaffee S, Koro-Johnson L, Mary A, Smith AA, Short SA. Use of site-directed mutagenesis to enhance the epitope-shielding effect of covalent modification of proteins with polyethylene glycol. Proc Natl Acad Sci USA. 1991;88:7185–7189.

33 Sherman MR, Saifer MG, Perez-Ruiz F. PEG-uricase in the management of treatment-resistant gout and hyperuricemia. Adv Drug Deliv Rev. 2008;60:59–68.

34 Wu X, Wakamiya M, Vaishnav S, Geske R, Montgomery CM Jr, Jones P, Bradley A, Caskey CT. Hyperuricemia and urate nephropathy in urate oxidase-deficient mice. Proc Natl Acad Sci USA. 1994;91:742–746.

35 Kelly SJ, Delnomdedieu M, Oliverio MI, Williams LD, Saifer MGP, Sherman MR, Coffman TM, Johnson GA, Hershfield MS. Diabetes insipidus in uricase-deficient mice: A model for evaluating therapy with poly(ethylene glycol)-modified uricase. J Am Soc Nephrol. 2001;12:1001–1009.

36 Ganson NJ, Kelly SJ, Scarlett E, Sundy JS, Hershfield MS. Control of hyperuricemia in subjects with refractory gout, and induction of antibody against poly(ethylene glycol) (PEG), in a phase I trial of subcutaneous PEGylated urate oxidase. Arthritis Res Ther. 2006;8:R12.

37 Sundy JS, Ganson NJ, Kelly SJ, Scarlett E, Rehrig CD, Huang W, Hershfield MS. Pharmacokinetics and pharmacodynamics of intravenous PEGylated recombinant mammalian urate oxidase in patients with refractory gout. Arthritis Rheum. 2007;56:1021–1028.

38 Sundy JS, Becker M, Baraf HS, Barkuizen A, Moreland LW, Huang B, Waltrip RW, Maroli AN, Horowitz Z. Reduction of plasma urate following multiple doses of pegloticase (PEG-Uricase) in subjects with treatment failure gout. Arthritis Rheum. 2008; 58:2882–2891.

39 Sundy JS, Baraf HS, Becker M, Edwards NL, Gutierrez-Urena SR, Treadwell EL, Vazquez-Mellado J, Yood RA, Horowitz Z, Huang B, Maroli AN, Waltrip RW. Efficacy and safety of intravenous (IV) pegloticase (PGL) in subjects with treatment failure gout (TFG): Phase 3 results from GOUT1 and GOUT2. Abstract 635, American College of Rheumatology Scientific Meeting, San Francisco CA, November 2008.

40 Baraf HS, Becker M, Edwards NL, Gutierrez-Urena SR, Sundy JS, Treadwell EL, Vazquez-Mellado J, Yood RA, Horowitz Z, Huang B, Maroli AN, Waltrip RW. Tophus response to pegloticase (PGL) therapy: pooled results from GOUT1 and GOUT2, PGL phase 3 randomized, double

blind, placebo-controlled trials. Abstract 22, American College of Rheumatology Scientific Meeting, San Francisco CA, November 2008.

41 Edwards NL, Baraf HS, Becker M, Gutierrez-Urena SR, Sundy JS, Treadwell EL, Vazquez-Mellado J, Yood RA, Kawata AK, Benjamin KL, Horowitz Z, Huang B, Maroli AN, Waltrip RW. Improvement in health-related quality of life (HRQL) and disability index in treatment failure gout (TFG) after peglotidase (PGL) therapy: pooled results from GOUT1 and GOUT2, phase 3 randomized, double blind, placebo (PBO)-controlled trials. Abstract 27, American College of Rheumatology Scientific Meeting, San Francisco CA, November 2008.

42 Becker M, Treadwell EL, Baraf HS, Edwards NL, Gutierrez-Urena SR, Sundy JS, Vazquez-Mellado J, Yood RA, Horowitz Z, Huang B, Maroli AN, Waltrip RW, Wright D. Immunoreactivity and clinical response to pegloticase (PGL): pooled data from GOUT1 and GOUT2, PGL phase 3 randomized, double blind, placebo-controlled trials. Abstract 1945, American College of Rheumatology Scientific Meeting, San Francisco CA, November 2008.

Certolizumab pegol: a PEGylated anti-tumour necrosis factor alpha biological agent

Andrew M. Nesbitt[1], Sue Stephens[2] and Elliot K. Chartash[3]

[1] *Inflammation Research, UCB Celltech, Slough SL1 3WE, UK*
[2] *Non-Clinical Development, UCB Celltech, Slough SL1 3WE, UK*
[3] *Clinical Development, UCB Inc, Atlanta, GA, USA*

Abstract

Tumour necrosis factor (TNF)α is a proinflammatory cytokine involved in systemic inflammation that mediates chronic inflammatory diseases such as rheumatoid arthritis (RA), Crohn's disease (CD) and psoriasis. Recognition of TNFα as a primary mediator of inflammatory disease has driven the development of monoclonal antibodies (mAbs) against TNFα as potential novel therapies for these disorders. Certolizumab pegol is a novel, polyethylene glycol (PEG)-conjugated, humanised, antigen-binding fragment (Fab') of an anti-TNFα mAb that does not mediate apoptosis or neutrophil degranulation. Preclinical studies have shown excellent bioavailability, with preferential distribution and retention in inflamed tissue, which could be due to the low diffusion rate of PEGylated molecules and/or the lack of an Fc, which prevents FcRn-mediated transport. Pharmacokinetics are linear and predictable. Certolizumab pegol is a potentially valuable new treatment option for several inflammatory diseases. It has shown promising efficacy and tolerability results in Phase II and III trials for RA, CD and psoriasis.

History of anti-tumour necrosis factor agents in autoimmune inflammatory disease states

Tumour necrosis factor α, its structure and function and biological roles

Tumour necrosis factor (TNF)α is a proinflammatory cytokine involved in systemic inflammation that is known to be a mediator of chronic inflammatory diseases such as Crohn's disease (CD), rheumatoid arthritis (RA) and psoriasis [1]. The existence of lymphotoxin (LT), a cytotoxic factor produced by lymphocytes, was first described at the University of California in 1968 [2]. TNF itself was subsequently isolated by researchers at the Memorial Sloan-Kettering Cancer Center in New York from macrophages in 1975 [3].

Recognition of the sequential and functional homology of TNF and LT led to the renaming of these two compounds as TNFα and TNFβ, respectively. TNFα was then recognised as having a key role in cachexia and as a principal mediator of septic shock in patients with infection [4, 5]. This molecule was ultimately found to be the prototype for the large family of TNF cytokines whose members are involved in the control of cell differentiation, proliferation and apoptosis, most notably in the immune and haematopoietic systems. Human TNFα is a nonglycosyated protein consisting of 157 amino acids that exists in both soluble and membrane-bound forms and is secreted by a variety of cell types, including macrophages, monocytes, neutrophils and T cells [6].

TNFα binds to two receptors, the 55 kDa TNFR1 (CD120a or p55, widely expressed on virtually all nucleated cell types) and the 75 kDa TNFR2 (CD120b or p75, expressed mainly by activated white blood cells and endothelial cells) [7]. The extracellular domains of TNFR1 and TNFR2 bind to the cleft between the subunits of the TNFα molecule, which initiates signalling [8]. The presence of two receptors allows for a wide diversity of signalling functions. Activation of TNFR1 can have a number of outcomes, depending on the availability of accessory proteins in differing cell types: the cytoplasmic domain of TNFR1 includes a death domain motif that initiates apoptosis after activation of caspases 3 and 8 (Fig. 1). Alternatively, TNF receptor–associated factor 2 (TRAF2) can recruit cellular inhibitors of apoptosis and activate pathways leading to nuclear translocation of antiapoptotic transcription factors such as nuclear factor-κB (NF-κB) and activator protein-1 (AP-1) (Fig. 1).

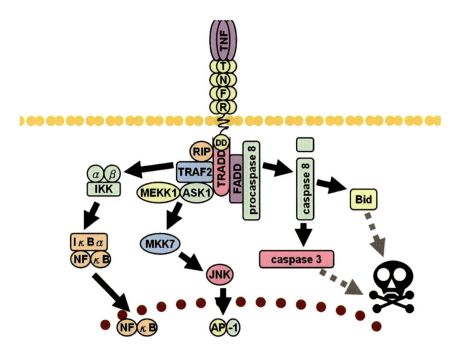

Figure 1. Simplified signaling pathways of TNFα. Dashed lines represent multiple steps. AP-1, activator protein-1; ASK1, apoptosis signal-regulating kinase 1; FADD, Fas-associated death domain; IKK, I κB kinase; JNK, c-Jun N-terminal kinase; MEK, mitogen-activated protein kinase kinase; MEKK1, MEK kinase 1; MKK, mitogen-activated protein kinase kinase 7; NF, nuclear factor-κB; RIP, receptor-interacting protein; TNFR, TNF receptor; TRADD, TNF receptor-associated death domain; TRAF2, TNF receptor-associated factor 2. Bid is a pro-apoptotic member of the Bcl-2 family.

These regulate the expression of genes blocking apoptosis, increasing cell proliferation and increasing expression of proinflammatory proteins [9]. Indeed, the NF-κB pathway alone activates more than 200 proinflammatory genes [10]. TNFR2 has specific signalling functions in T cells [11]. Although this receptor lacks the death domain found on TNFR1, it can mediate apoptosis via another, currently unknown, pathway [12]. TNFR2 can also activate NF-κB and AP-1.

TNFα in the pathogenesis of CD, RA and psoriasis

Normally, expression of TNFα and other proinflammatory cytokines is held in balance by anti-inflammatory factors, but this balance is shifted in inflammatory disease. Uncontrolled or excessive activity of TNFα leads among other effects to the chronic inflammation that characterises diseases such as CD, RA and psoriasis [13]. Through the pathways summarised above, TNFα upregulates adhesion molecules in endothelial tissue, stimulates fibroblast proliferation and recruits leukocytes into synovial fluid [14]. TNFα stimulates production of other cytokines and chemokines, reactive oxygen species, nitric oxide and prostaglandins, and increases rates of protease-mediated tissue remodelling [15, 16]. It promotes angiogenesis and osteoclast differentiation and activates bone-resorbing osteoclasts, which leads to joint erosion, particularly at marginal surfaces [17, 18]. TNFα also directly mediates pain, fever and cachexia.

CD

CD is mediated by T cells and is characterised by relapsing inflammation of the gut with extraintestinal manifestations typically involving the skin, eyes and joints. The causative factors remain unknown, although bacteria appear to play a major role in the inflammatory process [19]. The significance of the disease in terms of healthcare resource consumption is underlined by a prevalence of around 0.1% across the developed world [20] and the presence of complications such as perianal fistulae (which may require surgery) in as many as 43% of patients [21].

An imbalance in cytokine expression is central to the pathogenesis of CD. Inflammation of the gut is marked by mucosal infiltration by neutrophils and macrophages (Fig. 2), which in turn activates T cells [22]. T-helper (Th) type 1 cells subsequently stimulate the production of proinflammatory cytokines, which amplify the immune response and promote tissue destruction [23]. Indeed, the immunopathogenesis of CD involves interactions between a number of cytokines which include not only TNFα but also a number of interleukins (ILs) (Fig. 2), as shown by studies in lamina propria mononuclear cells isolated from colonic biopsies from patients with untreated inflammatory bowel disease [23]. These cells were noted to produce IL-1β and IL-6 as well as TNFα. In addition, lamina propria T cells have been reported to resist apoptosis in patients with inflammatory bowel disease [24, 25].

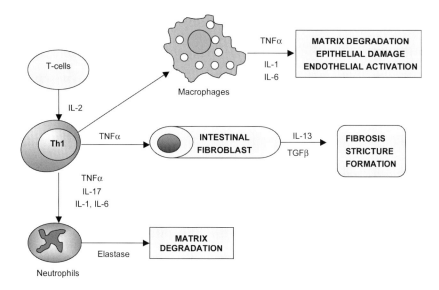

Figure 2. Extracellular molecular mediators of CD. IL, interleukin; Th1, T-helper cell type 1; TGFβ, transforming growth factor β; TNFα, tumour necrosis factor α.

This resistance is one of the best described disturbances to the immune system described in patients with this type of disorder; the beneficial effect of sulfasalazine in patients with CD appears to be at least partially attributable to the drug's proapoptotic effects in lamina propria T cells [26].

RA

As with CD, the precise aetiology of RA is unknown. However, it has become clear that the key drivers of inflammation in this disorder include most notably TNFα, in addition to IL-1 and IL-6 (Fig. 3) [14]. As in CD, Th1 cells stimulate the production of a variety of proinflammatory cytokines and destructive proteinases. This is seen after the activation of Th1 cells by antigen-presenting cells (APCs) and costimulatory pathways, and their infiltration of the synovium [14].

Cartilage explants treated with recombinant human TNFα show signs of tissue destruction, as demonstrated by enhanced resorption and inhibition of proteoglycan synthesis in experiments carried out in the 1980s [27]. Further experiments showed clinical and histologic changes indicative of RA in genetically engineered mice constitutively expressing TNFα [28]. Confirmation of these effects has come from results from three laboratories that show reduction of disease activity by anti-TNF antibodies as indicated by a standard collagen-induced arthritis model [29–31].

Psoriasis

Psoriasis, a chronic inflammatory disease of uncertain origin that affects approximately 2% of the population [32], can cause debilitating arthropathy in

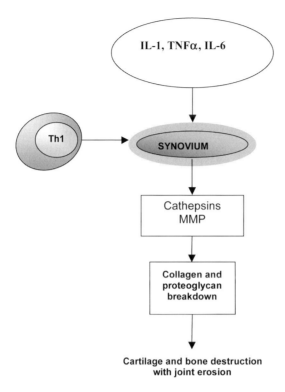

Figure 3. Extracellular molecular mediators of RA. IL, interleukin; MMP, matrix metalloproteinase; Th1, T-helper cell type 1; TNFα, tumour necrosis factor α.

as many as 30% of patients [33], and is characterised by the growth of scaly erythematous plaques. Similarly to CD and RA, psoriasis and psoriatic arthritis are mediated by T cells, which are found in abundance together with increased vascularity in skin plaques and inflamed synovial tissue. The inflammatory cascade is believed to be triggered by activation of CD4+ T cells, which generate in turn a number of proinflammatory cytokines that include TNFα. These cytokines activate CD8+ T cells, the main effectors in this disorder [34]. The centrality of the role of TNFα has been demonstrated by high concentrations of this factor in psoriatic skin lesions and the synovium of affected joints, and by its apparent importance in the perpetuation of inflammation in addition to the stimulation of angiogenesis and proliferation of keratinocytes [35].

Definitive work showing the key role played by TNFα in the pathogenesis of psoriasis has been carried out using a model in which human pre-psoriatic skin was grafted on to immunodeficient mice [36]. Approximately 6–8 weeks after engraftment, clinical and histologic features of psoriasis appeared, with increased expression of Ki-67 protein, major histocompatibility complex (MHC) Class II antigen, TNFα, IL-12, keratin 16, intracellular adhesion molecule (ICAM)-1 and platelet/endothelial cell adhesion molecule (PECAM)-1, which are all associated with inflammation. Neutralisation of TNFα significantly reduced indices of papillomatosis and acanthosis, and was linked to reduced numbers of T cells in grafts. This implies that the development of psoriasis and associated proliferation of T cells depends on the expression of TNFα.

The development of anti-TNFα agents

The monoclonal antibodies: infliximab (Remicade®) and adalimumab (Humira®)

Evolving understanding of the role of TNFα as the primary mediator of the inflammatory diseases, together with advances in recombinant gene technology, has driven the development of monoclonal antibodies (mAbs) against TNFα as potential novel therapies for CD, RA and psoriasis. The first steps in mAb therapy were taken in the mid-1980s with the development of a murine mAb (muromonab-CD3) that recognised the CD3 antigen found on human T cells, and this proved useful in the management of renal transplant rejection. This compound was suitable for short-term use only, however, because of the development of an immune response in most patients [37, 38], and further research was needed to reduce the immunogenicity of these antibodies [39]. Further progress in this field resulted in the introduction of two mAbs against TNFα for use in patients with immunologic inflammatory disease; these are the chimeric mAb infliximab, and, more recently, the human mAb adalimumab.

Pharmacology
In the chimeric mAb infliximab (Tab. 1), the first anti-TNFα agent to be introduced for the management of CD, RA and psoriasis, the human regions consist of immunoglobulin G1κ (IgG1κ) constant regions, while the variable regions are murine. As reviewed by Wong et al. [40], infliximab binds to both soluble and membrane-bound TNFα with high affinity and specificity, but does not bind to TNFβ. Its binding to transmembrane TNFα can mediate apoptosis, and its specificity limits potential for unwanted effects on other biological pathways. However, the presence of the murine variable region can lead to the production of human anti-chimeric antibodies, which may limit the therapeutic applications of infliximab [41].

Administration is by intravenous infusion, typically at a dose of 3–5 mg/kg every eight weeks, increasing dosing to 10 mg/kg or increasing the dosing interval frequency to every 4–5 weeks, if efficacy weakens [42].

Adalimumab is a humanized IgG1 anti-TNFα mAb generated by antibody guided selection using phage display technology. Similarly to infliximab, it binds soluble and transmembrane TNFα with high affinity, but unlike infliximab is not chimeric and has entirely human amino acid sequences.

Table 1. Comparative characteristics and pharmacology of anti-TNFα agents [96, 42, 97]

Characteristics and parameters	Agent		
	Infliximab	Adalumimab	Certolizumab pegol
Structure	Chimeric mouse-human IgG1 mAb	Recombinant human IgG1 mAb	PEGylated human-ized anti-Fab' fragment of an anti-TNFα mAb No Fc region
Conjugate	None	None	PEG2[*]: 2×20 kDa chains
Route of administration	Intravenous	Subcutaneous	Subcutaneous
Dose	3–5 mg/kg at weeks 0, 2 and 6, then 8-weekly. Can be increased to 10 mg/kg or frequency increased	40 mg every second week (or weekly if necessary) (RA). An induction of up to 160 mg followed by 80 mg 2 weeks later is required for CD. An 80-mg initial dose followed by 40 mg 1 week later is required for psoriasis	400 mg every 2 weeks for first 4 weeks, then every 4 weeks
Half-life (days)	7.7–9.5	10–20	14
Major bio-markers affected	CRP, inflammatory cytokines, anti-CCP, RF, MMPs, bone and carti-lage markers, regulatory T cells	CRP, inflammatory cyto-kines, anti-CCP, RF, MMPs	CRP

CCP, cyclic citrullinated peptide; CD, Crohn's disease; CRP, C-reactive protein; Ig, immunoglobulin; mAb, monoclonal antibodies; MMPs, matrix metalloproteinases; PEG, polyethylene glycol; RA, rheumatoid arthritis; RF, rheumatoid factor; TNF, tumour necrosis factor.
[*] PEG2 is linked to the Fab cysteine thiol group.

Nevertheless, immune responses to adalimumab in patients are observed, and have been found to weaken the clinical response to treatment [43, 44]. The formation of antibodies to infliximab and adalimumab can be attenuated by the use of immunosuppressants such as methotrexate (MTX) or azathio-prine/6-MP. In contrast to infliximab, adalimumab is given subcutaneously at a dose of 40 mg every two weeks, although this can be escalated to 40 mg once a week if necessary (Tab. 1).

Clinical development
The efficacy of infliximab in CD has been documented in randomised and controlled clinical studies. An early trial in 108 patients with moderate to severe disease showed an 81% response to a single dose of 5 mg/kg; this was compared with a 17% response rate in patients receiving placebo. Clinical remission was achieved by half of all patients on active treatment [45]. In the later ACCENT I study in 573 patients with a score of at least 220 on the Crohn's

Disease Activity Index (CDAI), 58% of patients responded to a single infusion of infliximab within two weeks [46]. Patients who responded to an initial dose of infliximab were found to be more likely to be in remission after 30 and 54 weeks, to discontinue corticosteroids and to maintain their response for longer periods, if active therapy was maintained every eight weeks.

Infliximab has also been shown to be effective for the management of fistulising disease, and for long-term therapy in patients with active CD not responding to conventional treatments [47, 48]. Approval for infliximab maintenance therapy in patients with fistulising CD was based on the results of the ACCENT II study in 306 patients [49]. The time to loss of response was significantly longer for patients who received infliximab maintenance therapy than for those who took placebo (>40 weeks *versus* 14 weeks; $P < 0.001$). At week 54, 19% of patients in the placebo group and 36% of those taking infliximab had complete absence of draining fistulae ($P = 0.009$).

Infliximab is approved for the treatment of active RA in patients failing disease-modifying therapy, and for severe progressive RA in previously untreated patients. In the ATTRACT trial in 428 patients taking MTX, patients given infliximab showed significant improvements in all indices of response [50]. Median changes from baseline to week 102 in the total radiographic score were 4.25 for MTX plus placebo and 0.50 for MTX plus infliximab. Quality of life was improved for up to two years in the MTX plus infliximab group. Benefit of early aggressive therapy for RA with infliximab was shown in the ASPIRE trial in patients not previously treated with MTX [51]. In this trial in 1049 randomised patients, infliximab plus MTX virtually halted radiographic progression of disease, with improvement noted in all American College of Rheumatology (ACR) indices.

Infliximab was approved for the management of psoriasis on the basis of two randomised and controlled trials. In the SPIRIT study in 249 patients with severe plaque psoriasis, 88% of patients treated with infliximab 5 mg/kg and 72% of those receiving 3 mg/kg achieved a 75% or greater improvement from baseline at week 10 in Psoriasis Area and Severity Index (PASI) score, compared with 6% of patients receiving placebo [52]. These changes were accompanied by improvements in health-related quality of life. In the subsequent EXPRESS study in 378 patients with moderate to severe disease [53], 80% of patients treated with infliximab achieved an improvement from baseline of at least 75% in PASI score, compared with 3% of placebo patients. High percentages of patients who received infliximab maintained their PASI responses for one year.

Adalimumab has been shown to be effective in the induction and maintenance of clinical efficacy in CD in the CLASSIC [54, 55], CHARM [56] and GAIN [57] double-blind, placebo-controlled trials. In CLASSIC I [54], 299 patients with moderate to severe CD received at weeks 0 and 2 subcutaneous injections of adalimumab 40 mg/20 mg, 80 mg/40 mg or 160/80 mg or placebo. Rates of remission at week 4 ranged from 18–36% in the adalimumab groups, compared with 12% in the placebo group. Maintenance of clinical remission for up to 56 weeks with adalimumab was shown in the subsequent

CLASSIC II trial [55]. In the CHARM trial [56], open-label induction therapy with adalimumab 80 mg was followed by 40 mg at week 2; randomisation stratified by response took place at week 4. Percentages of randomised responders in remission were significantly ($P < 0.001$) greater with adalimumab 40 mg every other week or 40 mg weekly than with placebo at week 26 (40%, 47% and 17%, respectively) and at week 56 (36%, 41% and 12%, respectively). In the four-week, placebo-controlled GAIN trial, 325 adult patients with moderate to severe CD who had symptoms despite infliximab therapy or who could not take infliximab because of intolerance were randomly assigned to receive induction doses of adalimumab, 160 mg and 80 mg, at weeks 0 and 2, respectively, or placebo at the same time points. 21% of patients in the adalimumab group *versus* 7% of those in the placebo group achieved remission at week 4 [57].

Adalimumab has been assessed in more than 2000 patients with RA [42, 58, 59]. In study DE019, 63, 39 and 21% of patients receiving adalimumab 40 mg every other week plus MTX achieved ACR20, 50 and 70 responses respectively at week 24 [59]. All response rates were statistically significantly better than those in patients receiving placebo plus MTX ($P < 0.001$). Recent data from a 52-week study in 1,212 patients have shown efficacy of adalimumab in moderate to severe psoriasis [60]. At week 16, an improvement of at least 75% in PASI score was seen in 71% of patients receiving adalimumab 40 mg every other week and in 7% of placebo recipients. Continuing treatment with adalimumab was associated with greater proportions of patients maintaining response for up to a year than with placebo.

Infliximab and adalimumab are both licensed to treat adults with moderate to severely active CD, whilst infliximab, the older of the two mAbs, is also indicated for the treatment of paediatric CD patients and in adults with moderately to severely active ulcerative colitis. Both agents are also approved to treat moderately to severely active RA, psoriatic arthritis, ankylosing spondylitis as well as severe plaque psoriasis. Adalimumab is further indicted in juvenile idiopathic arthritis.

Certolizumab pegol (CIMZIA®): a PEGylated antigen-binding fragment of a TNFα monoclonal antibody

Certolizumab pegol represents a further technological advance in the development of therapeutic mAbs by combining the science of pharmaceutical PEGylation with recombinant gene technology.

Certolizumab pegol is a polyethylene glycol (PEG)-conjugated, humanised, antigen-binding fragment (Fab') of an anti-TNFα mAb that is suitable for subcutaneous administration (Tab. 1). This agent has been developed to bind to and neutralise both soluble and membrane-bound forms of TNFα, and is formed by grafting the short hypervariable complementarity-determining regions derived from the murine mAb HTNF40 into the framework of a human Ig Fab' fragment.

The pharmacokinetic properties of Fab' fragments *in vivo* usually result in a short half-life, but attachment of a 40 kDa PEG moiety markedly increases the half-life of the molecule to approximately 14 days, making it suitable for every-other-week, or monthly, dosing. Engineering of the Fab' fragment with a single free cysteine residue in the hinge region enables site-specific attachment of PEG without affecting the ability of the Fab' fragment to bind and neutralize TNFα [61]. Importantly, certolizumab pegol is devoid of the Fc region, therefore, unlike infliximab and adalimumab, there is no complement fixation and cell lysis is not observed (Fig. 4) [62]. In addition, there is no neonatal Fc

a.

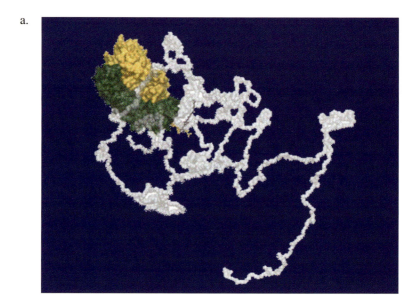

b.

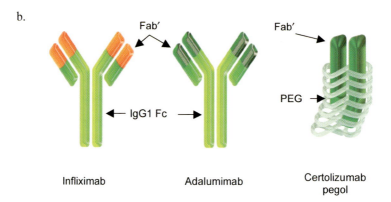

Figure 4. a. Molecular structure of certolizumab pegol. The PEG moiety is shown in grey. b. Comparative schematic representations of the three anti-TNFα monoclonal antibodies.

receptor(FcRn)-mediated transport, which could be the reason for preferential distribution and retention in inflamed tissue [63].

The FcRn molecule is involved in protecting Ig from catabolism and transporting it around the body as well as mediating transport across the placenta to provide immunity for the foetus. Certolizumab pegol is also distinguished within anti-TNFα mAbs by its valency: this compound is univalent, thus reducing the potential for large immune complex formation [64] whereas infliximab and adalumimab are divalent.

Affinities for TNFα and neutralisation properties of the anti-TNFα mAbs are essentially similar [62], but PEGylation does confer distinct characteristics. PEG is a branched polyether of variable chain length that is heavily hydrated, which increases the haemodynamic radius of proteins. PEGylated proteins tend not to diffuse well and are therefore retained with high bioavailability in the bloodstream [65]. The detailed structure of certolizumab pegol has been determined by x-ray crystallography [66], and the molecule found to be highly flexible and asymmetrical (Fig. 4). The PEG moiety is attached to Fab' at a specific point at the Fab' C-terminus, but it does not interact directly with the surface of the Fab', which remains unaffected by any interaction, rearrangement or modification by the PEG moiety.

Mechanism of action

The mechanism of action of certolizumab pegol has been investigated and contrasted with those of infliximab and adalimumab in a series of *in vitro* studies examining affinity for and neutralisation of TNFα; induction of antibody-dependent cell-mediated cytotoxicity (ADCC), complement-dependent cytotoxicity (CDC), apoptosis, degranulation of neutrophils and inhibition of lipopolysaccharide (LPS)-induced cytokine production [62]. All agents tested neutralised soluble TNFα and bound to and neutralised membrane-bound TNFα. Neutralisation of membrane-bound TNFα signalling was found to be mediated via both p55 and p75 receptors [67]. Affinity of certolizumab pegol for TNFα is high: measurement of the relative affinities of infliximab, adalimumab and certolizumab pegol was carried out using surface plasmon resonance. The affinity of certolizumab pegol for soluble TNFα was higher than that of either of the other two agents (Tab. 2) [62].

Resistance of cells such as T cells to apoptosis has been proposed to play an important role in inflammatory bowel disease [68] for example, and TNFα is a known survival factor in certain cell types. When compared with control human IgG1, treatment with infliximab and adalimumab caused CDC and ADCC of a TNFα transfected cell line with high levels of TNFα on the cell surface. As expected, certolizumab pegol did not mediate cell killing either directly through fixation of complement or via recruitment of effector cells; this is attributable to its lack of an Fc region. Additional experiments illustrated that certolizumab pegol did not mediate apoptosis of activated lymphocytes and monocytes by signalling through membrane TNFα *in vitro*, whereas infliximab and adalimumab both mediated this effect [62]. Therefore, direct

Table 2. Affinities of infliximab, adalimumab, and certolizumab pegol for soluble TNFα [62]

Parameter	Agent		
	Infliximab	Adalumimab	Certolizumab pegol
Association rate constant, k_a ($M^{-1} s^{-1}$)	$1.01 \pm 0.06 \times 10^6$	$0.724 \pm 0.30 \times 10^6$	$1.22 \pm 0.09 \times 10^6$
Dissociation rate constant, k_d (s^{-1})	$2.30 \pm 0.34 \times 10^{-4}$	$1.14 \pm 0.12 \times 10^{-4}$	$1.09 \pm 0.13 \times 10^{-4}$
Equilibrium constant, K_D (k_d/k_a, pM)	227.2	157.4	89.3

Data are means ± SD.

induction of apoptosis via binding of anti-TNFα agents to membrane-bound TNFα may not necessarily be required for clinical efficacy in autoimmune inflammatory disease. The reason why certolizumab pegol does not induce apoptosis in cells bearing membrane-bound TNFα is not clear; one hypothesis is that this agent binds to a different epitope to the other mAbs, which may lead to a different signalling pattern inside the cell. Indeed, it has recently been shown that all three clinically effective anti-TNFα agents are able to induce *in vitro* apoptosis in co-cultivated CD4+/CD14+ cells cultivated from lamina propria mononuclear cells derived from gut specimens of patients with CD. However, this is believed to be due to an indirect mechanism which is not related to signalling through membrane-bound TNFα [69].

Neutrophils are commonly found at sites of inflammation and are able to produce TNFα upon stimulation [70]. Incubation with certolizumab pegol *in vitro* did not result in changes to the integrity of polymorphonuclear neutrophilic leukocytes (PMNs), whereas degranulation and loss of cell membrane integrity were markedly increased after incubation with infliximab or adalimumab. Thus, infliximab and adalimumab induce loss of cell viability and release of intracellular granular material into the surrounding medium, whereas certolizumab pegol does not [62].

Preincubation of human monocytes with certolizumab pegol at concentrations of 1 µg/mL and above completely inhibited production of IL-1β in response to stimulation with LPS. Although all three agents caused complete inhibition of this cytokine at concentrations of 1 µg/mL, certolizumab pegol appeared to be the most potent compound in this respect, showing a slower titration of the attenuation of IL-1β expression at concentrations below 1 µg/mL [62]. LPS stimulation of cells of the immune system represents a powerful signalling mechanism, which suggests that the inhibition mediated by anti-TNFα agents may be an important anti-inflammatory process.

Stoichiometry and complex formation
The univalent nature of certolizumab pegol may be at least in part responsible

for some of the differences in mechanistic behaviour of this molecule when compared with infliximab and adalumimab. Binding of biological agents to TNFα can result in the formation of immune complexes, which can have unwanted destructive effects such as the PMN degranulation and superoxide production. Dynamic light-scattering studies have shown that, at an anti-TNFα:TNFα molar ratio of 1:1, adalimumab and infliximab form very large complexes of diameter exceeding 30 nm. In contrast, certolizumab pegol does not form these large complexes (Tab. 3). Isothermal titration calorimetry data show also that certolizumab pegol binds to 2.9 monomers in a TNFα trimer at saturation compared with 2.63 and 2.44 monomers in a TNFα trimer for inflix-imab and adalimumab, respectively [64]. Taken together, the dynamic light scattering and isothermal titration calorimetry data show that infliximab and adalimumab are able to cross-link TNFα trimers by binding to monomers in different trimers (Fig. 5). Certolizumab pegol cannot cross-link in this way due to its univalent structure. The large complexes produced by infliximab and adalimumab were noted to have proinflammatory effects on PMNs *in vitro* (see previous), which caused these cells to degranulate and produce superoxide ions [64].

Table 3. Approximate molecular weights (kDa) of anti-TNFα/TNFα immune complexes at differing molar ratios [64]

Parameter	Agent		
	Infliximab	Adalumimab	Certolizumab pegol
TNFα alone	48	33	34
Antibody alone	289	286	417
Antibody:TNFα ratio 1:1	2727	2304	613
Antibody:TNFα ratio 4:1	559	277	581

PEGylation and tissue penetration

PEGylated molecules tend to diffuse slowly from blood because of the haemo-dynamic properties of PEG [65]. This gives rise to the possibility that such molecules might have distribution patterns that differ from intact IgG1 in terms of their exposure times in inflamed and non-inflamed tissues. The disposition of certolizumab pegol, infliximab and adalumimab has therefore been investi-gated using a biofluorescence method in healthy and inflamed murine tissue in two experiments [63]. All agents were conjugated with Alexa680, a low mole-cular weight fluorescent dye, and administered intravenously to healthy DBA/1 mice and to DBA/1 mice with collagen-induced arthritis. Levels of agents in hind paws were then measured.

All three agents penetrated inflamed tissue more effectively than non-inflamed tissue, but certolizumab pegol penetrated inflamed arthritic paws

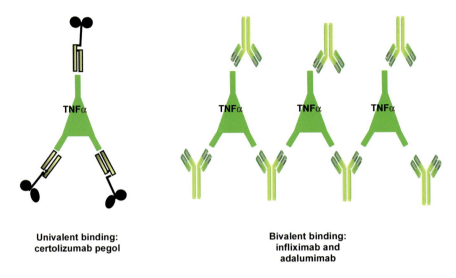

Figure 5. Formation of immune complexes via TNFα trimer cross-linking [64].

most effectively and was retained for longer than infliximab or adalimumab. The longer retention time could be a result of the low diffusion rate of PEGylated molecules and/or the lack of an Fc, which leads to a lack of recycling by the FcRn. Therefore, it is possible that, due to PEGylation, certolizumab pegol will have a higher inflamed to normal tissue ratio than a mAb. Indeed, macromolecules such as PEG have been suggested as possible vehicles for delivery of drugs to rheumatoid joints because of this very property [71].

As discussed earlier, certolizumab pegol has no Fc region. Adalimumab and infliximab are whole IgG1 molecules and possess an Fc region, which potentially allows passage across the placenta via specific neonatal FcRn receptors [72]. In an experiment to test for the likelihood of placental transfer of PEGylated Fab's [73], rats were injected during gestation with either a complete chimeric anti-murine TNFα antibody (cTN3 IgG1) or a PEGylated Fab' fragment (cTN3 PF). In foetal samples from five rats given cTN3 PF while pregnant, PEGylated Fab' was found at very low concentrations in only two of the samples and was entirely undetectable in the others. This suggests that the PEGylated fragment does not undergo FcRn-mediated placental transfer, but that there may be a low level of passive transfer as with all proteins. However, biologically relevant levels of intact IgG1 were found in rats given whole immunoglobulin, which indicates that this molecule had crossed the placenta. Concentrations in milk of PEGylated Fab' were also much lower than those of IgG1 eight days postpartum.

Thus, on the basis of these findings, certolizumab pegol, as an Fc-free molecule, would not be expected to cross the placenta via FcRn receptors. As the number of pregnant women whom have been exposed to these drugs is low, this area of study remains unresolved and requires further research.

In vivo *pharmacology*

In vivo pharmacological testing of certolizumab pegol was originally carried out in animal models, with initial research focusing on the fate of the PEG moiety after administration. Radiolabelled PEG is not suitable for quantification because it is unstable [74]. Therefore, distribution and elimination studies have been carried out using ^{1}H nuclear magnetic resonance spectroscopy (NMR) and sodium dodecyl sulfate-polyacrylamide gel electrophoresis (SDS-PAGE).

In rat studies [75], nine animals were given one subcutaneous injection of certolizumab pegol at a dose of 400 mg/kg, with daily urine and faeces samples being collected for 84 days; a further group of rats received 100 mg/kg subcutaneously of either certolizumab pegol or cTN3 PF (cTN3 PF is a Fab' conjugated with 40 kDa PEG, which acts as a homologous reagent binding to and neutralising rat TNFα and was used to show any tissue targeting due to TNFα binding). These rats were killed at prespecified intervals in groups of four, and their major organs removed for NMR examination. In an additional study, certolizumab pegol was injected into the tail veins of male Lewis rats at doses of 10 or 100 mg/kg, and urine collected over 28 days for analysis by SDS-PAGE [75]. The data showed overall that PEG derived from certolizumab pegol is distributed to major organs (excluding the brain), it has a plasma half-life of approximately three weeks, it is not influenced by target site binding and it is excreted largely unchanged by the kidneys with first-order kinetics. Moreover, urinary data showed the rate of excretion essentially to match the rate of administration, with no evidence of accumulation.

NMR analysis showed PEG to be detectable in urine for up to 84 days after administration, with the peak concentration of 198 µg/mL being observed on day 4. PEG concentrations then declined in a first-order manner to 14 µg/mL during week 12. Mean daily excretion reached 1.9% on day 6 and declined to 0.21% during week 12. The mean cumulative dose excreted in urine was 65% after 84 days. Extrapolation to infinity projected a total urinary excretion of 73%, with a half-life for elimination of the PEG component of 23.4 days, much longer than the two-day plasma elimination half-life of certolizumab pegol. Total faecal excretion in rats was 18%, with possible ingestion of a portion of the PEG dose from the injection site and some urinary contamination accounting for approximately 11% of faecal recovery. Published murine data [76] suggest that approximately 1% of the dose of 40 kDa PEG might be excreted via the biliary route, although this remains unconfirmed. Overall, NMR analysis indicated the mean urinary and faecal excretion of PEG in rats to be 83% of an administered dose after 84 days, with extrapolation to a final total of >90% [75]. There was no evidence of accumulation in any major organ, with the spleen having the highest level of exposure, and the highest proportion of the dose being found in the liver (the largest organ). Distribution of certolizumab pegol was found to be similar to that for cTN3 PF, which implies that TNFα binding is not a significant mediator of PEG distribution in tissue [75]. SDS-PAGE analysis showed that the PEG excreted in urine was 40 kDa, suggesting that after the Fab' was cleaved from the PEG, the PEG was cleared by the kidney with no further metabolism [75].

Clinical development

The safety profile and pharmacological properties of certolizumab pegol have been assessed in primates and in healthy human volunteers. Several clinical trials have also been conducted to examine the efficacy and tolerability of certolizumab pegol in patients with CD, RA and psoriasis.

Pharmacology in primates and humans

The disposition of certolizumab pegol has been characterised over a range of doses in primates and in healthy human volunteers [77]. Human males aged 18–50 years were recruited in two studies: a double-blind, placebo-controlled, ascending single-dose trial in 16 patients of certolizumab pegol 0.3–10.0 mg/kg given by intravenous infusion over 60 min, and a double-blind, double-dummy, ascending single-dose study in 24 patients who received a subcutaneous bolus of 20, 60 or 200 mg or an intravenous infusion of 1 mg/kg. In a study in non-human primates, 20 cynomolgus monkeys were divided into five groups and given certolizumab pegol by intravenous infusion over 60 min (50, 100 or 400 mg) or by subcutaneous bolus injection (3.0 or 31 mg/kg). Plasma concentrations of certolizumab pegol were determined using an enzyme-linked immunosorbent assay (ELISA).

Certolizumab pegol showed linear pharmacokinetics over the dose ranges studied in both species, with biexponential disposition after intravenous dosing and monoexponential disposition after subcutaneous injection (Tab. 4). Volumes of distribution were equivalent in terms of species-relevant plasma volumes, and subcutaneous absorption was sustained, with low clearances and extended plasma elimination half-lives (approximately two weeks).

Table 4. Pharmacokinetic parameters (modeled estimates with 95% CIs) of certolizumab pegol in humans and primates (cynomolgus monkeys) after intravenous or subcutaneous administration [77]

Parameter	Humans	Primates
Central volume of distribution, V_1 (mL/kg)	45 (44–47)	46 (44–48)
Total body clearance, CL (mL/h/kg)	0.17 (0.16–0.18)	0.29 (0.27–0.31)
Peripheral volume of distribution, V_2 (mL/kg)	29 (25–33)	31 (26–37)
Intercompartmental clearance, CLd (mL/h/kg)	0.69 (0.52–0.90)	1.1 (0.8–1.5)
Elimination half-life, $t_{1/2}$ (h)	313 (284–345)	192 (174–212)
Bioavailability, F	1 (0.9–1.1)	1.1 (1.0–1.3)
Absorption rate constant, Ka (h^{-1})	0.018 (0.015–0.020)	0.11 (0.08–0.16)

Pharmacokinetics were shown to be predictable, with extended or equivalent exposure relative to anti-TNFα agents already available for clinical use. Certolizumab pegol had bioavailability of approximately 80% (ranging from 76–88%) following subcutaneous administration in humans [78].

Evaluation of 10,275 plasma concentration *versus* time records from 1,580 persons (80% with CD and 15% with RA, and 5% of whom were healthy individuals) showed the clearance of certolizumab pegol to be typically 0.428 L/d, with volume of distribution 4.0 L in a 70 kg person [79]. The disposition of certolizumab pegol was found to be unaffected by age, gender, creatinine clearance, leukocyte count and concomitant use of drugs such as corticosteroids, aminosalicylic acid and its analogues and anti-infectives. Of other covariates tested, only the presence of antibodies to certolizumab pegol had an effect of more than 30% on peak plasma concentrations and areas under curves of plasma concentration *versus* time, but such antibodies were found in only 8% of individuals tested [79]. A doubling in body weight was found to be the second most influential covariate. Overall, however, none of the covariates tested had a clinically relevant effect. Notably, other recent data show that the PEG component of certolizumab pegol is not immunogenic and therefore does not provoke the production of antibodies [80]. A small proportion of patients have been noted to produce antibodies to the protein part of certolizumab pegol, but these antibodies have been shown not to cross-react with other anti-TNFα mAbs [81].

CD

The efficacy and safety of subcutaneously administered certolizumab pegol was examined in two Phase II studies in patients with moderate to severely active CD [82, 83]. In the first, 92 patients with moderate to severely active disease were randomised to a single intravenous infusion of certolizumab pegol 1.25 mg/kg, 5 mg/kg, 10 mg/kg or 20 mg/kg or placebo. Although the primary endpoint of this study, a difference in the rate of clinical response or remission at week 4 between either of the active treatment arms and placebo, was not attained, a significantly different remission rate at week 2 was observed for the 10 mg/kg dose compared with placebo [82]. Certolizumab pegol was well tolerated by the patients in this study, with the majority of adverse events being of mild to moderate intensity and with no infusion reactions being reported.

In a larger, Phase II dose-ranging study, patients received subcutaneous injections of certolizumab pegol 100 mg (n = 74), 200 mg (n = 72), 400 mg (n = 72) or placebo (n = 73) at weeks 0, 4 and 8 with no induction dose, and efficacy assessments were made every two weeks until week 12. The highest subcutaneous dose of 400 mg was the most effective in inducing clinical response and remission relative to all time points [83]. Although the primary efficacy endpoint of clinical response at week 12 was not met, statistically significant differences between certolizumab pegol 400 mg and placebo were noted in favour of the active drug as early as week 2, and also at week 4, week

8 and week 12. Certolizumab pegol 400 mg had a favourable safety profile, with a similar percentage of patients reporting adverse events as in the placebo group over a 12-week period.

Data from these Phase II studies facilitated the design of the two Phase III trials (PRECiSE 1 and 2), which were the next stage in the development of certolizumab pegol for the treatment of patients with moderate to severely active CD. Patients received an induction dose consisting of certolizumab pegol 400 mg at weeks 0, 2 and 4 followed by a four-weekly 400-mg maintenance dose. To further investigate influence of CRP, patients in PRECiSE 1 and 2 were stratified according to baseline CRP levels ≥10 mg/L or <10 mg/L. In the PRECiSE 1 trial, 662 patients were randomised to double-blind induction and maintenance treatment with certolizumab pegol or placebo for 26 weeks [84]. In the 26-week PRECiSE 2 trial, 668 patients received open-label induction with certolizumab pegol and those (n = 428; 64%) with a clinical response (≥100 points decrease in baseline CDAI score) at week 6 were stratified according to CRP level and randomised into the double-blind, placebo-controlled maintenance phase [85].

The efficacy endpoints in PRECiSE 1, of clinical response at week 6 and clinical response at both week 6 and week 26, were met both in the primary population of patients with CRP levels ≥10 mg/L and in the overall patient population [84]. Similarly, CRP levels also had no influence over the efficacy outcomes in the PRECiSE 2 study [85].

Patients who completed PRECiSE 1 and 2 were eligible for entry into the long-term open-label extension study, PRECiSE 3, where the long-term efficacy and safety outcomes of scheduled certolizumab pegol treatment can be assessed. Patients withdrawing from PRECiSE 1 or 2 due to exacerbation of symptoms of CD were eligible for entry into the open-label study, PRECiSE 4, designed to evaluate the efficacy and safety of reinduction. Interim results from these two studies have shown that remission rates after 18 months of continued treatment in PRECiSE 3, and after 12 months of treatment following reinduction in PRECiSE 4, have remained stable. Furthermore, no new safety signals were identified following re-exposure of placebo patients from PRECiSE 2 to certolizumab pegol [86].

Certolizumab pegol has a good tolerability profile in patients with CD. Safety data for 1,328 patients who took part in the PRECiSE 1 and 2 studies illustrated that the majority of adverse events were rated as mild to moderate in intensity, with headache, nasopharyngitis, abdominal pain and cough being the most common adverse events in the certolizumab pegol treatment groups [87]. Local injection reactions were rare with certolizumab pegol (<3%) and autoantibody seroconversion rates were low (anti-nuclear: 8.3% or less and anti-double-stranded DNA: 1.4% or less) [87].

The clinical development programme of certolizumab pegol in CD continues with Phase IIIb studies which have been designed to examine aspects such as the efficacy of certolizumab pegol in infliximab-refractory patients (WELCOME) and in other measures of efficacy such as mucosal healing (MUSIC).

RA

Several Phase II and III trials have assessed the safety and efficacy of certolizumab pegol in patients with RA. In a Phase II double-blind, randomised trial (n = 36), certolizumab pegol demonstrated significant efficacy in reducing inflammation and improving symptoms of RA [88]. In this trial, 75% of patients who received an intravenous infusion of certolizumab pegol at 5 or 20 mg/kg achieved ACR 20 scores after eight weeks of treatment.

In a Phase III trial, the efficacy and safety of subcutaneous certolizumab pegol monotherapy (400 mg every four weeks) was assessed in patients with active RA who had previously failed therapy with at least one disease-modifying antirheumatic drug (DMARD). At week 24, the ACR20 responder rate was 45.5% in the certolizumab pegol group (n = 111) compared with 9.3% in the placebo group (n = 109). The effect of certolizumab pegol treatment was evident early after treatment, with 80.6% of ACR20 responders having achieved this response by week 1 [89]. In addition, certolizumab pegol treatment provided rapid and sustained improvements in physical function and pain relief in these patients who had previously failed DMARD therapy [90].

The efficacy and safety of certolizumab pegol as add-on therapy to methotrexate (MTX) were evaluated in two Phase III double-blind, placebo-controlled, randomised trials (RAPID 1 and 2) [91, 92]. In these trials, adults with active RA (n = 982 and n = 619 in RAPID 1 and 2, respectively) were randomised 2:2:1 to receive subcutaneous certolizumab pegol, lyophilized formulation (RAPID 1) or liquid formulation (RAPID 2), 200 mg or 400 mg plus MTX, or placebo plus MTX, every two weeks for 52 weeks (RAPID 1) or 24 weeks (RAPID 2). In the 200-mg active treatment groups, a loading dose of 400 mg of certolizumab pegol at weeks 0, 2 and 4 was also implemented. Enrolled patients were maintained on a stable dose of MTX for the duration of the trial. Primary study endpoints were ACR20 response rate at week 24 (RAPID 1 and 2) and change from baseline in modified Total Sharp Score (mTSS) at week 52 (RAPID 1). Secondary endpoints included ACR50 and ACR70 response rates at weeks 24 and 52 in RAPID 1 and week 24 in RAPID 2. Patient-reported outcomes included physical function, health-related quality of life, pain, fatigue and RA-related work and household productivity.

In the RAPID 1 trial, the ACR20, ACR50 and ACR70 response rate for patients receiving certolizumab pegol plus MTX were significantly higher than that for patients receiving placebo plus MTX [91]. Comparable results for ACR20/50/70 response rates were observed in the RAPID 2 trial [92]. Certolizumab pegol had a fast onset of action, with statistically significant ACR20 and ACR50 improvements after the first dose [92].

In both trials, the radiographic progression of structural joint damage was significantly inhibited by certolizumab pegol plus MTX. In RAPID 1, after 52 weeks of therapy and RAPID 2, after 24 weeks of therapy, the changes from baseline in mTSS, erosion scores and joint space narrowing were significantly lower in patients receiving certolizumab pegol plus MTX compared with patients receiving placebo plus MTX. In RAPID 1, the effects of certolizumab pegol plus MTX on the retardation of structural joint damage were observed as early as week 16 [93].

Results from RAPID 1 and 2 revealed that in addition to significantly improving the signs and symptoms of RA, certolizumab pegol plus MTX significantly improved all aspects of patients' quality of life, including physical function, SF-36 physical and mental component summary scores (a widely used health questionnaire containing 36 questions to assess quality of life), pain, fatigue and work and home productivity [94]. The rapid onset of action was also reflected in these patient reported outcomes, with improvements in physical function, pain and fatigue showing statistical significance as early as Week 1 [94].

Certolizumab pegol was well tolerated by RA patients, both when administered as monotherapy treatment and when administered as add-on therapy to MTX [90–92]. The majority of adverse events were mild to moderate, and discontinuation due to adverse events was low in all groups (<6%). In all studies there was a low incidence of injection-site pain or reactions. The incidence of infections was also minimal, and was comparable to that seen with other anti-TNFα agents [90–92].

These data further support a role for certolizumab pegol monotherapy, or combination therapy with MTX, as an effective and well-tolerated treatment option for patients with RA. Certolizumab pegol has an acceptable safety profile, significantly reduces the signs and symptoms of RA and inhibits progression of structural damage. Onset of benefit of certolizumab pegol treatment is rapid, and the combination of certolizumab pegol plus MTX causes sustained and long-term reduction in the signs and symptoms of RA.

Psoriasis

Results from the first study to evaluate the efficacy and safety of certolizumab pegol in patients with moderate to severe psoriasis illustrated that both the 200-mg and 400-mg liquid formulations of certolizumab pegol, given every two weeks over a period of 12 weeks, significantly reduced the redness, thickness and scaliness of lesions when compared with placebo treatment [95]. In this Phase II, randomised, double-blind, placebo-controlled dose-ranging study, 176 patients with moderate to severe chronic plaque psoriasis, who were candidates for systemic therapy and/or phototherapy or photochemotherapy, were randomised. 60 patients (34.1%) had previously used biologicals, including 41 (23.3%) with prior anti-TNFα use [95].

By week 12, 74.6% and 82.8% of patients in the certolizumab pegol 200- and 400-mg groups, respectively, achieved response rates on the PASI 75, compared with 6.8% in the placebo group ($P<0.001$). Furthermore, 52.5% and 72.4% of patients in the certolizumab pegol 200- and 400-mg groups, respectively, gave the rating 'clear' or 'almost clear' on the Psoriasis Global Assesment (PGA) scale, compared with 1.7% in the placebo group ($P < 0.001$) [95].

Certolizumab pegol was well tolerated in patients with psoriasis, with a safety profile consistent with what is expected from this class of therapy. The most common adverse events were headache, pruritus and cough. Administration of certolizumab pegol was well tolerated with <3% of the study population reporting injection-site reactions and <1% of patients reporting injection-site pain [95].

The results from this study suggest that certolizumab pegol has the potential for further development as a valuable new treatment option for this difficult-to-treat disease.

Closing statement

The involvement of the cytokine TNFα in immune-mediated inflammatory diseases was little known just two decades ago. In a short period, basic scientific researchers utilising cutting-edge advancements in recombinant gene technology developed mAbs-targeting TNFα, which, over the years, has been proven to be effective in bringing relief to thousands of patients with RA, CD, ulcerative colitis and psoriasis. Certolizumab pegol utilises the technique of PEGylation to enhance the pharmacokinetic properties of the antigen-binding fragment of an anti-TNFα mAb without detrimental effect to the pharmacodynamic properties. These advances have led to a compound that can be administered via subcutaneous injection every-other-week or once monthly, with a rapid onset of action, a low level of injection-site pain, preferential distribution in inflamed tissue, and, low levels of immunogenicity [84, 83]. Certolizumab pegol was approved for use in the United States in adults with moderate to severe Crohn's disease in April 2008. The continuation of the clinical development of CIMZIA® (certolizumab pegol) may lead to a significant advance in the treatment of CD, RA and psoriasis and improve the lives of patients with these diseases.

References

1 Bradley JR (2008) TNF-mediated inflammatory disease. *J Pathol* 214: 149–160
2 Kolb WP, Granger GA (1968) Lymphocyte *in vitro* cytotoxicity: characterization of human lymphotoxin. *Proc Natl Acad Sci U S A* 61: 1250–1255
3 Carswell EA, Old LJ, Kassel RL, Green S, Fiore N, Williamson B (1975) An endotoxin-induced serum factor that causes necrosis of tumors. *Proc Natl Acad Sci U S A* 72: 3666–3670
4 Beutler B, Greenwald D, Hulmes JD, Chang M, Pan YC, Mathison J, Ulevitch R, Cerami A (1985) Identity of tumour necrosis factor and the macrophage-secreted factor cachectin. *Nature* 316: 552–554
5 Beutler B, Milsark IW, Cerami AC (1985) Passive immunization against cachectin/tumor necrosis factor protects mice from lethal effect of endotoxin. *Science* 229: 869–871
6 Chatzantoni K, Mouzaki A (2006) Anti-TNF-α antibody therapies in autoimmune diseases. *Curr Top Med Chem* 6: 1707–1714
7 Aggarwal BB (2003) Signalling pathways of the TNF superfamily: a double-edged sword. *Nat Rev Immunol* 3: 745–756

8 Bazzoni F, Beutler B (1996) The tumor necrosis factor ligand and receptor families. *N Engl J Med*
   334: 1717–1725
9 Baud V, Karin M (2001) Signal transduction by tumor necrosis factor and its relatives. *Trends Cell
   Biol* 11: 372–377
10 Kumar A, Takada Y, Boriek AM, Aggarwal BB (2004) Nuclear factor-κB: its role in health and
   disease. *J Mol Med* 82: 434–488
11 Grell M, Becke FM, Wajant H, Mannel DN, Scheurich P (1998) TNF receptor type 2 mediates
   thymocyte proliferation independently of TNF receptor type 1. *Eur J Immunol* 28: 257–263
12 Haridas V, Darnay BG, Natarajan K, Heller R, Aggarwal BB (1998) Overexpression of the p80
   TNF receptor leads to TNF-dependent apoptosis, nuclear factor-κB activation, and c-Jun kinase
   activation. *J Immunol* 160: 3152–3162
13 Feldmann M, Brennan FM, Paleolog E, Cope A, Taylor P, Williams R, Woody J, Maini RN (2004)
   Anti-TNFalpha therapy of rheumatoid arthritis: what can we learn about chronic disease? *Novartis
   Found Symp* 256: 53–69
14 Choy EH, Panayi GS (2001) Cytokine pathways and joint inflammation in rheumatoid arthritis. *N
   Engl J Med* 344: 907–916
15 Paleolog EM, Hunt M, Elliott MJ, Feldmann M, Maini RN, Woody JN (1996) Deactivation of vascular endothelium by monoclonal anti-tumor necrosis factor alpha antibody in rheumatoid arthritis. *Arthritis Rheum* 39: 1082–1091
16 Tak PP, Taylor PC, Breedveld FC, Smeets TJ, Daha MR, Kluin PM, Meinders AE, Maini RN
   (1996) Decrease in cellularity and expression of adhesion molecules by anti-tumor necrosis factor alpha monoclonal antibody treatment in patients with rheumatoid arthritis. *Arthritis Rheum* 39:
   1077–1081
17 Goldring SR, Gravallese EM (2002) Pathogenesis of bone lesions in rheumatoid arthritis. *Curr
   Rheumatol Rep* 4: 226–231
18 Dayer JM, Beutler B, Cerami A (1985) Cachectin/tumor necrosis factor stimulates collagenase
   and prostaglandin E2 production by human synovial cells and dermal fibroblasts. *J Exp Med* 162:
   2163–2168
19 De Hertogh G, Aerssens J, Geboes KP, Geboes K (2008) *World J Gastroenterol* 14: 845–852
20 Loftus EV Jr, Schoenfeld P, Sandborn WJ (2002) The epidemiology and natural history of Crohn's
   disease in population-based patient cohorts from North America: a systematic review. *Aliment
   Pharmacol Ther* 16: 51–60
21 Schwartz DA, Pemberton JH, Sandborn WJ (2001) Diagnosis and treatment of perianal fistulas in
   Crohn's disease. *Ann Intern Med* 135: 906–918
22 Hanauer SB (2006) Inflammatory bowel disease: epidemiology, pathogenesis, and therapeutic
   opportunities. *Inflamm Bowel Dis* 12(Suppl 1): S3–S9
23 Reinecker HC, Steffen M, Witthoeft T, Pflueger I, Schreiber S, MacDermott RP, Raedler A (1993)
   Enhanced secretion of tumour necrosis factor-alpha, IL-6, and IL-1β by isolated lamina propria
   mononuclear cells from patients with ulcerative colitis and Crohn's disease. *Clin Exp Immunol* 94:
   174–181
24 Boirivant M, Marini M, Di Felice G, Pronio AM, Montesani C, Tersigni R, Strober W (1999)
   Lamina propria T cells in Crohn's disease and other gastrointestinal inflammation show defective
   CD2 pathway-induced apoptosis. *Gastroenterology* 116: 557–565
25 Ina K, Itoh J, Fukushima K, Kusugami K, Yamaguchi T, Kyokane K, Imada A, Binion DG, Musso
   A, West GA et al. (1999) Resistance of Crohn's disease T cells to multiple apoptotic signals is
   associated with a Bcl-2/Bax mucosal imbalance. *J Immunol* 163: 1081–1090
26 Doering J, Begue B, Lentze MJ, Rieux-Laucat F, Goulet O, Schmitz J, Cerf-Bensussan N,
   Ruemmele FM (2004) Induction of T lymphocyte apoptosis by sulphasalazine in patients with
   Crohn's disease. *Gut* 53: 1632–1638
27 Saklatvala J (1986) Tumour necrosis factor alpha stimulates resorption and inhibits synthesis of
   proteoglycan in cartilage. *Nature* 322: 547–549
28 Keffer J, Probert L, Cazlaris H, Georgopoulos S, Kaslaris E, Kioussis D, Kollias G (1991)
   Transgenic mice expressing human tumour necrosis factor: a predictive genetic model of arthritis.
   EMBO J 10: 4025–4031
29 Piguet PF, Grau GE, Vesin C, Loetscher H, Gentz R, Lesslauer W (1992) Evolution of collagen
   arthritis in mice is arrested by treatment with anti-tumour necrosis factor (TNF) antibody or a
   recombinant soluble TNF receptor. *Immunology* 77: 510–514
30 Thorbecke GJ, Shah R, Leu CH, Kuruvilla AP, Hardison AM, Palladino MA (1992) Involvement

of endogenous tumor necrosis factor alpha and transforming growth factor beta during induction of collagen type II arthritis in mice. *Proc Natl Acad Sci U S A* 89: 7375–7379

31 Williams RO, Feldmann M, Maini RN (1992) Anti-tumor necrosis factor ameliorates joint disease in murine collagen-induced arthritis. *Proc Natl Acad Sci U S A* 89: 9784–9788

32 Gelfand JM, Weinstein R, Porter SB, Neimann AL, Berlin JA, Margolis DJ (2005) Prevalence and treatment of psoriasis in the United Kingdom: a population-based study. *Arch Dermatol* 141: 1537–1541

33 Zachariae H (2003) Prevalence of joint disease in patients with psoriasis: implications for therapy. *Am J Clin Dermatol* 4: 441–447

34 Gudjonsson JE, Johnston A, Sigmundsdottir H, Valdimarsson H (2004) Immunopathogenic mechanisms in psoriasis. *Clin Exp Immunol* 135: 1–8

35 Veale DJ, Ritchlin C, FitzGerald O (2005) Immunopathology of psoriasis and psoriatic arthritis. *Rheum Dis* 64(Suppl 2): ii26–ii29

36 Boyman O, Hefti HP, Conrad C, Nickoloff BJ, Suter M, Nestle FO (2004) Spontaneous development of psoriasis in a new animal model shows an essential role for resident T cells and tumor necrosis factor-alpha. *J Exp Med* 199: 731–736

37 Goldstein G, Fuccello AJ, Norman DJ, Shield CF 3rd, Colvin RB, Cosimi AB (1986) OKT3 monoclonal antibody plasma levels during therapy and the subsequent development of host antibodies to OKT3. *Transplantation* 42: 507–511

38 Bach JF, Chatenoud L (1987) Immunologic monitoring of orthoclone OKT3-treated patients: the problem of antimonoclonal immune response. *Transplant Proc* 19(2 Suppl 1): 17–20

39 Knight DM, Wagner C, Jordan R, McAleer MF, DeRita R, Fass DN, Coller BS, Weisman HF, Ghrayeb J (1995) The immunogenicity of the 7E3 murine monoclonal Fab antibody fragment variable region is dramatically reduced in humans by substitution of human for murine constant regions. *Mol Immunol* 32: 1271–1281

40 Wong M, Ziring D, Korin Y, Desai S, Kim S, Lin J, Gjertson D, Braun J, Reed E, Singh RR (2008) TNFα blockade in human diseases: mechanisms and future directions. *Clin Immunol* 126: 121–136

41 Harriman G, Harper LK, Schaible TF (1999) Summary of clinical trials in rheumatoid arthritis using infliximab, an anti-TNFα treatment. *Ann Rheum Dis* 58(Suppl 1): i61–i64

42 Kuek A, Hazleman BL, Ostör AJ (2007) Immune-mediated inflammatory diseases (IMIDs) and biologic therapy: a medical revolution. *Postgrad Med J* 83: 251–260

43 Bartelds GM, Wijbrandts CA, Nurmohamed MT, Stapel S, Lems WF, Aarden L, Dijkmans BA, Tak PP, Wolbink GJ (2007) Clinical response to adalimumab: relationship to anti-adalimumab antibodies and serum adalimumab concentrations in rheumatoid arthritis. *Ann Rheum Dis* 66: 921–926

44 West RL, Zelinkova Z, Wolbink GJ, Kuipers EJ, Stokkers PCF, van der Woude CJ (2008) Does immunogenicity play a role in adalimumab treatment for Crohn's disease? *J Crohn's Colitis Suppl* 2: 17 (P039)

45 Targan SR, Hanauer SB, van Deventer SJ, Mayer L, Present DH, Braakman T, DeWoody KL, Schaible TF, Rutgeerts PJ (1997) A short-term study of chimeric monoclonal antibody cA2 to tumor necrosis factor alpha for Crohn's disease. Crohn's Disease cA2 Study Group. *N Engl J Med* 337: 1029–1035

46 Hanauer SB, Feagan BG, Lichtenstein GR, Mayer LF, Schreiber S, Colombel JF, Rachmilewitz D, Wolf DC, Olson A, Bao W et al; ACCENT I Study Group (2002) Maintenance infliximab for Crohn's disease: the ACCENT I randomised trial. *Lancet* 359: 1541–1549

47 Present DH, Rutgeerts P, Targan S, Hanauer SB, Mayer L, van Hogezand RA, Podolsky DK, Sands BE, Braakman T, DeWoody KL et al. (1999) Infliximab for the treatment of fistulas in patients with Crohn's disease. *N Engl J Med* 340: 1398–1405

48 Rutgeerts P, D'Haens G, Targan S, Vasiliauskas E, Hanauer SB, Present DH, Mayer L, Van Hogezand RA, Braakman T, DeWoody KL et al. (1999) Efficacy and safety of retreatment with anti-tumor necrosis factor antibody (infliximab) to maintain remission in Crohn's disease. *Gastroenterology* 117: 761–769

49 Sands BE, Anderson FH, Bernstein CN, Chey WY, Feagan BG, Fedorak RN, Kamm MA, Korzenik JR, Lashner BA, Onken JE et al. (2004) Infliximab maintenance therapy for fistulizing Crohn's disease. *N Engl J Med* 350: 876–885

50 Maini RN, Breedveld FC, Kalden JR, Smolen JS, Furst D, Weisman MH, St Clair EW, Keenan GF, van der Heijde D, Marsters PA et al; Anti-Tumor Necrosis Factor Trial in Rheumatoid

Arthritis with Concomitant Therapy Study Group (2004) Sustained improvement over two years in physical function, structural damage, and signs and symptoms among patients with rheumatoid arthritis treated with infliximab and methotrexate. *Arthritis Rheum* 50: 1051–1065

51 St Clair EW, van der Heijde DM, Smolen JS, Maini RN, Bathon JM, Emery P, Keystone E, Schiff M, Kalden JR, Wang B et al; Active-Controlled Study of Patients Receiving Infliximab for the Treatment of Rheumatoid Arthritis of Early Onset Study Group (2004) Combination of infliximab and methotrexate therapy for early rheumatoid arthritis: a randomised, controlled trial. *Arthritis Rheum* 50: 3432–3443

52 Gottlieb AB, Evans R, Li S, Dooley LT, Guzzo CA, Baker D, Bala M, Marano CW, Menter A (2004) Infliximab induction therapy for patients with severe plaque-type psoriasis: a randomised, double-blind, placebo-controlled trial. *J Am Acad Dermatol* 51: 534–542

53 Reich K, Nestle FO, Papp K, Ortonne JP, Evans R, Guzzo C, Li S, Dooley LT, Griffiths CE; EXPRESS study investigators (2005) Infliximab induction and maintenance therapy for moderate-to-severe psoriasis: a phase III, multicentre, double-blind trial. *Lancet* 366: 1367–1374

54 Hanauer SB, Sandborn WJ, Rutgeerts P, Fedorak RN, Lukas M, MacIntosh D, Panaccione R, Wolf D, Pollack P (2006) Human anti-tumor necrosis factor monoclonal antibody (adalimumab) in Crohn's disease: the CLASSIC-I trial. *Gastroenterology* 130: 323–333

55 Sandborn WJ, Hanauer SB, Rutgeerts P, Fedorak RN, Lukas M, MacIntosh DG, Panaccione R, Wolf D, Kent JD, Bittle B et al. (2007a) Adalimumab for maintenance treatment of Crohn's disease: results of the CLASSIC II trial. *Gut* 56: 1232–1239

56 Colombel JF, Sandborn WJ, Rutgeerts P, Enns R, Hanauer SB, Panaccione R, Schreiber S, Byczkowski D, Li J, Kent JD et al. (2007) Adalimumab for maintenance of clinical response and remission in patients with Crohn's disease: the CHARM trial. *Gastroenterology* 132: 52–65

57 Sandborn WJ, Rutgeerts P, Enns R, Hanauer SB, Colombel JF, Panaccione R, D'Haens G, Li J, Rosenfeld MR, Kent JD et al. (2007b) Adalimumab Induction Therapy for Crohn's Disease Previously Treated with Infliximab: A Randomized Trial. *Ann Intern Med* 146: 829–838

58 Weinblatt ME, Keystone EC, Furst DE, Moreland LW, Weisman MH, Birbara CA, Teoh LA, Fischkoff SA, Chartash EK (2003) Adalimumab, a fully human anti-tumor necrosis factor alpha monoclonal antibody, for the treatment of rheumatoid arthritis in patients taking concomitant methotrexate: the ARMADA trial. *Arthritis Rheum* 48: 35–45

59 Keystone EC, Kavanaugh AF, Sharp JT, Tannenbaum H, Hua Y, Teoh LS, Fischkoff SA, Chartash EK (2004) Radiographic, clinical, and functional outcomes of treatment with adalimumab (a human anti-tumor necrosis factor monoclonal antibody) in patients with active rheumatoid arthritis receiving concomitant methotrexate therapy: a randomized, placebo-controlled, 52-week trial. *Arthritis Rheum* 50: 1400–1411

60 Menter A, Tyring SK, Gordon K, Kimball AB, Leonardi CL, Langley RG, Strober BE, Kaul M, Gu Y, Okun M et al. (2008) Adalimumab therapy for moderate to severe psoriasis: A randomised, controlled phase III trial. *J Am Acad Dermatol* 58: 106–115

61 Weir N, Athwal D, Brown D, Foulkes R, Kollias G, Nesbitt A, Popplewell A, Spitali M, Stephens S (2006) A new generation of high-affinity humanized PEGylated Fab' fragment anti-tumor necrosis factor-α monoclonal antibodies. *Therapy* 3: 535–545

62 Nesbitt A, Fossati G, Bergin M, Stephens P, Stephens S, Foulkes R, Brown D, Robinson M, Bourne T (2007) Mechanism of action of certolizumab pegol (CDP870): *in vitro* comparison with other anti-tumor necrosis factor alpha agents. *Inflamm Bowel Dis* 13: 1323–1332

63 Palframan R, Vugler A, Moore A, Nesbitt A, Foulkes R (2008) Differing distribution of certolizumab pegol, adalimumab and infliximab in the inflamed paws of mice with collagen-induced arthritis compared with normal mice. *Ann Rheum Dis* 67(Suppl II): 595

64 Henry AJ, Kennedy J, Fossati G, Nesbitt AM (2007) Stoichiometry of binding to and complex formation with TNF by certolizumab pegol, adalimumab, and infliximab, and the biologic effects of these complexes. *Gastroenterology* 132 (4, Suppl 2): A231

65 Chapman AP (2002) PEGylated antibodies and antibody fragments for improved therapy: a review. *Adv Drug Deliv Rev* 54: 531–545

66 Smith B, Ceska T, Henry A, Heads J, Turner A, King M, Krebs M, Heywood S, O'Hara J, Nesbitt A (2008) Detailing the novel structure of the biopharmaceutical certolizumab pegol. *J Crohn's Colitis Suppl* 2: 50

67 Gramlick G, Fossati G, Nesbitt A (2006) Neutralization of soluble and membrane tumor necrosis factor-α (TNF-α) by infliximab, adalimumab, or certolizumab pegol using P55 or P75 TNF-α receptor-specific bioassays. *Gastroenterology* 130(Suppl 2): A697

68 Peppelenbosch MP, van Deventer SJ (2004) T cell apoptosis and inflammatory bowel disease. *Gut* 53: 1556–1558

69 Atreya R, Bartsch B, Galle PR, Neurath MF (2008) Anti-TNF agents target the mucosal intercellular signaling in inflammatory bowel diseases: A common molecular mechanism of action of clinically effective anti-TNF agents. *Gastroenterology* 134(4, Suppl 1): A644 (W1153)

70 Balazovich KJ, Suchard SJ, Remick DG, Boxer LA (1996) Tumor necrosis factor-alpha and FMLP receptors are functionally linked during FMLP-stimulated activation of adherent human neutrophils. *Blood* 88: 690–696

71 Wang D, Miller SC, Sima M, Parker D, Buswell H, Goodrich KC, Kopecková P, Kopecek J (2004) The arthrotropism of macromolecules in adjuvant-induced arthritis rat model: a preliminary study. *Pharm Res* 21: 1741–1749

72 Baert F, Noman M, Vermeire S, Van Assche G, D'Haens G, Carbonez A, Rutgeerts P (2003) Influence of immunogenicity on the long-term efficacy of infliximab in Crohn's disease. *N Engl J Med* 348: 601–608

73 Nesbitt A, Brown D, Stephens S, Foulkes R (2006) Placental transfer and accumulation in milk of the anti-TNF antibody TNF in rats: immunoglobulin G1 *versus* pegylated Fab'. *Am J Gastroenterol* 101: S438

74 Webster R, Didier E, Harris P, Siegel N, Stadler J, Tilbury L, Smith D (2007) PEGylated proteins: evaluation of their safety in the absence of definitive metabolism studies. *Drug Metab Dispos* 35: 9–16

75 Parton T, King L, van Asperen J, Heywood S, Nesbitt A (2008) Investigation of the distribution and elimination of the PEG component of certolizumab pegol in rats. *J Crohn's Colitis Suppl* 2: 26

76 Yamaoka T, Tabata Y, Ikada Y (1994) Distribution and tissue uptake of poly(ethylene glycol) with different molecular weights after intravenous administration to mice. *J Pharm Sci* 83: 601–606

77 Baker M, Stringer F, Stephens S (2006) Pharmacokinetic properties of the anti-TNF agent certolizumab pegol. *Gut* 55 (Suppl V): A122

78 CIMZIA Prescribing Information. 2008. Available at: http://www.fda.gov/cder/foi/label/2008/125160s000lbl.pdf Accessed: 09 July 2008

79 Pigeolet E, Jacqmin ML, Sargentini-Maier ML, Parker G, Stockis A (2007) Population pharmacokinetics of certolizumab pegol. *Clin Pharmacol Ther* 81(Suppl 1): S72

80 Vetterlein O, Kopotsha T, Nesbitt A, Stephens S (2008) No antibodies to PEG detected in patients treated with certolizumab pegol. *Ann Rheum Dis* 67(Suppl II): 326

81 Vetterlein O, Kopotsha T, Nesbitt A, Brown D, Stephens S (2007) In patients with rheumatoid arthritis treated with the anti-TNF certolizumab pegol or infliximab, antibodies produced do not cross react with other agents. *Ann Rheum Dis* 66(Suppl II): 161

82 Winter TA, Wright J, Ghosh S, Jahnsen J, Innes A, Round P (2004) Intravenous CDP870, a PEGylated Fab' fragment of a humanized antitumour necrosis factor antibody, in patients with moderate-to-severe Crohn's disease: an exploratory study. *Aliment Pharmacol Ther* 20: 1337–1346

83 Schreiber S, Rutgeerts P, Fedorak RN, Khaliq-Kareemi M, Kamm MA, Boivin M, Bernstein CN, Staun M, Thomsen OØ, Innes A; CDP870 Crohn's Disease Study Group (2005) A randomised, placebo-controlled trial of certolizumab pegol (CDP870) for treatment of Crohn's disease. *Gastroenterology* 129: 807–818

84 Sandborn WJ, Feagan BG, Stoinov S, Honiball PJ, Rutgeerts P, Mason D, Bloomfield R, Schreiber S; PRECISE 1 Study Investigators (2007c) Certolizumab pegol for the treatment of Crohn's disease. *N Engl J Med* 357: 228–238

85 Schreiber S, Khaliq-Kareemi M, Lawrance IC, Thomsen OØ, Hanauer SB, McColm J, Bloomfield R, Sandborn WJ; PRECISE 2 Study Investigators (2007a) Maintenance therapy with certolizumab pegol for Crohn's disease. *N Engl J Med* 357: 239–250

86 Schreiber S, Panes J, Mason D, Lichtenstein G, Dandborn WJ (2008) Efficacy and tolerability of certolizumab pegol are sustained over 18 months: data from PRECiSE 2 and its extension studies (PRECiSE 3 and 4). *Gastroenterology* 134(4, Suppl 1): A490 (T1133)

87 Schreiber S, Feagan B, Hanauer S, Rutgeerts PJ, Sandborn WJ (2007) Subcutaneous certolizumab pegol is well tolerated by patients with active Crohn's disease: Results from two phase III studies (PRECiSE 1 and 2). *J Crohn's Colitis Suppl* 1: 43

88 Choy EHS, Hazleman B, Smith M, Moss K, Lisi L, Scott DG, Patel J, Sopwith M, Isenberg DA (2002) Efficacy of a novel PEGylated humanized anti-TNF fragment (CDP870) in patients with rheumatoid arthritis: a phase II double-blinded, randomised, dose-escalating trial. *Rheumatology*

41: 1133–1137

89 Fleischmann R, Mason D, Cohen S (2007) Efficacy and safety of certolizumab pegol monotherapy in patients with rheumatoid arthritis failing previous DMARD therapy. *Ann Rheum Dis*
66(Suppl II): 169

90 Fleischmann R, Keininger DL, Tahiri-Fitzgerald E, Mease P (2007) Certolizumab pegol
monotherapy 400 mg every 4 weeks improves physical functioning and reduces pain in patients
with rheumatoid arthritis who have previously failed DMARD therapy. *Ann Rheum Dis* 66(Suppl
II): 170

91 Keystone E, Mason D, Combe B (2007) The anti-TNF certolizumab pegol in combination with
methotrexate is significantly more effective than methotrexate alone in the treatment of patients
with active rheumatoid arthritis: 1-year results from the RAPID 1 study. *Ann Rheum Dis* 66(Suppl
II): 55

92 Smolen J, Brzezicki J, Mason D, Kavanaugh A (2007) Efficacy and safety of certolizumab pegol
in combination with methotrexate (MTX) in patients with active rheumatoid arthritis despite MTX
therapy: Results from the RAPID 2 study. *Ann Rheum Dis* 66(Suppl II): 187

93 Van der Heijde D, Strand V, Keystone E, Landewe R (2007) Inhibition of radiographic progression by lyophilized certolizumab pegol added to methotrexate in comparison with methotrexate
alone in patients with rheumatoid arthritis: The RAPID 1 trial. *Arthritis Rheum* 56(suppl):
390–391

94 Strand V, Keininger DL, Tahiri-Fitzgerald E (2007) Certolizumab pegol results in clinically mean-
ingful improvements in physical function and health-related quality of life in patients with active
rheumatoid arthritis despite treatment with methotrexate. *Arthritis Rheum* 56(Suppl): 393

95 Reich K, Tasset C, Ortonne J (2007) Efficacy and safety of certolizumab pegol in patients with
chronic plaque psoriasis: Preliminary results of a randomised, double-blind, placebo-controlled
trial. *Ann Rheum Dis* 66 (Suppl II): 251

96 Shealy DJ, Visvanathan S (2008) Anti-TNF antibodies: lessons from the past, roadmap for the
future. *Handb Exp Pharmacol* 181: 101–129

97 Blick SKA, Curran MP (2007) Certolizumab pegol: in Crohn's disease. *BioDrugs* 21: 195–201

PEGylated Protein Drugs: Basic Science and Clinical Applications
Edited by F.M. Veronese
© 2009 Birkhäuser Verlag/Switzerland

PEG: a useful technology in anticancer therapy

Anna Mero, Gianfranco Pasut and Francesco M. Veronese

Department of Pharmaceutical Sciences, University of Padua, via Marzolo 5, 35100 Padova, Italy

Abstract

Cancer chemotherapy dates back to the 1940s with the first use of nitrogen mustards and antifolate drugs. The use of small molecule and biopharmaceutical drugs is today the acceptable approach to cancer treatment in both ambulatory and in patient care. Drug delivery of these drugs has given rise to safer and more efficacious options. Today, the use of polymers for sustained and targeted delivery has allowed oncologists to deal with the earlier limitations of chemotherapy. In this chapter the focus is on polymer conjugation of anticancer drugs, such as high molecular weight proteins and low molecular weight compounds. Examples will be presented to demonstrate an increase in the pharmacological therapeutic index by targeting the drug molecules to the diseased sites with corresponding reduction in drug related side effects. The focus will be on the attachment of polyethylene glycol (PEG) to oncolytic drugs, a process referred to as PEGylation. This technology has been completely validated in the area of protein modification, but is very much in its infancy in the modification of small molecular weight drugs. However, increasing and encouraging efforts have recently been made and will be presented. This chapter will also discuss recent achievements in PEGylation processes with a particular emphasis on the application of PEG to non-conventional therapies such as oxidation therapy, photodynamic therapy and radiopharmaceutical therapy.

Introduction

The modern era of chemotherapy can be dated back to 1942, when Louis Goodman and Alfred Gilman, both pharmacologists at the Yale School of Medicine, found that the nitrogen mustard, a simple but highly reactive molecule, regressed the growth of non-Hodgkin's lymphoma in patients, forming a covalent bond with DNA and inducing apoptosis [1]. A great number of chemotherapeutic agents were developed in the following years, but unfortunately most of them are limited by undesirable low pharmacological therapeutic indices, owing to toxic side effects. It is also known that antitumor agents might be more active on cancer tissues than on healthy normal cells, but no one chemotherapeutic drug used in traditional cancer therapy is selective only on cancer cells. They act on proliferating normal cells as well, resulting in known side effects.

In the past decade, molecular and genetic approaches allowed for a better knowledge of the tumor biology and in particular demonstrated that the mechanisms, regulating cellular activities such as proliferation and survival, are largely altered in cancer cells [2]. The capacity to address and repair these molecular defects in cancer cells paved the way to the era of 'targeted therapy'. The first drug belonging to this class has been imatinib mesylate (gleevec®), approved by the US Food and Drug Administration (FDA) in 2001. Imatinib mesylate is a moderately potent inhibitor of the kinase BCR-ABL, the fusion protein product of a chromosomal translocation involved in the pathogenesis of chronic myeloid leukemia (CML) [3, 4].

Unfortunately, the discovery of highly selective drugs, such as imatinib mesylate, is rare. Therefore, the need to design drug delivery systems (DDS), properly tailored to carry potent antitumor drugs has evolved, resulting in medications with vastly improved therapeutic indices. The aim of these DDS is to increase the amount of drug delivered to the tumor tissue, reducing undesirable side effects and frequency of dosing.

PEGylated small organic molecules

In the mid-1970s Ringsdorf proposed for the first time the idea of covalently attached chemotherapeutic agents to a water-soluble polymer in order to increase the solubility and the stability of hydrophobic drugs and, possibly, to enhance the tumor specificity by the addition of a targeting residue [5]. So far, several chemistries have been investigated, some of which include polymers with covalent links to amino acid residues of antibodies, hormones and peptides; and polymers linked to small molecules in order to alter size and hydrophobic-hydrophilic balance (HLB). In particular, the size of the conjugate can permit tumor tissue accumulation by exploiting the known enhanced permeability and retention (EPR) effect [6]. This is a passive tumor-targeting mechanism, based on the observation that tumor vessels have increased endothelial fenestrations and architectural anarchy, allowing the extravasation and protracted retention in the tumor tissue of macromolecules or particulates [7].

Among the several biomaterials that are used as drug carriers, human albumin (HA), polyglutamic acid (PGA), N-hydroxy methyl methacrylamide copolymer (HPMA) and polyethylene glycol (PEG) have been widely tested. Abraxane® is a nanoparticle formulation that is used to treat metastatic breast cancer. It is co-solvent free complex of human protein albumin bound to the paclitaxel [8]. The other polymers possess several functional groups along the polymeric backbone, allowing an increase drug loading. For instance, OPAX-IO™, a paclitaxel-PGA conjugate (formerly Xyotax®) presents an increased drug payload whereas maintaining a higher water solubility and *in vivo* biodegradability. The product has been tested in Phase III clinical trials for the treatment of lung, ovarian, colorectal, breast and oesophageal cancer [9].

PEG is very successful in protein conjugation for its unique characteristics of biocompatibility, high solubility and non-toxicity, but on the other hand in the case of low molecular weight drugs it presents the one limitation of a low drug loading when compared to PGA. This is because there are only one or two hydroxyl groups per polymer chain available for drug attachment. Although there are several PEGylated proteins and one PEGylated aptamer on the market, there are no PEGylated small molecular weight drugs commercially available [10]. The first attempt at clinical PEG oncology was conducted by Enzon Pharmaceuticals, Inc. PEG-camphotecin or Pegamotecan was prepared by the coupling of two molecules of camptothecin to a diol PEG having a molecular weight of 40 kDa (Fig. 1). This prodrug approach was accomplished by using an amino acid spacer between the drug molecule and the PEG reagent. Unfortunately the drug loading was only of 1.7 wt% and despite the good results obtained in Phase I and II [11, 12], the company announced the discontinuation of further clinical testing. New PEG morphologies have emerged such as dendrimer PEG [13], multi-arm PEG and PEG dendrons and early clinical study results have shown a lot of promise. Exploiting this technology, Nektar Therapeutics proposed a new compound called PEG-irinotecan (NKTR-102), where the topoisomerase-I inhibitor, Irinotecan, is covalently bound to four arms of PEG. In preclinical studies NKTR-102 showed a plasma half-life in a mouse model, evaluated on the basis of the released active metabolite SN-38, of 15 days when compared to 4 h with irinotecan only [14]. In Phase I clinical trials, safety, pharmacokinetics and anti-tumor activity of the compound was evaluated in 57 patients with advanced solid tumors, such as breast, ovarian, cervical, or non-small cell lung cancer that had failed prior treatments with irinotecan or with standard treatment. Only 13 reported significant anti-tumor activity while seven had a response in a reduction of tumor size ranging from 40% to 58%. Six patients showed minor response only [15]. Multiple Phase II studies are underway with NKTR-102 for the treatment of ovarian, breast, colorectal and cervical cancer. As a follow on, Enzon introduced EZN-2208, a conjugate of SN-38 and a 40 kDa PEG with four arms (see Fig. 2). This compound showed activity in a panel of human tumor xenografts [16, 17]. A Phase I study to evaluate the safety and tolerability of intravenous EZN-2208 in patients with advanced solid tumor or lymphoma is currently ongoing [18]. The PEGylation of docetaxel, NKTR-105 is another product has shown good preclinical activity in colon and lung cancer xenograft models [19]. This product has just entered Phase I clinical studies.

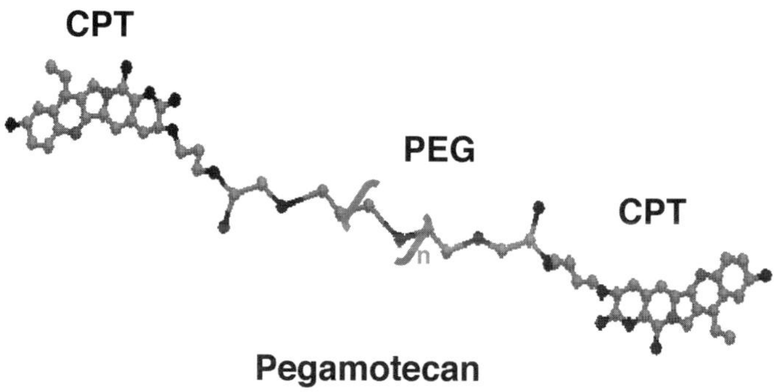

Figure 1. Schematic representation of Pegamotecan

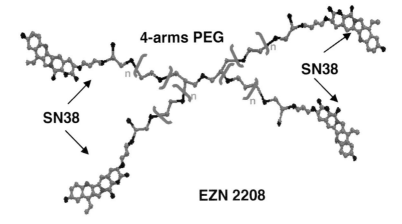

Figure 2. Schematic representation of EZN 2208

PEG small molecule conjugates for combination therapy

A general problem related with the classical protocols of chemotherapy is the onset of drug resistances, mainly due to increase drug efflux from cells or mutations in cell-death pathways. To overcome resistance to single agents, combination therapy is essential for tumor eradication and cure. In fact, combinations of drugs have proved to be more effective than single agents against both metastatic cancers and in patients at high risk of relapse after primary surgical treatment [20].

The idea of combination therapy has been extended also to polymer-drug conjugates basically by three ways: i) the delivery of two different drugs on the same polymer, exploiting the multivalency of many carriers, ii) the co-administration of a polymer-drug conjugate and a classic low molecular weight drug, or iii) the co-administration of two conjugates, each one bearing a different drug.

The last approach was studied by Minko and co-workers testing free CPT (camptothecin), CPT-PEG, CPT-PEG-BH3 (BCL2 homology 3) or CPT-PEG-LHRH (luteinizing hormone-releasing hormone) conjugates and the mixture of CPT-PEG-BH3 and CPT-PEG-LHRH conjugates in human ovarian carcinoma cells [21]. It was demonstrated that conjugation of CPT to PEG increased its pro-apoptotic activity and that a further enhancement was achieved by using the peptide BH3 to yield CPT-PEG-BH3 or the peptide LHRH in a CPT-PEG-LHRH conjugate. The mixture of these two last derivatives increased the therapeutic response acting as a suppressor of cell anti-apoptotic defence [22]. Other groups also explored the effectiveness of combination therapy, using either HPMA or PGA as carriers. A phase III clinical trial compared PGA-paclitaxel in association with standard chemotherapy (carboplatin *versus* paclitaxel + carboplatin [23]). The same conjugate has also been tested in combination with radiotherapy in a Phase I trial for esophageal and gastric cancer and four complete clinical responses (33%) were observed in this study [24].

A conjugate obtained by coupling one epirubicin (EPI) and several nitric oxide (NO) releasing moieties to a heterobifunctional PEG-dendron, was studied as a different concept of combination therapy (Fig. 3). The combination of NO and EPI on the same carrier has presented several advantages: i) NO has been shown to protect cardiomyocytes against doxorubicin-induced apoptosis [25], thus preventing the drawback of a heavy cardiomyopathy, ii) NO enhances the antitumor activity of free doxorubicin only in cancer cells [26] iii) the contemporaneous conjugation of EPI and NO to the same polymeric carrier ensures that the two active molecules will undergo the same body distribution and cell internalization. The first studies confirmed that the conjugate EPI-PEG-(NO)$_8$ (Fig. 3) has a different cytotoxic profile compared to free EPI against tumor and normal cells. In particular, this conjugate was able to induce a higher degree of apoptosis in Caco-2 cells while reducing EPI toxicity in embryonic rat heart-derived myoblast (H9c2). In addition, its efficacy was confirmed *in vivo* showing a 95% tumor reduction in human colon carcinoma (Caco-2) cells and human ovarian cancer (SKOV-2) cells tumor-bearing mice [27].

The potentials of hetero-bifunctional PEGs were exploited also for the preparation of PEG-gemcitabine conjugate bearing folic acid as a targeting moiety. In mice the polymer conjugation was found to increase the plasma half-life of the drug, by reducing its kidney clearance and by decreasing the

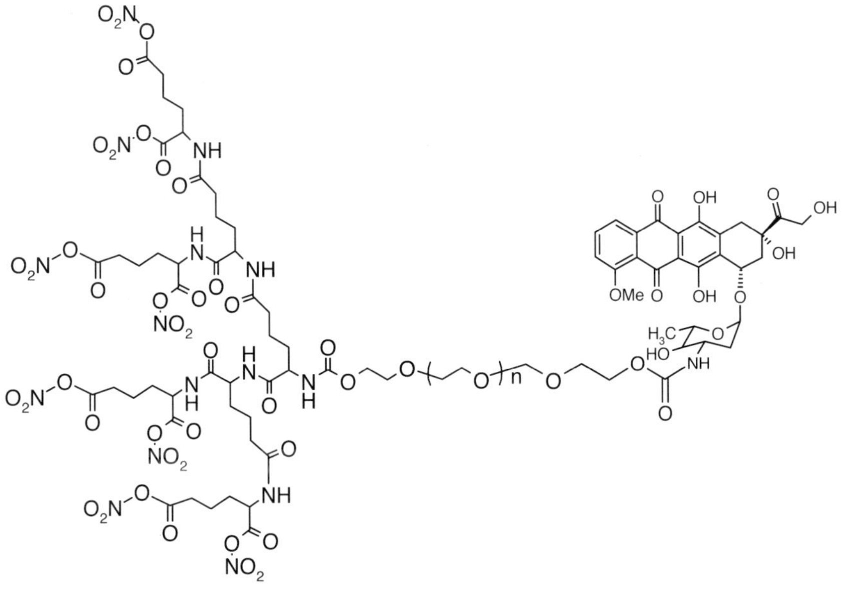

Figure 3. Schematic representation of dendron PEG structure, EPI-PEG-(NO)$_8$

degradation through the cytidine-deaminase enzyme. The targeted conjugates showed a higher antiproliferative activity and higher selectivity than the non-targeted ones when studied against KB-3-1 cell line, which over-expresses the folic acid receptor. The decreased cytotoxicity of these targeted conjugates in cell lines that do not over-express the folic acid receptor (HT-29) can indicate that they require a receptor-mediated endocytosis mechanism for cell penetration [28].

PEG-proteins as valuable anticancer drugs

To avoid repetitions with other chapters of this book, we will review here only the amino acids degrading enzymes as PEGylated proteins used in chemotherapy.

The rational for the basis of this therapy is in the observation that some tumors cells are not able to synthesize *de novo* some amino acids because they have a deficient gene or poor protein metabolism. These cells are therefore dependent on external supply of enzymatic therapy in order to allow for a steady depletion of these essential amino acids, thereby resulting in starvation of cancer cells [29]. Hence, in order to be therapeutically effective the deficient enzymes need to be steadily infused into the body in order to last for several days thereby resulting in amino acid depletion and eventual tumor cell death. PEGylation has been therefore investigated as a suitable procedure that would allow for serum steady-state concentrations of these enzymes [10]. A high level of enzyme surface shielding could be reached by randomly PEGylating the amines to reduce the risk of immunogenicity and rapid blood clearance. Moreover, it was also thought that a high molecular weight PEG enzyme complex may accumulate in tumor tissue by EPR effect.

Among the several enzymes of potential antitumour interest, PEG-asparaginase was the only one that was commercialized. Its PEGylation represented a milestone in protein conjugation because it opened the way to additional PEG protein products. Other enzymes, such as arginine deiminase, arginase and methioninase have been PEGylated and well-characterized and have been tested in a clinical setting.

PEG-asparaginase

The L-asparaginase (L-ASNase) that catalyses the conversion of asparagines to aspartic acid was isolated for the first time from *E. coli* (Fig. 4).

It is a tetramer composed of four identical subunits with an active site on each. For the last 25 years, L-ASNase has been an important component in the treatment of acute lymphoblastic leukemia (ALL). Indeed, while normal cells can produce asparagine, the leukemic ones are unable to produce enough asparagine to survive. L-ASNase is given to ALL patients to ensure depletion

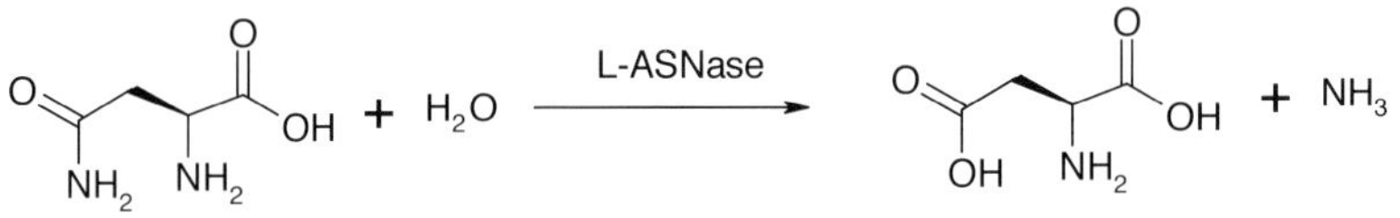

Figure 4. Conversion of L-asparagine to L-aspartic acid mediated by ASNase

of asparagine circulating in the blood that ultimately results in leukemic cell death. Elspar®, isolated from *E. coli*, and *Erwinia* L-asparaginase, isolated from *Erwinia chrysanthemi* are presently available for clinical use by Merck & Co. and by Ogden Bioservices Pharmaceutical respectively. The main limitation in their use is clinical hypersensitivity, which develops in 3–78% of patients through acute allergic reactions or silent hypersensitivities. L-asparaginase was randomly PEGylated at the ε-amino residues using an acetylating PEG 5 kDa to obtain a mixture of multi-PEG-asparaginase conjugates with different degree of modification [30, 31]. The first clinical studies, started in 1984, showed soon the great potentials of the derivative: i) decreased immunogenic response, ii) enhanced circulation in blood and iii) resistance to plasma proteases [32]. The conjugate, brand name ONCASPAR®, developed by Enzon Pharmaceuticals inc., was approved by the FDA in 1994. ONCASPAR® is indicated as a component of a multiagent chemotherapeutic regimen for the first-line treatment of children with acute lymphoblastic leukemia (ALL) and for the treatment of patients with acute lymphoblastic leukemia and hypersensitivity to native forms of L-asparaginase [33]. It is administered through intramuscular injection or intravenous infusion, one dose achieved similar levels of asparagine depletion as nine doses of native L-asparaginase. The use of ONCASPAR® for the treatment of ALL continues to be explored even to date to evaluate the optimal dose and duration of treatment. It is contraindicated in patients with a history of serious allergic reactions to ONCASPAR®, and in patients with a history of serious thrombosis, pancreatitis, or serious hemorrhagic events with prior L-asparaginase therapy. It should be discontinued in the case of anaphylaxis or serious allergic reactions, thrombosis, or pancreatitis. The most common adverse reactions with ONCASPAR (>2%), monitored by several Phase IV clinical studies, are allergic reactions (including anaphylaxis), hyperglycemia, pancreatitis, central nervous system (CNS) thrombosis, coagulopathy, hyperbilirubinemia, and elevated transaminases. In study II (n = 2770), the per-patient incidence for Grades 3 and 4 non-hematologic toxicities were: elevated transaminases (11%), coagulopathy (7%), hyperglycemia (5%), CNS thrombosis/hemorrhage (2%), pancreatitis (2%), clinical allergic reaction (1%), and hyperbilirubinemia (1%). There were three deaths due to pancreatitis [34].

Worth of noting is that, during the numerous studies carried out with this conjugate, there were some reports claiming the presence of antibodies against polyethylene glycol in the sera of patients accompanied by undetectable asparaginase activity. Most probably the absence of the PEG-asparaginase activity depends on antibodies against the polymer that enhance the elimination rate of the conjugate. This occurrence needs further investigations to explain why a polymer, known to be not-immunogenic, can elicit antibodies preventing the therapeutic efficacy of the conjugate. An explanation can be that the highly immunogenicity of the heterologous asparaginase may acts as an adjuvant stimulating the immunogenic response against PEG. Some authors claim that this phenomenon, discovered also in the case of PEG-uricase, is now occurring more frequently than in the past [35, 36]. This is probably due to the increased use of this polymer, for therapeutic or non-therapeutic applications, in cosmetics and foods (see also chapter by Armstrong in this book).

PEG-arginine deiminase

Arginine is synthesised *de novo* in healthy cells from citrulline by argininosuccinate synthetase (ASS) and arginine lyase (ASL). It was found that some cell lines, which often possess an incomplete enzymatic pool, lack the ASS enzyme being therefore auxothophic for arginine [37]. Arginine deiminase (ADI) degrades arginine into citrulline and ammonia (Fig. 5) leading to tumor cell starvation and consequently to inhibition of tumor growth and angiogenesis [38].

ADI, expressed in *E. coli*, has been demonstrated to be a potent enzyme in killing human leukemia cells *in vitro*. It led to cell apoptosis and anti-angiogenic effects in some human melanoma and human hepatocellular carcinoma cell lines also. The IC_{50} is from <0.01 to 0.3 µg/mL for the former and <0.01 µg/mL for the latter, respectively [39]. The main limits of its therapeutic application have been found to reside in its short half-life that does not allow the sufficient time of starvation of arginine in blood that is needed for cell death and in the immunological response due to heterologous origin.

PEGs of different molecular weight, shape and linkers for the conjugation were used [40]. It was observed a loss of activity of increasing the number of linked PEG chains. Two derivatives, obtained using PEG of 5 and 20 kDa linked to about 40% of the amine groups of the proteins, were chosen for the *in vivo* experiments. The 20 kDa derivative showed a longer pharmacokinetics than the 5 kDa one, independently to the route of administration and animal gender [40]. After a single i.m. injection in mice a single dose of PEG_{20kDa}-ADI achieved similar levels of arginine depletion as 14 doses of native enzyme with an half-life of seven days against the 5 h of the native one.

Figure 5. Conversion of L-arginine to citrulline mediated by ADI

PEG$_{20kDa}$-ADI was evaluated in humans in two different clinical trials in patients with metastatic melanoma and hepatocellular carcinoma (HCC) [41, 42]. In the first trial 19 patients were treated with 160 IU/m^2 weekly administration of the conjugate. Among these, two patients showed a complete response, seven a partial one, seven remained stable and in three cases the disease was progressive. In all of these cases, the drug was well-tolerated without significant side-effects. In the second trial the conjugate was administered to patients with metastatic melanoma with three cycles of i.m. injections at different concentrations ranging from 40–640 IU/m^2. The dose of 160 IU/m^2 decreased the concentration of arginine from 130 µmmol/L to 2 µmmol/L and six out of 24 patients responded to the treatment with a prolonged survival [42].

PEG-arginase

Arginase activity determines the depletion of arginine by the hydrolysis of this amino acid to ornithine and urea (Fig. 6). The enzyme has not been considered for a long time as a potential therapeutic drug, because the enzymes, initially extracted from bovine and murine source, were not effective *in vivo*. The reason was the high K_m and the high pH (9.6) for optimum activity [43].

Recently a human arginase has been obtained by recombinant techniques possessing a lower K_m (1.9 mM) and optimum pH activity at physiological conditions. Furthermore, this enzyme is worthy of consideration because, being normally expressed in humans, it presents less immunogenic problems with respect to the previously studied enzymes. Anyway, it suffers from very short half-life (few minutes) that reduces its efficacy.

Different types of PEGs and different extent of modification were used to overcome the problem. All of the prepared conjugates maintained the enzymatic activity, except the one obtained with mPEG-cyanuric chloride for which a drop in activity of about 70% was found. The derivative with mPEG-SPA, with a degree of amine group modification of about the 40–50% and a half-life of three days, showed to be the most promising candidate for therapeutic applications [44].

IC$_{50}$ values of the PEGylated specie was almost the same of that of the native protein evaluated in several HCC cell lines and encouraged for following *in vivo* studies. It was found that the PEGylated arginase, in combination therapy with 5-fluorouracil, inhibited in mice the tumor growth of OTC-deficient Hep3B (ornithine transcarbamylase), being ASS positive. These results

Figure 6. Conversion of L-arginine to ornithine mediated by Arginase

were in favor of PEG-arginase as a valuable anticancer drug especially towards some ADI resistant tumor [45]. The Bio-Cancer Company that developed this conjugate, coded PEG-BCT-100, announced the availability on the market in China since the 2008.

PEG-methioninase

It has been observed that in many cancer cell lines and cancer xenografts in animal models the methionine starvation leads to the inhibition of the growth and to death of the cells. Methioninase (METase), a pirydoxal-5-phosphate (PLP) depending enzyme, cleaves methionine into α-ketobutirrate, methanethiol and ammonia (Fig. 7).

METase was found in *Pseudomonas putida, Aeromonas* and *Clostridium*, but not in yeast, plants or mammals. The enzyme was cloned for the first time from *Pseudomonas putida* and produced in *Escherichia Coli*. For his specificity, this enzyme seems be a good candidate as anticancer drug and its efficacy was successfully evaluated in several cancer cell lines. METase alone or in combination with chemotherapeutic agents, such as cisplatin or 5-fluorouracil, has shown efficacy and synergy in several tumors. The findings from a pilot Phase I clinical trial showed that the plasma methionine levels were undetectable over a period of about 24 h after METase administration in patients with advanced cancer. Unfortunately, after repeated administrations in monkeys, a severe anaphylactic shock and a death of one animal occurred, indicating the very high immunogenic potential of the enzyme. A PEGylated derivative has been prepared using PEG_{5kDa} yielding a conjugate that retained 70% of the enzymatic activity. Pharmacokinetic studies in mice showed the enhanced half-life of the conjugates up to 20-fold and an increased methionine depletion time up to 12-fold compared with unmodified enzyme, without any immunogenic reactions. Furthermore, it was possible to still increase the half-life of the conjugate with co-administration of the PLP co-factor. In fact PLP dissociates *in vivo* with a subsequent loss of enzyme activity. A combined administration of the enzyme with the co-factor allowed for the reduction and the frequency of the protein administration. Antibodies against the conjugate were observed in primate models after repeated administrations, but their lev-

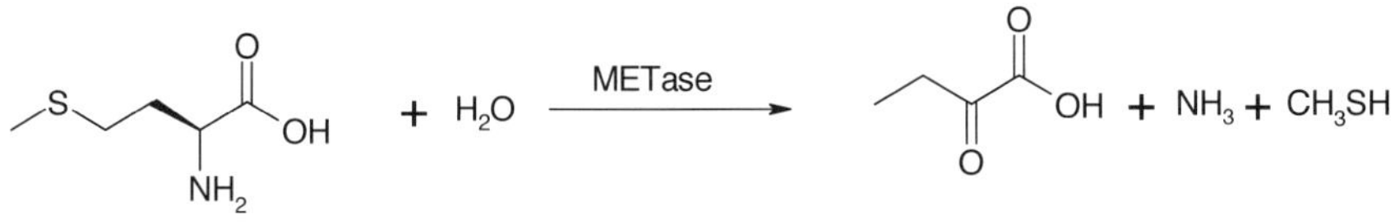

Figure 7. Degradation of L-methionine to α-ketobutirrate, methanethiol and ammonia mediated by METase

els were 100–1,000-fold lower than the native enzyme and interestingly they did not neutralize the enzyme activity [46–48].

Alternative approaches in the treatment of cancer

For years, the use of polymer-drug conjugates has been restricted to established and clinically used small organic molecules, such as doxorubicine, paclitaxel and camptothecin. The discovery of new molecular targets for cancer therapy and an increased knowledge of the biology of the tumor tissue prompted the synthesis of new conjugates using non-conventional drugs.

An interesting new approach is represented by the 'oxidation therapy' that exploits H_2O_2-generating enzymes or inhibitor of heme oxigenase 1 (HO-1) to kill cancer cells [49]. It is in fact known that high levels of oxidative stress causes cytotoxicity by inhibiting cell proliferation, finally leading to apoptotic/necrotic cell death.

To improve this strategy PEG was conjugated to xanthine oxidase (XO), a metalloflavoprotein that catalyses the oxidation of xanthine/hypoxanthine to generate superoxide anion (O_2^-) [50]. The conjugation reduces the affinity towards the blood vessels of the native enzyme, that is responsible of damage and hypertension, and allows to exploit the EPR effect thanks the increased hydrodynamic volume of the conjugate [51]. The *in vivo* studies demonstrated that after administration of PEG-XO followed by infusion of the substrate hypoxanthine, the tumor growth was remarkably suppressed up to at least 52 days and a complete regression of tumor was found in three out of seven tumor-bearing mice, while no apparent side-effects were observed [52].

Another approach takes advantage of PEGylation of D-amino acid oxidase (DAO), an enzyme that catalyses the stereoselective oxidative deamination of D-amino-acids to the corresponding α-ketoacids with the formation of H_2O_2 [53]. Since the substrate of the enzyme does not exist in physiological conditions, it is possible to control the H_2O_2 production by regulating the exogenous infusion of D-amino acids. Furthermore, the PEG conjugation could create a macromolecule that by EPR effect is selectively accumulated in tumor tissue. It was demonstrated that, after conjugate administration and subsequent tumor accumulation, the infusion of D-proline induced suppression of tumor growth and significant increase of survival rate in tumor-bearing mice [54].

A different approach is based on zinc protoporphyrin (ZnPP) that is an inhibitor of HO-1. For this compound the use is limited by the insolubility in physiological solutions. ZnPP solubility was increased by PEGylation, yielding macromolecular micelles of 200 nm in diameter. When administered intravenously the plasma residence time was 40 times higher than that of the native ZnPP. Again, the possibility of an EPR effect may play some role. The derivative inhibited the growth of S-180 and colon 38 tumors in ddY and BALB/c mice, without substantial side effects [55]. This conjugate can be potentially used also for photodynamic therapy (PDT), one of the most promising and less invasive antitumor approaches. In the conventional phototherapy, a photosensitizer, exposed to the laser beam at 630 nm, generates the highly reactive singlet oxygen [56]. ZnPP efficiently generates highly reactive singlet oxygen also under a tungsten-xenon light, an irradiation source usually employed for endoscopic purposes. Under these conditions, it was found that the conjugate induced marked tumor regression, both in mouse tumor xenograft model and chemical induced rat cancer [57]. The lower cost and easy availability of tungsten-xenon light with respect to the laser apparatus for conventional PDT makes an easier application of this therapeutic strategy. In general, PDT is receiving increasing attention for the treatment of superficial cancers, such as lung, gastric, bladder and cervical cancer or of other cavities reachable by endoscopic means. The advantage over other therapeutic treatment is the specificity and localization, this reducing the occurrence of systematic toxicity [58].

Psoralens, a photosensitizer agent that under UV-A light specifically binds to DNA, was also PEGylated (Fig. 8). With this product it is possible to reach high specificity: the activity only in tissues irradiated by UV-A, a tumor accumulation by EPR effect and a very precise mechanism of action [59].

In a recent work, a bi-functional PEG was used to link a magnetic resonance imaging (MRI) contrast agent and a photosensitizer to reduce non-specific uptake, particularly liver uptake and to improve tumor targeting in MRI-guided photodynamic therapy with a very low invasive treatment [60]. The conjugates revealed a prolonged blood residence time due to the reduced recognition by liver macrophages and a very high accumulation in tumor tissue after 18 hours. Worth noting is the use of the technique of three-dimensional high-resolution dynamic contrast enhanced MRI that was effective for non-invasive visualization of the real-time pharmacokinetics and biodistribution of the polymer conjugates in mouse tumor models. This allowed for identifying the specific uptake of the conjugate in tumor tissue [61].

Polymer conjugation presents further advantages of radiotherapy, offering the possibility of delivering low selectivity radiopharmaceuticals. PEG might

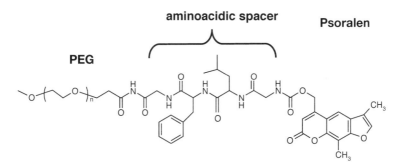

Figure 8. Schematic representation of mPEG-psoralen. The drug is conjugated to PEG through an aminoacidic spacer cleavable in lysosomes by cathepsin-B, a protease present and over-expressed in tumors.

convey also new properties to conjugates. An interesting case is the PEGylated PN$_2$S (N-(N-(3-diphenylphosphinopropionyl)glycyl)cysteine) a ligand for ^{99m}Tc labeling. The presence of PEG allowed to simplify the labeling procedure thanks to the supramolecular assembly of the PN$_2$S-PEG conjugates. The micelles, spontaneously formed in aqueous solution, favored the reduction and coordination of technetium [62]. The labeling did not modify the polymer behavior in water; and at the same time PEG was able to contribute towards improved pharmacokinetic properties and biodistribution. The method presents an useful strategy to further improve the potential of ^{99m}Tc in diagnosis. It may also be applied to rhenium-186/188 for radiotherapy through the labeling of targeted biomolecules of conjugated PN$_2$S set with a bi-functional PEG [63]. PEG conjugation has been also investigated to improve the pharmacokinetic and biodistribution of monoclonal antibodies bearing radiopharmaceuticals. An example is the PEGylated anti-CD45 monoclonal antibody coupled with AHN-12, which can be labeled with the high-energy beta-particle-emitting isotope ^{90}Y. The complex was tested *in vivo* in CD45+ B cell Daudi lymphoma, grown as flank tumors in athymic nude mice, displaying a significantly better antitumor effect than the non-PEGylated antibody. The study evaluated PEGs of different sizes and as expected, that rats treated with 43 kDa PEGylated AHN-12 had higher radiolabeled antibody blood levels and improved pharmacokinetics, than those treated with the 5 kDa derivative or the non-PEGylated AHN-12 moiety. The surviving mice revealed no signs of kidney, liver, or gastrointestinal damage after histological examination. Notably, *in vitro* studies also indicated that PEGylation did not have major effects on the labeling efficiency and the binding of labeled antibody [64].

Conclusions

Human malignancies fall into very diverse groups of diseases, even within histological classifications, and they must be treated in different manners. The application of classical chemotherapy in the period from 1950 to 1980 did not lead to the expected results due to the insurgence of common cancer resistance mechanisms and to several toxic side effects in patients. So far a huge effort was dedicated to find new approaches to overcome the limits of the classical anticancer drugs and to improve the knowledge of the tumor tissue retention. Nanotechnology approaches and polymer conjugation appears to be suitable options in the delivery of anti-cancer drugs to patients seeking a better and safer quality of life.

From the results reported in this and other chapters of the book, PEG can be considered an effective polymer that has reached great potential in drug delivery. Its benefit in delivering proteins and small molecules have been realized in the clinic. Lessons learnt from these products will give rise to PEG constructs that are beyond the simple linear, dendronic or multi-arm morphologies. A newer generation of PEG or PEG-like polymers are just around the corner [65] and it will be interesting to see how these polymers will be available to address some of the limitations of PEG and PEG conjugation.

Acknowledgments
The authors thank Dr. Tacey Viegas for the useful suggestions in the preparation of the chapter.

References

1. Gilman, A. (1946) The biological actions and therapeutic applications of the B-chloroethyl amines and sulphides. Science 103, 409–436.
2. Chabner, B.A., Roberts, T.G. (2005) Chemotherapy and the war on cancer. Nat. Rev. Cancer. 5, 65–72.
3. Druker, B.J. (2001) Efficacy and safety of a specific inhibitor of the BCR–ABL tyrosine kinase in chronic myeloid leukemia. N. Engl. J. Med. 344, 1031–1037.
4. Kantarjian, H. (2002) Hematologic and cytogenetic responses to imatinib mesylate in chronic myelogenous leukemia. N. Engl. J. Med. 346, 645–652.
5. Ringsdorf, H. (1975) Structure and properties of pharmacologically active polymers. J. Polym. Sci. Symp. 51, 135–153.
6. Matsumura, Y., Maeda, H. (1986) A new concept for macromolecular therapeutics in cancer chemotherapy: mechanism of tumoritropic accumulation of proteins and the antitumour agent, SMANCS. Cancer Res. 6, 6387–6392.
7. Maeda, H., Wu, J., Sawa, T., Matsumura, Y., Hori, K. (2000) Tumour vascular permeability and the EPR effect in macromolecular therapeutics: a review. J. Control. Release 65, 271–284.
8. Hawkins, M.J., Soon-Shiong, P., Desai, N. (2008) Protein nanoparticles as drug carriers in clinical medicine. Adv. Drug Deliv. Rev. 60, 876–885.
9. Singer, J.W., Shaffer, S., Baker, B., Bernareggi, A., Stromatt, S., Nienstedt, D., Besman, M. (2005) Paclitaxel poliglumex (XYOTAX; CT-2103): an intracellularly targeted taxane. Anticancer Drugs 16, 243–254.
10. Pasut, G., Veronese, F.M. (2007) Progress in polymer science, Polymer–drug conjugation, recent achievements and general strategies. Progress in Polymer Science 32, 933–961.
11. Rowinsky, E.K., Rizzo, J., Ochoa, L., Takimoto, C.H., Forouzesh, B., Schwartz, G. (2003) A phase I and pharmacokinetic study of pegylated camptothecin as a 1-hour infusion every 3 weeks in patients with advanced solid malignancies. J. Clin. Oncol. 21, 148–157.
12. Scott, L.C., Yao, J.C., Benson A.B., Thomas, A.L., Falk, S., Mena, R.R., Picus, J., Wright, J., Mulcahy, M.F., Ajani, J.A., Evans, T.R. (2008) A phase II study of pegylated-camptothecin (pegamotecan) in the treatment of locally advanced and metastatic gastric and gastro-oesophageal junction adenocarcinoma. Cancer Chemother. Pharmacol. 63, 363–370.
13. Berna, M., Dalzoppo, D., Pasut, G., Manunta, M., Izzo, L., Jones, A.T., Duncan, R., Veronese, F.M. (2006) Novel monodisperse PEG-dendrons as new tools for targeted drug delivery: synthesis, characterization and cellular uptake. Biomacromolecules 7, 146–153.
14. Eldon, M.A., Staschen CM., Viegas, T., Bentley, M. (2007) NKTR-102, a novel PEGylated-irinotecan conjugate, results in sustained tumor growth inhibition in mouse models of human colorectal and lung tumors that is associated with increased and sustained tumor SN38 exposure. 2007 AACR-NCI-EORTC International Conference, San Francisco, Poster C157.
15. Von Hoff, D.D., Jameson, G.S., Borad, M.J., Rosen, L.S., Utz, J., Basche, M., Alemany, C., Dhar, S., Acosta, L., Barker, T., Walling, J., Hamm, J.T. (2008) First phase I trial of NKTR-102 (PEG-Irinotecan) reveals early evidence of broad anti-tumour activity in three different schedules. 20th EORTC-NCI-AACR Symposium on Molecular Targets and Cancer Therapeutics, Geneva, Switzerland, Poster 595.
16. Zhao, H., Rubio, B., Sapra, P., Wu, D., Reddy, P., Sai, P., Martinez, A., Gao, Y., Lozanguiez, Y., Longley, C., Greenberger, L.M., Horak, I.D. (2008) Novel prodrugs of SN38 using multiarm poly(ethylene glycol) linkers. Bioconjug Chem. 19, 849–859.
17. Sapra, P., Zhao, H., Mehlig, M., Malaby, J., Kraft, P., Longley, C., Greenberger, L.M., Horak, I.D. (2008) Novel Delivery of SN38 Markedly Inhibits Tumour Growth in Xenografts, Including a Camptothecin-11–Refractory Model. Clin. Can. Res. 14, 1888–1896.

18. Guo, Z., Wheler, J.J., Naing, A., Mani, S., Goel, S., Mulcahy, M., Gamza, F., Longley, C., Buchbinder, A., Kurzrock, R. (2008) Clinical pharmacokinetics (PK) of EZN-2208, a novel anticancer agent, in patients (pts) with advanced malignancies: A phase I, first-in-human, dose-escalation study. J. Clin. Oncol. 26 (abstr 2556).

19. Wolff, R., Routt, S., Hartsook, R., Riggs, J., Zhang W., Persson, H., Johnson, R. (2008) NKTR-105, a novel PEGylated-docetaxel demonstrates superior anti-tumor activity compared to docetaxel in human non-small cell lung and colon cancer xenografts. 20th EORTC-NCI-AACR Symposium on Molecular Targets and Cancer Therapeutics, Geneva, Switzerland, Poster 448.

20. Greco, F., Vicent, M.J. (2008) Polymer-drug conjugates: current status and future trends, Front Biosci. 13, 2744–2756.

21. Dharap, S.S., Wang, Y., Chandna, P., Khandare, J.J., Qiu, B., Gunaseelan, S., Sinko, P.J., Stein, S., Farmanfarmaian, A., Minko, T. (2005) Tumour-specific targeting of an anticancer drug delivery system by LHRH peptide, Proc. Natl. Acad. Sci. USA. 102, 12962–12967.

22. Khandare, J.J., Chandna, P., Wang, Y., Pozharov, V.P., Minko, T. (2006) Novel polymeric prodrug with multivalent components for cancer therapy. J. Pharmacol. Exp. Ther. 317, 929–937.

23. Nemunaitis, J., Cunningham, C., Senzer, N., Gray, M., Oldham, F., Pippen, J., Mennel, R., Eisenfeld, A. (2005) Phase I study of CT-2103, a polymer-conjugated paclitaxel, and carboplatin in patients with advanced solid tumours. Cancer Invest. 23, 671–676.

24. Dipetrillo, T., Milas, L., Evans, D., Akerman, P., Ng, T., Miner, T., Cruff, D., Chauhan, B., Iannitti, D., Harrington, D., Safran, H. (2006) Paclitaxel poliglumex (PPX-xyotax) and concurrent radiation for esophageal and gastric cancer – A phase I study. Am. J. Clin. Oncol.–Canc. 29, 376–379.

25. Santucci, L., Mencarelli, A., Renga, B., Pasut, G., Veronese, F.M., Zacheo, A., Germani, A., Fiorucci, S. (2006) Nitric oxide modulates proapoptotic and antiapoptotic properties of chemotherapy agents: the case of NO-pegylated epirubicin. Faseb J. 20, 765–767.

26. Fogli, S., Nieri, P., Breschi, M.C. (2004) The role of nitric oxide in anthracycline toxicity and prospects for pharmacologic prevention of cardiac damage. Faseb J. 18, 664–675.

27. Cantucci, L., Mencarelli, A., Renga, B., Ceccobelli, D., Pasut, G., Veronese, F.M., Distrutti, E., Fiorucci, S. (2007) Cardiac safety and antitumoral activity of a new nitric oxide derivative of pegylated epirubicin in mice. Anticancer Drugs. 18, 1081–1091.

28. Pasut, G., Canal, F., Dalla Via, L., Arpicco, S., Veronese, F.M., Schiavon, O. (2008) Antitumoral activity of PEG-gemcitabine prodrugs targeted by folic acid. Journal Control. Rel. 127, 239–248.

29. Vellard, M. (2003) The enzymes as a drug: application of enzymes as pharmaceuticals. Curr. Opin. Biotechnol. 14, 444–450.

30. Abuchowski, A., Van Es, T., Palczuk, N.C., McCoy, J.R., Davis, F.F. (1979) Treatment of L5178Y tumour-bearing BDF1 mice with a nonimmunogenic L-glutaminase-L-asparaginase. Cancer Treat. Rep. 63, 1127–1132.

31. Abuchowski, A., Kazo, G.M., Verhoest, J.R., Van Es, T., Kafkewitz, D., Nucci, M.L., Viau, A.T., Davis, F.F. (1984) Cancer therapy with chemically modified enzymes. I. Antitumour properties of polyethylene glycol-asparaginase conjugates. Cancer Biochem. Biophys. 7, 175–186.

32. Kurtzberg, J. (2000) Cancer Medicines. 5th ed. Gansler T., Decker Inc, Canada.

33. Dinndorf, P.A., Gootenberg, J., Cohen, M.H., Keegan, P., Pazdur, R. (2007) FDA drug approval summary: Pegaspargase (Oncaspar®) for the first-line treatment of children with acute lymphoblastic leukemia (ALL). The Oncologist 12, 991–998.

34. Apostolidou, E., Swords, R., Alvarado, Y., Giles, F.J. (2007) Treatment of acute lymphoblastic leukemia (ALL): a new era. Drugs 67, 2153–2171.

35. Sherman, M.R., Saifer, M.G., Perez-Ruiz, F. (2008) PEG-uricase in the management of treatment-resistant gout and hyperuricemia. Adv. Drug Deliv. Rev. 60, 59–68.

36. Armstrong, J.K., Hempel, G., Koling, S., Chan, L.S., Fisher, T., Meiselman, H.J., Garratty, G. (2007) Antibody against poly(ethylene glycol) adversely affects PEG-asparaginase therapy in acute lymphoblastic leukemia patients. Cancer 110, 103–111.

37. Sugimura, K., Ohno, T., Kusuyama, T., Azuma, I. (1992) High sensitivity of human melanoma cell lines to the growth inhibitory activity of mycoplasmal arginine deiminase in vitro. Melanoma Res. 2, 191–196.

38. Cheng, P.N., Leung, Y.C., Lo, W.H., Tsui, S.M., Lam, K.C. (2005) Remission of hepatocellular carcinoma with arginine depletion induced by systemic release of endogenous hepatic arginase due to transhepatic arterial embolisation, augmented by high-dose insulin: arginase as a potential drug candidate for hepatocellular carcinoma. Cancer Lett. 224, 67–80.

39. Gong, H., Zolzer, F., von Recklinghausen, G., Havers, W., Schweigerer, L. (2000) Arginine deim-

inase inhibits proliferation of human leukemia cells more potently than asparaginase by inducing cell cycle arrest and apoptosis. Leukemia 14, 826–829.

40. Holtsberg, F.W., Ensor, C.M., Steiner, M.R., Bomalaski, J.S., Clark, M.A. (2002) Poly(ethylene glycol) (PEG) conjugated arginine deiminase: Effects of PEG formulations on its pharmacological properties. J. Control. Rel. 80, 259–271.

41. Izzo, F., Marra, P., Beneduce, G., Castello, G., Vallone, P., De Rosa, V., Cremona, F., Ensor, C.M., Holtsberg, F.W., Bomalaski, J.S., Clark, M.A., Ng, C., Curley, S.A. (2004) PEGylated arginine deiminase treatment of patients with unresectable hepatocellular carcinoma: Results from phase I/II studies. J. Clin. Oncol. 22, 1815–1822.

42. Ascierto, P.A., Scala, S., Castello, G., Daponte, A., Simeone, E., Ottaiano, A. (2005) Pegylated arginine deiminase treatment of patients with metastatic melanoma: results from phase I and II studies. J. Clin. Oncol. 23, 7660–7668.

43. Savoca, K.V., Davis, F.F., Van Es, T., McCoy, J.R., Palczuk, N.C. (1984) Cancer therapy with chemically modified enzymes. II. The therapeutic effectiveness of arginase, and arginase modified by the covalent attachment of polyethylene glycol, on the taper liver tumour and the L5178Y murine leukaemia. Cancer Biochem. Biophys. 7, 261–268.

44. Tsui, S.M., Lam, W.M., Lam, T.L., Chong, H.C., So, P.K., Kwok, S.Y., Arnold, S., Cheng, P.N.M., Wheatley, D.N., Lo, W.H., Leung, Y.C. (2009) PEGylated derivatives of recombinant human arginase I (rhArg1) for sustained in vivo activity in cancer therapy.1. Preparation and characterization. J. Cell Mol. Med. 9, 9–22.

45. Cheng, P.N., Lam, T., Lam, W., Tsui, S., Cheng, A.W., Lo, W., Leung, Y. (2007) Pegylated recombinant human arginase (rhArg-peg5,000mw) inhibits the in vitro and in vivo proliferation of human hepatocellular carcinoma through arginine depletion. Cancer Res. 67, 309–317.

46. Yang, Z., Sun, X., Li, S., Tan, Y., Wang, X., Zhang, N., Yagi, S., Takakura, T., Kobayashi, Y., Takimoto, A., Yoshioka, T., Suginaka, A., Frenkel, E.P., Hoffman, R.M. (2004) Circulating half-life of PEGylated recombinant methioninase holoenzyme is highly dose dependent on cofactor pyridoxal-5-phosphate. Cancer Res. 64, 5775–5778.

47. Yang, Z., Wang, J., Yoshioka, T., Li, B., Lu, Q., Li, S., Sun, X., Tan, Y., Yagi, S., Frenkel, E.P., Hoffman, R.M. (2004) Pharmacokinetics, methionine depletion, and antigenicity of recombinant methioninase in primates. Clin. Cancer Res. 10, 2131–2138.

48. Yang, Z., Wang, J., Lu, Q., Xu, J., Kobayashi, Y., Takakura, T., Takimoto, A., Yoshioka, T., Lian, C., Chen, C., Zhang, D., Zhang, Y., Li, S., Sun, X., Tan, Y., Yagi, S., Frenkel, E.P., Hoffman, R.M. (2004) PEGylation confers greatly extended half-life and attenuated immunogenicity to recombinant methioninase in primates. Cancer Res. 64, 6673–6678.

49. Fang, J., Seki, T., Maeda, H. (2009) Therapeutic strategies by modulating oxygen stress in cancer and inflammation. Adv. Drug Deliv. Rev. 61, 290–302.

50. Yoshikawa, T., Kokura, S., Tanaka, K., Naito, Y., Kondo, M. (1995) A novel cancer therapy based on oxygen radicals. Cancer Res. 55,1617–1620.

51. Ben-Yoseph, O., Ross, B. D. (1994) oxidation therapy: the use of a reactive oxygen species generating enzyme system for tumour treatment. Br. J. Cancer 70, 1131–1135.

52. Sawa, T., Wu, J., Akaike, T., Maeda, H. (2002) Tumour-targeting chemotherapy by a xanthine oxidase-polymer conjugate that generates oxygen-free radicals in tumour tissue. Cancer Res. 60, 666–671.

53. Yagi, K. (1971) Reaction mechanism of D-amino acids oxidase. Adv. Enzymol. Relat. Areas Mol. Biol. 34, 41–78.

54. Fang, J., Nakamura, H., Deng, D.W., Akuta, T., Greish, K., Iyer, A.K., Maeda, H. (2008) Oxystress inducing antitumour therapeutics via targeted delivery of PEG-conjugated D-amino acid oxidase. Int. J. Cancer 122, 1135–1144.

55. Fang, J., Sawa, T., Akaike, T., Akuta, T., Sahoo, S.K., Greish, K., Hamada, A., Maeda, H. (2003) In vivo antitumour activity of pegylated zinc protoporphyrin: targeted inhibition of heme oxygenase in solid tumour. Cancer. Res. 63, 3567–3674.

56. Regehly, M., Greish, K., Rancan, F., Maeda, H., Bohm, F., Roder, B. (2007) Water-soluble polymer conjugates of ZnPP for photodynamic tumour therapy. Bioconjug. Chem. 18, 494–499.

57. Iyer, A.K., Greish, K., Seki, T., Okazaki, S., Fang, J., Takeshita, K., Maeda, H. (2007) Polymeric micelles of zinc protoporphyrin for tumour targeted delivery based on EPR effect and singlet oxygen generation. J. Drug Target. 15, 496–506.

58. Schiwon, K., Brauer, H.D., Gerlach, B., Muller, C.M., Montforts, F.P. (1994) Potential photosensitizers for photodynamic therapy. IV. Photophysical and photochemical properties of azapor-

phirin and azachlorin derivatives. J. Photochem. Photobiol. B. Biol. 23, 239–243.

59. Bettio, F., Canevari, M., Marzano, C., Bordin, F., Guiotto, A., Greco, F., Duncan, R., Veronese, F.M. (2006) Synthesis and biological *in vitro* evaluation of novel PEG-psoralen conjugates, Biomacromolecules 7, 3534–3541.

60. Vaidya, A., Sun, Y., Feng, Y., Emerson, L., Jeong, E., Zheng-Rong L. (2008) Contrast-enhanced MRI-guided photodynamic cancer therapy with a pegylated bifunctional polymer conjugate. Pharm. Res., 25, 2002–2011.

61. Wang, Y., Ye, F., Jeong, E.K., Sun, Y., Parker, D.L., Lu, Z.R. (2007) Non-invasive visualization of pharmacokinetics, biodistribution and tumour targeting of poly[N-(2-hydroxypropyl)methacrylamide] in mice using contrast enhanced MRI. Pharm. Res. 24, 1208–1216.

62. Visentin, R., Pasut, G., Veronese, F.M., Mazzi U. (2004) Highly efficient Technetium-99 m labeling procedure based on the conjugation of N-[N-(3-Diphenylphosphinopropionyl)glycyl]cysteine ligand with poly(ethylene glycol). Bioconjug. Chem., 15, 1046–1054.

63. Meléndez-Alafort, L., Nadali, A., Pasut, G., Zangoni, E., De Caro, R., Cariolato, L., Giron, M.C., Castagliuolo, I., Veronese, F.M., Mazzi, U. (2009) Detection of sites of infection in mice using 99mTc-labeled PN(2)S-PEG conjugated to UBI and 99mTc-UBI: a comparative biodistribution study. Nucl. Med. Biol. 36, 57–64.

64. Vallera, D.A., Sicheneder, A.R., Taras, E.P.,Brechbiel, M.W.,Vallera, J.A., Panoskaltsis-Mortari, A., Burns, L.J. (2007) Radiotherapy of CD45-expressing Daudi tumours in nude mice with yttrium-90-labeled, PEGylated anti-CD45 antibody. Cancer Biother. Radiopharm. 22, 488–500.

65. Mero, A., Pasut, G., Via, L.D., Fijten, M.W.M., Schubert, U.S., Hoogenboom, R., Veronese, F.M. (2008) Synthesis and characterization of poly(2-ethyl 2-oxazoline) conjugates with proteins and drugs: Suitable alternatives to PEG-conjugates? J. Control. Release, 125, 87–95.

Regulatory strategy and approval processes considered for PEG-drug conjugates and other nanomedicines

Tacey X. Viegas[1] and Francesco M. Veronese[2]

[1] *Serina Therapeutics, Inc., 601 Genome Way, Huntsville, AL 35806, USA*
[2] *Department of Pharmaceutical Sciences, University of Padua, Via F.Marzolo 5, 35131 Padua, Italy*

Abstract

Therapeutic products prepared by PEG conjugation currently undergo the same regulatory scrutiny as small molecule drugs. A brief review of historical and current regulatory submission strategies is discussed in this chapter. In addition, some forward looking suggestions and considerations are made for nanomedicines that employ PEG and other polymers.

Introduction

In the different chapters of this book, the authors reported on the chemistry and biology of PEG-drug conjugates. In addition, case studies were presented to show how a PEG drug can be tested in the clinic with favourable patient treatment outcomes. The obvious question drug developers need to ask is "what was the regulatory process used to gain approval for these PEG-drugs and what is the best regulatory pathway that one can adopt for future polymer mediated drug products?" This is a question that is often ignored in early research and development.

The acceptance of polyethylene glycol (PEG) in the biotechnology and pharmaceutical industry is well known with the approval of nine drugs and three devices. This would suggest that the various international regulatory agencies are well versed in the use of this technology in drug delivery. Each product is reviewed for its merits and demerits on a case by case basis because not every molecule is suited for PEGylation. So where are PEGs normally used in nanomedicines? A review of literature will show that PEG is used in attachment to proteins, peptides, aptamers and small molecules. In addition, PEG also enables other technologies such as stealth liposomes, targeted liposomes, microcapsules, nanocapsules, dendrimers, dendrons and multicomponent aggregates. All these structures can be grouped under the category 'nanomedicine'. The '2005 European Science Foundation Forward Look' defined nanomedicine as: 'systems working from the molecular level using engineering devices and nanostructures to achieve a medical benefit'. In this context, nanoscale includes active components or objects ranging in size from 1 nm to 100 nm [1]. In another analogy, Ruth Duncan introduced 'Polymer Therapeutics' as nano-sized (5–100 nm) polymer-based pharmaceuticals, which include rationally designed macromolecular drugs, polymer-drug and polymer-protein conjugates, polymeric micelles containing covalently bound drug, and polyplexes for DNA delivery [2]. Both reviews make recommendations regarding the clinical potential of these molecules, the need for new tools to study them, their risk-benefit assessments in acute and long-term exposure, and the need for further study in this field. While the nanotechnology field is on its growth spur, the use of PEG in direct attachment to drugs may be at its full potential at this time. On the other hand, the combination of PEG to polysaccharides, lipids and water insoluble materials is still under early development, and because of the complex nature of these drug delivery constructs, the regulatory process may require a different scrutiny.

Historical approval processes

PEGylated adenosine deaminase (Adagen®) and L-asparaginase (Oncaspar®) developed by Enzon, Inc., Piscataway, NJ, in early 1990s were the first modified protein products. They are indicated for severe combined immunodeficiency disease (SCID) and acute lymphoblastic leukaemia (ALL), respectively. These two rare diseases allowed the agency to review these molecules as 'orphan drugs', a process that allows for an expedited review prior to approval. Both compounds contained numerous strands of PEG 5000 covalently attached to the enzyme [3, 4]. During the early regulatory process, PEG was being viewed as an excipient that contributed to the pharmacokinetic properties of the molecule. It was comparable to a methylcellulose derivative that provided a sustained release or controlled release delivery of a small molecule drug in an oral tablet formulation. The distribution and fate of the metabolised PEG enzymes was not a major concern at that time. But, now with improved analytical tools and the methods to pin point the amino acid residues directly involved with PEG conjugation, it is a requirement to measure not only the number of PEG strands attached to each protein molecule, but also the critical chemical and biological properties. This was the case when the technique was applied to two PEGylated interferon products used in the treatment of hepatitis C. Both, PEGylated interferon alpha2b (PEG Intron®, Schering-Plough) and PEGylated interferon alpha2a (PEGASYS®, Hoffman-LaRoche Ltd) were more thoroughly characterised. The number of PEG strands attached to the protein, the specific sites of conjugation, the description of the individual isomers, their relative biological activity, the stability of the conjugates and their immunogenicity were characterised and reported during the drug approval process. Interferon alpha2b (Intron A, Schering-Plough) was conjugated to a 12,000 Da monomethoxypolyethylene glycol polymer [5]. It was approximately 95% monopegylated and the primary, the secondary, and the tertiary structures were unaltered. Though pegylation appeared to decrease the specific activity of the interferon alpha-2b protein, the potency of the conjugate was comparable to the native protein at both the molecular and cellular level. In addition, pegylation did not affect the epitope recognition of antibodies used for Intron A quantitation. An extensive analysis of the pegylated positional isomers revealed that approximately 50% of the PEG was attached to the Histidine-34 residue and this isomer had the highest antiviral activity.

Interferon alpha2a (Roferon-A, Hoffman-LaRoche) is a potent drug used to treat various types of cancer and viral diseases including Hepatitis B/C infections. To improve the pharmacological properties of the drug, a branched 40,000 Da monomethoxypolyethylene glycol polymer was covalently bound to a lysine side chain of the protein [6]. The drug substance was described as a mixture of mainly six monopegylated positional isomers modified at lysine residues K31, K134, K131, K121, K164, and K70. Each isomer was identified and characterised and it was shown that the PEG-K31 and PEG-K134 isomers had higher antiviral activity than the other isomers listed above [7]. These observations were included in the drug substance characterisation and non-clinical biology sections of the regulatory submission document.

Current approval processes

With the merger of small molecule drugs and biological products under the wing of the Center for Drug Evaluation and Research (CDER) at the US Food and Drug Administration (FDA), sponsors are traditionally required to submit an Investigational New Drug (IND) application in accordance with 21 Code of Federal Regulations (CFR) Part 312. This must occur prior to conducting human clinical studies. INDs are required to contain detailed information about the investigational new drug and this includes sections such as:

1. Chemistry, manufacturing, and controls which will include details about the active ingredient and its formulation, the methods of manufacturing, impurity profiles, specifications, analytical methods, packaging and stability.
2. Pre-clinical pharmacological and toxicological results from studies of the drug in animals. This will include biological activity, pharmacokinetics, toxicology in rodents and non-rodents after single and repeated doses, histopathology, genotoxicity and carcinotoxicity.
3. Clinical pharmacology which is the clinical trial protocol specifically catered to the safety of the drug substance in humans. The study design typically includes considerations for using normal human subjects or patients from a target disease population; the starting dose and justifications for dose escalation; a pharmacokinetic evaluation section; and an adverse event section that explains when the 'maximum tolerated dose (MTD)' has been reached.

During the IND review period, the FDA may identify additional information necessary to assure the safety of subjects and assure that the study design is adequate to permit an evaluation of the drug's safety or effectiveness in humans.

After the drug has been adequately studied in different phases of clinical trials, the applicant is required to submit a new drug application (NDA) as part of 21 CFR 314 or a biologics license application (BLA) as part of 21 CFR 601, in order to obtain US approval to market the drug. NDAs and BLAs will typically contain the same sections listed above, but additional information about product manufacturing and analytical validation is required, long-term stability in the proposed container closure and package, long-term toxicology studies in animals which will include reproductive and developmental data, extensive data generated from the safety and efficacy clinical evaluations in humans which may include drug combination and drug interaction information. Pharmacovigilance, pharmacoeconomics and environmental risk assessments are also submitted at this time.

A recent paper by Roger Gaspar and Ruth Duncan lists the additional considerations required when polymers are used [8]. Since polymers are not monodisperse, the polydispersity index (PDI) is normally reported for the PEG reagent used. It is known that as the molecular weight of the polymer is increased, the PDI will also increase. A large variation of PDI is not desired, as this will result in PEG-drug conjugates having a larger range in molecular weight and in turn this may affect the pharmacokinetic profile of the PEG-drug. In this scenario, reproducibility and validation become critical process parameters in the Chemistry Manufacturing and Controls (CMC). In the case of PEG-biological drug conjugates it is important to report additional information in the drug impurity section. Limits are set for residual unconjugated PEG, native protein; high molecular weight conjugates resulting from multi PEGylated isomers; and cross-linked products resulting from free diol in some PEG reagents. The potential of interaction of PEG-biological drugs with the container closure system has to be evaluated. When small molecules are attached to PEG or other polymers, the chemistry of attachment will differ from PEG-protein drugs. PEGylation of small molecules typically results in the loss of biological activity because the conjugate is too bulky to bind to target ligands and cell membranes receptors or to be internalised into the cells to reach the site of action. So in these cases, the covalent attachment is reversible thus allowing the small molecule to be released in the body over time. Besides the drug and the polymer, the linker chemistry becomes critical. Formulations of these kinds are classified as 'Sustained and Controlled Release Parenterals'. A joint workshop on 'Assuring Quality and Performance of Sustained and Controlled Release Parenterals' was conducted in 2002 between the American Association of Pharmaceutical Scientists, the Food and Drug Administration, and the United States Pharmacopoeia. Experts from industry, the regulatory agencies and academia debated on processes and analytical tools required to correctly measure the performance of these parenteral formulations. In their report [9], they identified a number of *in vitro* and *in vivo* tests that were required to evaluate the performance of these parenterals. A number of issues need to be addressed for PEG-small molecule drugs, one of which is the stability of the hydrolysable bond between the linker and the drug. It is essential that this bond be stable in the container during shelf life stability. It is also essential that this bond not immediately hydrolyze *in vivo* after injection, in order to avoid dose dumping. Low pH stable and lyophilized formulations were discussed as part of the formulation options and *in vitro* dissolution tests were recommended as part of the specifications.

The higher the molecular weight of the PEG selected, the better the pharmacokinetics of the biological or small molecule. The *in vivo* half-life is extended and the clearance rate of the drug is reduced. However, because PEG is non biodegradable, there are questions about how it traverses from the circulation into extravascular tissue and how it is eliminated from the body. There are two pathways through which PEGs and PEG conjugated compounds can be excreted. One pathway is through kidney glomeruli and this is for molecular weights that are less than 50 kDa. Another pathway is through the hepatic bile duct system and this is for molecular weights that are greater than 50 kDa [10]. Two published studies address the presence of renal vacuoles observed in rats after chronic administration of PEG and PEG conjugates. In the first study PEG-haemoglobin (Hb) was infused into the tail veins of rats at a dose of about 1.5 g/kg [11]. Minimal to moderate vacuolation was apparent in the kidneys of the PEG-Hb treated animals seven days post infusion. No vacuoles were observed when either bovine Hb or M-PEG 5 kDa was infused. In the second study PEG 20 kDa TNF-bp was injected to rats every other day for 3 months at doses of 40, 20 or 10 mg/kg. Despite the presence of marked kidney vacuolation, there were no changes in blood urea nitrogen (BUN), blood creatinine, urinalysis parameters such as urinary N-acetyl-β-D-glucosaminidase (NAG), urinary microglobulin, or sodium excretion. Equivalent doses of PEG alone did not cause light microscopic evidence of vacuolation [12]. In both these studies, the distension of lysosomes was attributed to the hygroscopic nature of PEG. These studies demonstrated that PEG linked proteins have the capacity to induce renal tubular vacuoles and the approval process will require that PEG drugs be monitored for this pathological observation. More recently, a markedly higher occurrence (22–25%) of anti-PEG antibodies has been observed in a healthy blood donor population (350 donors tested), and they were of the IgG and IgM types. The presence of anti-PEG antibodies was very closely associated with rapid clearance of PEG asparaginase from the body [13]. Although the advantages of PEG-conjugation are apparent, it may be advisable to screen patients for pre-existing anti-PEG antibodies. With the usage of low and medium sized PEGs in liquid medications, parenteral injections, dermal creams and shampoos, food and beverages, there is a high likelihood that a particular subset of the human population will test positive for anti-PEG antibodies. Low molecular weight and nearly monodisperse PEGs have also been recently reported to activate the classical and alternative pathways of the complement system. Increased levels of C3, C4d, Bb, C3a-desArg and SC5b-9 were detected *in vitro*, in serum during a time course study [14]. But in another study by the same authors, it was reported that polydisperse PEG 400 Da and 1960 Da and their corresponding non-ionic liposomal compositions did not increase the levels of the complement proteins described above [15]. In any case, studies may need to be conducted to address the observations noted above, and as part of the pharmacology and toxicology review process.

Biogenerics

The regulatory approval process for follow-on protein products (biogenerics) is scientifically more challenging than for small molecule generic drugs. For the latter, the abbreviated new drug application (ANDA) format is extensively used. The sponsor of such an application needs to demonstrate that the product is chemically similar and is bioequivalent. The agencies rely on the safety and efficacy of the innovator product in order to accept a pathway for approval of the generic product. In the case of biogenerics, both the US and the European regulatory agencies have experience in dealing with follow-on products. Protein products in the US are approved as drugs under the Food and Drug Cosmetic (FDC) Act or licensed as biological products under the Public Health Service (PHS) Act. In the case of follow-on proteins, an abbreviated pathway is described in section 505(b)(2) of the FDC Act, which permits a sponsor to rely on published literature or on the Agency's finding of safety and effectiveness for a referenced approved drug product to support approval of a proposed product. In November 2004, the European Medicine Agencies (EMEA)/Committee for Human Medicinal Products (CHMP) issued a set of guidelines for Biosimilars which addressed general, quality-relevant and pre-clinical/clinical requirements for specific products such as somatropin, epoetin, filgrastim, and insulin [16]. A Biosimilar is defined as "a product having highly similar quality attributes before and after manufacturing process changes and that no adverse impact on the safety or efficacy, including immunogenicity, of the drug product occurs". The first biosimilars approved and marketed were Omnitrope® (recombinant somatropin) in the US, and Omnitrope® and Binocrit® (epoetin alpha) in Europe [17]. The FDA and the EMEA used the above case studies to recommend requirements for a biosimilar submission:

1. Chemistry, Manufacturing and Controls: Physicochemical testing to prove that the structure of the biogeneric is similar to the approved product.
   a. Primary structure: A combination of the amino acid sequence and structural investigations will demonstrate that the biogeneric has the same functional characteristics as the approved product, e.g., Edman sequencing and peptide mapping by mass spectroscopy.

   b. Mass analysis: Matrix Assisted Laser Desorption/Ionization Time-of-Flight (MALDI-TOF) and Electrospray Ionisation Mass Spectroscopy (ESI-MS).

   c. Spatial structure: Circular Dichroism (CD) and Nuclear Magnetic Resonance (NMR) spectroscopy. In addition, post-translational modification of the protein such as glycosylation, acetylation, or phosphorylation is required. Where applicable, the assembly of protein molecules into aggregates needs to be demonstrated.

   d. Polarity: Reverse Phase High Performance Liquid Chromatography (HPLC)

   e. Charge: Capillary Electrophoresis and Isoelectric Focusing

   f. Size: Size Exclusion Chromatography (SEC) and Gel Electrophoresis (GE)

2. Non-clinical pharmacology: *In vitro* cell proliferation activity and *in vivo* bioassays are recommended in Europe as part of a standardised monograph. An example of a bioassay for interferon is referenced in the European Pharmacopoeia [18].

3. Immunogenicity is the ability of a therapeutic protein to stimulate an immune response. It can range from development of detectable but not clinically significant antibodies, to an immune response that will impact on safety or effectiveness by the creation of neutralising antibodies. The ability to predict immunogenicity of a protein product, particularly the more complex proteins, is extremely limited. Some clinical assessment is needed where the product is to be administered chronically and the immunogenic potential is measured. The possibility of generating a cross-reaction with similar endogenous proteins is assessed.

4. Preclinical toxicology: Sub-chronic toxicity studies in rats or dogs.

5. Clinical Pharmacokinetics and Pharmacodynamics (PK/PD) in a Phase I format to demonstrate bioequivalence.

6. Clinical efficacy and safety: Phase III studies in a patient population to demonstrate that the biosimilar has similar efficacy as the approved product.

7. Supportive literature on the clinical experience and long-term safety of the approved product.

Most of the PEGylated proteins and aptamers currently enjoy market exclusivity. But once their patents expire, these biogeneric versions may need to be developed using the same procedures described above. These products may then be sub-classified as 'follow-on polymer conjugated biologics' or simply as 'PEG biosimilars'.

Conclusions

Even though the polymer or biological carrier is devoid of biological activity it is no longer considered as a simple excipient. This fact implies that after a covalent coupling the conjugate must be considered as part of a new composition of matter, i.e., a new chemical entity. It is the protein that provides the biological activity for the conjugate and it is the polymer that redefines the pharmacokinetics and distribution of the protein. Hence, by default, PEG becomes part of the process that will undergo all the approval steps needed for new drugs. In the CMC section, PEG could be considered as a synthetic intermediate for the final conjugate for which the chemistry, analytical measurements, stability, toxicology and biological fate must be known. With the field of PEGylation rapidly expanding to oligonucleotides, small molecule drugs and newer nanomedicines, it is likely that longer and expensive approval processes will slow down the ability to bring safer and therapeutically more advantageous products to market. The cost may be so high that only the big pharmaceutical companies could afford to develop them [19]. But the expectations are high, as is demonstrated by the rapidly increasing number of studies in this field and the growing number of researchers in new biotechnology dedicated companies.

The guidelines and recommendations in the field of nanomedicines come mainly from EMEA in Europe, FDA in USA and MHRA in UK [20, 21]. It is accepted that no drug may be devoid of any risk, and it is responsibility of the drug sponsor to guarantee an acceptable risk/benefit balance. The researchers are therefore encouraged to refer to the guidelines presented by these agencies in the initial stages of any product development venture. As mentioned before, each product in nanomedicine will face different questions and will need new answers. A consortium from the different agencies may be required to better and uniformly regulate in this arena.

References

1. ESF Scientific Forward Look on Nanomedicines, 23 February 2005, (www.esf.org)
2. Duncan R (2003) The dawning era of polymer therapeutics. *Nat Rev Drug Discov* 2: 347–360
3. Adagen®, Physician Desk Reference, Thomson Medical Publishing
4. Oncaspar®, Physician Desk Reference, Thomson Medical Publishing
5. Grace M, Youngster S, Gitlin G, Sydor W, Xie L, Westreich L, Jacobs S, Brassard D, Bausch J, Bordens R (2001) Structural and biologic characterization of pegylated recombinant IFN-alpha 2b. *J Interferon Cytokine Res* 12: 1103–1115
6. Dhalluin C, Ross A, Leuthold LA, Foser S, Gsell B, Müller F, Senn H (2005) Structural and biophysical characterization of the 40 kDa PEG-interferon- alpha2a and its individual positional isomers. *Bioconjug Chem* 16: 504–517
7. Foser S, Schacher A, Weyer KA, Brugger D, Dietel E, Marti S, Schreitmüller T (2003) Isolation, structural characterization, and antiviral activity of positional isomers of monopegylated interferon alpha-2a (PEGASYS). *Protein Expr Purif* 30: 78–87
8. Gaspar R, Duncan R (2009) Polymer carriers: Preclinical safety and regulatory implications for design and development of polymer therapeutics. ADDR Theme issue: Polymer Therapeutics: Clinical applications and Challenges for Development (in press)
9. Burgess DJ, Hussain AS, Ingallinera TS, Chen ML (2002) Assuring quality and performance of sustained and controlled release parenterals: AAPS workshop report, co-sponsored by FDA and USP. *Pharm Res* 19: 1761–1768
10. Yamaoka T, Tabata Y, Ikada Y (1994) Distribution and tissue uptake of Poly(ethylene glycol) with

different molecular weights after intravenous administration to mice. *J Pharm Sci* 83: 601–606

11. Conover C, Lejeune L, Linberg R, Shum K, Shorr RGL (1996) Transitional vacuole formation following a bolus infusion of PEG-hemoglobin in the rat. *Art Cells, Blood Subs and Immob Biotech* 24: 599–611

12. Bendele B, Seely J, Richey C, Sennello G, Shopp G (1998) Short communication: renal tubular vacuolation in animals treated with polyethylene-glycol-conjugated proteins. *Toxicological Sciences* 42: 152–15 7

13. Armstrong JK, Hempel G, Koling S, Chan LS, Fisher T, Meiselman HJ, Garratty G (2007) Antibody against poly(ethylene glycol) adversely affects PEG-asparaginase therapy in acute lymphoblastic leukemia patients. *Cancer* 110: 1103–1111

14. Hamad I, Hunter AC, Szebeni J, Moghimi SM (2008) Poly(ethylene glycol)s generate complement activation products in human serum through increased alternative pathway turnover and a MASP-2-dependent process. *Mol Immuno* 46: 225–232

15. Moghimi SM, Hamad I, Andresen TL, Jørgensen K, Szebeni J (2006) Methylation of the phosphate oxygen moiety of phospholipid-methoxy(polyethylene glycol) conjugate prevents PEGylated liposome-mediated complement activation and anaphylatoxin production. *FASEB J* 20: 2591–2593

16. The EU Directive 2001/83/EC, Article 10, Paragraph 4, published in EU Directive 2004/27/EC (implemented in EU at the end of October 2005)

17. Nachtmann F (2008) The regulatory situation of biosimilars/follow-on-biologics: a pioneer's perspective. AAPS Annual Meeting and Exposition, Atlanta, GA, November 1817.

18. European Pharmacopoeia monograph 01/2005: No. 1110, as 'Interferon alfa-2 concentrated solution' and chapter 5.6, 'Assay of interferon'

19. Faunce T, Shats K (2007) Researching safety and cost-effectiveness in the life cycle of nanomedicine. *J Low Med* 15: 128–135

20. Federal Food, Drug, and Cosmetic Act (FDandC Act), Chapter 5, Research Into Pediatric Uses for Drugs and Biological Products

21. EMEA, Reflection paper on nanotechnology-based medicinal products for human use, (EMEA/CHMP/79769/2006), 29 June 2006

Index

The MDT-Series
Milestones in Drug Therapy

The discovery of drugs is still an unpredictable process. Breakthroughs are often the result of a combination of factors, including serendipidity, rational strategies and a few individuals with novel ideas. *Milestones in Drug Therapy* highlights new therapeutic developments that have provided significant steps forward in the fight against disease. Each book deals with an individual drug or drug class that has altered the approach to therapy. Emphasis is placed on the scientific background to the discoveries and the development of the therapy, with an overview of the current state of knowledge provided by experts in the field, revealing also the personal stories behind these milestone developments. The series is aimed at a broad readership, covering biotechnology, biochemistry, pharmacology and clinical therapy.

Forthcoming titles

Bortezomib in the Treatment of Multiple Myeloma, K.C. Anderson, P.G. Richardson, I. Ghobrial (Editors), 2009
Influenza Virus Sialidase – A Drug Discovery Target, M. von Itzstein (Editor), 2010
Drugs for HER2-positive Breast Cancer, C.C. Zielinski, M. Sibilia, T. Grunt, R. Bartsch (Editors), 2010

Published volumes

Erythropoietins, Erythropoietic Factors, and Erythropoiesis, 2nd Revised and Extended Edition, S. Elliott, M. Foote, G. Molineux (Editors), 2009
Bipolar Depression: Molecular Neurobiology, Clinical Diagnosis and Pharmacotherapy, C.A. Zarate, H.K. Manji (Editors), 2009
Treatment of Psoriasis, J.M. Weinberg (Editor), 2008
Aromatase Inhibitors, 2nd revised edition, B.J.A. Furr (Editor), 2008
Pharmacotherapy of Obesity, J.P.H. Wilding (Editor), 2008
Entry Inhibitors in HIV Therapy, J.D. Reeves, C.A. Derdeyn (Editors), 2007
Drugs affecting Growth of Tumours, H.M. Pinedo, C. Smorenburg (Editors), 2006
TNF-alpha Inhibitors, J.M. Weinberg, R. Buchholz (Editors), 2006
Aromatase Inhibitors, B.J.A. Furr (Editor), 2006
Cannabinoids as Therapeutics, R. Mechoulam (Editor), 2005
St. John's Wort and its Active Principles in Anxiety and Depression, W.E. Müller (Editor), 2005

MILESTONES IN DRUG THERAPY

Erythropoietins, Erythropoietic Factors, and Erythropoiesis

Molecular, Cellular, Preclinical, and Clinical Biology
2nd Revised and Extended Edition

BIRKHÄUSER

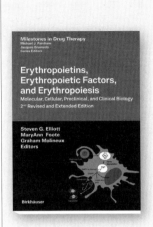

Elliott, S.G., Amgen Inc., Thousand Oaks, CA, USA / **Foote, M.A.,** MA Foote Associates, Westlake Village, CA, USA / **Molineux, G.,** Amgen Inc., Thousand Oaks, CA, USA **(Eds)**

Elliott, S.G. / Foote, M.A. /
Molineux, G. (Eds)
Erythropoietins, Erythropoietic Factors, and Erythropoiesis
2nd Revised and Extended Ed.
2009. 326 p. 56 illus., 11 in color.
Hardcover
ISBN 978-3-7643-8694-8
MDT — Milestones in Drug
Therapy

This second edition is a one-source guide to current information about red blood cell physiology and the action of native and recombinant human erythropoietic factors. Topics in the fields of erythropoiesis, recombinant protein discovery and production, and treatment of patients with anemia due to renal failure, cancer, or chronic diseases are covered. The newest theories in erythropoiesis (receptors, signaling), manufacturing, new formulations, and clinical research are discussed. This book is of interest to researchers and clinical investigators in academia and biotechnology and pharmaceutical companies, to clinical research associates, clinical monitors, and physician investigators.

From the Contents:
Erythropoiesis: an overview.- Regulation of endogenous erythropoietin production.- Biology of erythropoietin.- Erythropoiesis - genetic abnormalities.- Studies of erythropoiesis and the discovery and cloning of recombinant human erythropoietin.-Commercial production of recombinant erythropoietins.- Biosimilar epoetins.- New molecules and formulations.- Structural basis for the signal transduction of erythropoietin.- Intracellular signaling by the erythropoietin receptor.- Mechanism of erythropoietin receptor activation.- Pharmacokinetics of erythropoiesis-stimulating agents.- Use of erythropoietic stimulating agents in the setting of renal disease.- Abuse of recombinant erythropoietins and blood products by athletes.- Role and regulation of iron metabolism in erythropoiesis and disease.- Nonhematopoietic effects of erythropoiesis-stimulating agents.

www.birkhauser.ch